This is a comprehensive discussion of complexity as it arises in physical, chemical, and biological systems, as well as in mathematical models of nature. Common features of these apparently unrelated fields are emphasised and incorporated into a uniform mathematical description, with the support of a large number of detailed examples and illustrations.

The quantitative study of complexity is a rapidly developing subject with special impact in the fields of physics, mathematics, information science, and biology. Because of the variety of the approaches, no comprehensive discussion has previously been attempted. The aim of this book is to illustrate the ways in which complexity manifests itself and to introduce a sequence of increasingly sharp mathematical methods for the classification of complex behaviour. The authors offer a systematic, critical, ordering of traditional and novel complexity measures, relating them to well-established physical theories, such as statistical mechanics and ergodic theory, and to mathematical models, such as measure-preserving transformations and discrete automata. A large number of fully worked-out examples with new, unpublished results is presented. This study provides a classification of patterns of different origin and specifies the conditions under which various forms of complexity can arise and evolve. An even more important result than the definition of explicit complexity indicators is, however, the establishment of general criteria for the identification of anologies among seemingly unrelated fields and for the inference of effective mathematical models.

This book will be of interest to graduate students and researchers in physics (nonlinear dynamics, fluid dynamics, solid-state, cellular automata, stochastic processes, statistical mechanics and thermodynamics), mathematics (dynamical systems, ergodic and probability theory), information and computer science (coding, information theory and algorithmic complexity), electrical engineering and theoretical biology.

Cambridge Nonlinear Science Series 6

EDITORS

Professor Boris Chirikov
Budker Institute of Nuclear Physics, Novosibirsk

Professor Predrag Cvitanović
Niels Bohr Institute, Copenhagen

Professor Frank Moss
University of Missouri, St Louis

Professor Harry Swinney
Center for Nonlinear Dynamics,
The University of Texas at Austin

Complexity

TITLES IN PRINT IN THIS SERIES

Complexity

Hierarchical structures and scaling in physics

R. Badii
Paul Scherrer Institute, Villigen, Switzerland

A. Politi
Istituto Nazionale di Ottica, Florence, Italy

PUBLISHED BY THE PRESS SYNDICATE OF THE UNIVERSITY OF CAMBRIDGE
The Pitt Building, Trumpington Street, Cambridge, United Kingdom

CAMBRIDGE UNIVERSITY PRESS
The Edinburgh Building, Cambridge CB2 2RU, UK   www.cup.cam.ac.uk
40 West 20th Street, New York, NY 10011–4211, USA   www.cup.org
10 Stamford Road, Oakleigh, Melbourne 3166, Australia
Ruiz de Alarcón 13, 28014 Madrid, Spain

First published 1997
First paperback edition (with corrections) 1999

Typeface 10/13.5pt Times   *System* LaTeX   [UPH]

A catalogue record for this book is available from the British Library

Library of Congress Cataloguing in Publication data
Badii, R.
Complexity : hierarchical structures and scaling in physics / R. Badii, A. Politi.
p.   cm. – (Cambridge nonlinear science series : 6)
Includes bibliographical references and index.
ISBN 0 521 41890 9
1 Scaling (Statistical physics).   2 Complexity (Philosophy).
3 Mathematical physics.   I. Politi, A.   II. Title.   III. Series.
QC174.85.S34B33   1997
501$'$.17–DC20   96-18903   CIP

ISBN 0 521 41890 9 hardback
ISBN 0 521 66385 7 paperback

Transferred to digital printing 2003

Contents

Preface

The intuitive notion of complexity is well expressed by the usual dictionary definition: "a complex object is an arrangement of parts, so intricate as to be hard to understand or deal with" (Webster, 1986). A scientist, when confronted with a complex problem, feels a sensation of distress that is often not attributable to a definite cause: it is commonly associated with the inability to discriminate the fundamental constituents of the system or to describe their interrelations in a concise way. The behaviour is so involved that any specifically designed finite model eventually departs from the observation, either when time proceeds or when the spatial resolution is sharpened. This elusiveness is the main hindrance to the formulation of a "theory of complexity", in spite of the generality of the phenomenon.

The problem of characterizing complexity in a quantitative way is a vast and rapidly developing subject. Although various interpretations of the term have been advanced in different disciplines, no comprehensive discussion has yet been attempted. The fields in which most efforts have been originally concentrated are automata and information theories and computer science. More recently, research in this topic has received considerable impulse in the physics community, especially in connection with the study of phase transitions and chaotic dynamics. Further interest has been raised by the discovery of "glassy" behaviour and by the construction of the first mathematical models in evolutionary biology and neuroscience.

The aim of this book is to illustrate the ways in which complexity manifests itself in nature and to guide the reader through a sequence of increasingly sharp mathematical methods for the classification of complex behaviour. We propose a

systematic, critical, ordering of the available complexity measures, relating them
to well-established physical theories, such as statistical mechanics and ergodic
theory, and to mathematical models, such as measure-preserving transformations and discrete automata. The object (usually a pattern generated by some
unknown rule) is investigated in the infinite-time or in the infinite-resolution
limit, or in both, as appropriate. The difficulty of describing or reproducing its
scaling properties shall be interpreted as an evidence of complexity.

In Chapter 1, we introduce the scientific background in which the concept of
complexity arises, mentioning physical systems and models which will be more
thoroughly illustrated in the following two chapters. Our survey is intended
to discriminate those phenomena that are actually relevant to complexity. In
Chapter 4, we review the fundamentals of symbolic dynamics, the most convenient framework to achieve a common treatment of otherwise heterogeneous
systems. Chapters 5, 6, and 7 deal with probability, ergodic theory, information, thermodynamics, and automata theory. These fields form the basis for
the discussion of complexity. Although the concepts and quantities they deal
with may not all look modern or fashionable, they do provide, in a broad
sense, a classification of complexity. We stress the complementarity of different
indicators (power spectrum, degree of mixing, entropy, thermodynamic functions, automaton representation of a language) in the specification of a complex
system. Physical and mathematical aspects of this analysis are illustrated with
several paradigmatic examples, used throughout the book to compare various
complexity measures with each other. Chapters 5, 6, and 7 are therefore essential for the understanding of Chapter 8, where we introduce the principal
"classical" definitions of complexity and some of the most recent ones. We show
that only a few of them actually bring a novel and useful contribution to the
characterization of complexity. It will also appear that these definitions cannot
be ordered by elementary inclusion relations, in such a way that simplicity of an
object according to the most "liberal" measure implies simplicity according to
the strictest one and vice versa for a complex object. In many cases, the results
will take the form of existence or non-existence statements rather than being
expressed by numbers or functions.

Chapter 9 deals with complexity measures that explicitly refer to a hierarchical
organization of the dynamical rules underlying symbolic patterns produced by
unknown systems. In this context, complexity is related to a particularly
strong condition: namely, the lack of convergence of a hierarchical approach.
In Chapter 10, the main results presented in this book are summarized and
directions for future research are pointed out.

The study of complexity is a new subject in so far as it is just beginning to
encompass different fields of research. Therefore, we have necessarily neglected
a great number of interesting topics and mentioned others only in passing (for
all these, we refer the reader to the suggested bibliography).

We have tried to give all necessary definitions in as rigorous a way as allowed, without sacrificing clarity and readability. In the effort to be concise, we have concentrated on providing a guide through the areas of research that are essential for the understanding of the novel subjects treated in this book. We hope that the uniformity of notation and the relegation of technical mathematical notions to the appendices will help the reader follow the main course of the discussion without much need for consulting standard textbooks or the original research papers.

Our primary concern has been to stimulate the reader to explore the rich and beautiful world of complexity, to create and work out examples and, perhaps, to propose his/her own complexity measure.

We apologize for any errors and passages of weak or unclear exposition, as well as for misjudgment about the importance of experiments or mathematical tools for the characterization of complexity. We shall be happy to receive comments from all who care to note them.

This book has been written on the invitation of Simon Capelin, of Cambridge University Press, whose patience and constant encouragement we wish to acknowledge in a particular way. We received help in various forms from several colleagues. Suggestions for the improvement of the manuscript have been made by F. T. Arecchi, S. Ciliberto, P. Grassberger, P. Liò, R. Livi, S. Ruffo, and P. Talkner. Articles on subjects related to the arguments treated in the book have been provided by A. Arneodo, H. Atlan, M. Casartelli, G. Chaitin, A. De Luca, B. Derrida, J. D. Farmer, H. Fujisaka, J. P. Gollub, B. Goodwin, C. Grebogi, S. Großmann, H. A. Gutowitz, B. A. Huberman, R. Landauer, P. Meakin, S. C. Müller, A. S. Pikovsky, D. Ruelle, K. R. Sreenivasan, V. Steinberg, H. L. Swinney, T. Tél, and D. A. Weitz. Pictures contributed by G. Broggi, J. P. Gollub, S. C. Müller, E. Pampaloni, S. Residori, K. R. Sreenivasan, and D. A. Weitz have been included; others, although equally interesting, could not be accommodated in the available space. We have greatly benefitted from a long-time collaboration with F. T. Arecchi (R. B. and A. P.) and E. Brun (R. B.), as well as from discussions with D. Auerbach, C. Beck, G. Broggi, J. P. Crutchfield, P. Cvitanović, M. Droz, G. Eilenberger, M. J. Feigenbaum, M. Finardi, Z. Kovacs, G. Mantica, G. Parisi, C. Perez-Garcia, A. S. Pikovsky, I. Procaccia, G. P. Puccioni, M. Rasetti, P. Talkner, and C. Tresser. During the preparation of the manuscript, we enjoyed full support from our home institutions, the Paul Scherrer Institute (Villigen) and the Istituto Nazionale di Ottica (Florence). Part of the work has been carried out at the Institute for Scientific Interchange in Turin, where we have been invited by M. Rasetti. A. P. is grateful to I. Becchi for the access to the facilities of the PIN Center of the Engineering Department in Prato; R. B. wishes to thank the warm hospitality of A. and G. Pardi during his visits to Florence and acknowledges support by the Laboratorio FORUM-INFM within its research program on Nonlinear Physics.

Part I

Phenomenology and models

Chapter 1

Introduction

The scientific basis of the discussion about complexity is first exposed in general terms, with emphasis on the physical motivation for research on this topic. The genesis of the "classical" notion of complexity, born in the context of the early computer science, is then briefly reviewed with reference to the physical point of view. Finally, different methodological questions arising in the practical realization of effective complexity indicators are illustrated.

1.1 Statement of the problem

The success of modern science is the success of the experimental method. Measurements have reached an extreme accuracy and reproducibility, especially in some fields, thanks to the possibility of conducting experiments under well controlled conditions. Accordingly, the inferred physical laws have been designed so as to yield nonambiguous predictions. Whenever substantial disagreement is found between theory and experiment, this is attributed either to unforeseen external forces or to an incomplete knowledge of the state of the system. In the latter case, the procedure so far has followed a reductionist approach: the system has been observed with an increased resolution in the search for its "elementary" constituents. Matter has been split into molecules, atoms, nucleons, quarks, thus reducing reality to the assembly of a huge number of bricks, mediated by only three fundamental forces: nuclear, electro-weak and gravitational interactions.

The discovery that everything can be traced back to such a small number of different types of particles and dynamical laws is certainly gratifying. Can one thereby say, however, that one *understands* the origin of earthquakes, weather variations, the growing of trees, the evolution of life? Well, in principle, yes. One has just to fix the appropriate initial conditions for each of the elementary particles and insert them into the dynamical equations to determine the solution[1]. Without the need of giving realistic numbers, this undertaking evidently appears utterly vain, at least because of the immense size of the problem. An even more fundamental objection to this attitude is that a real understanding implies the achievement of a synthesis from the observed data, with the elimination of information about variables that are irrelevant for the "sufficient" description of the phenomenon. For example, the equilibrium state of a gas is accurately specified by the values of only three macroscopic observables (pressure, volume and temperature), linked by a closed equation. The gas is viewed as a collection of essentially independent subregions, where the "internal" degrees of freedom can be safely neglected. The change of descriptive level, from the microscopic to the macroscopic, allows recognition of the inherent simplicity of this system.

This extreme synthesis is no longer possible when it is necessary to study motion at a mesoscopic scale as determined, e.g., by an impurity. In fact, the trajectory of a Brownian particle (e.g., a pollen grain) in a fluid can be exactly accounted for only with the knowledge of the forces exerted by the surrounding molecules. Although the problem is once more intractable, in the sense mentioned above, a partial resolution in this case has been found in the passing from the description of *single items* to that of *ensembles*: instead of tracing an individual orbit, one evaluates the probability for the Brownian particle to be in a given state, which is equivalent to considering a family of orbits with the same initial conditions but experiencing different microscopic configurations of the fluid. Although less detailed, this new level of description in principle involves evaluation and manipulation of a much larger amount of information: namely, the time evolution of a continuous set of initial conditions. This difficulty has been overcome in equilibrium statistical mechanics by postulating the equiprobability of the microscopic states. This constitutes a powerful short-cut towards a compact model for Brownian motion in which knowledge of the macroscopic variables again suffices. The fluid is still at equilibrium but the Brownian particle constitutes an *open* subsystem that evolves in an erratic way, being subject to random fluctuations on one side and undergoing frictional damping on the other. In addition to this, deterministic drift may be present.

These examples introduce two fundamental problems concerning physical modelling: the practical feasibility of predictions, given the dynamical rules, and the relevance of a minute compilation of the system's features. The former question entails both the inanity of the effort of following the motion of a huge number of particles and the impossibility of keeping the errors under

1. Excluding the possible existence of other unknown forces.

control. In fact, as the study of nonlinear systems has revealed, arbitrarily small uncertainties about the initial conditions are exponentially amplified in time in the presence of deterministic chaos (as in the case of a fluid). This phenomenon may already occur in a system specified by three variables only. The resulting limitation on the power of predictions is not to be attributed to the inability of the observer but arises from an intrinsic property of the system. The second observation points out that the elimination of the particles' coordinates in favour of a few macroscopic variables does not imply, in many cases, a reduced ability to perform predictions for quantities of interest. The success of statistical mechanics in explaining, e.g., specific heats, electric conductivity, and magnetic susceptibility demonstrates the significance of this approach. As long as it affects just irrelevant degrees of freedom, chaotic motion does not downgrade coarse representations of the dynamics but may even accelerate their convergence.

Nature provides plenty of patterns in which coherent macroscopic structures develop at various scales and do not exhibit elementary interconnections: for instance, the often cited biological organisms or, more simply, vortices in the atmosphere or geological formations (sand dunes, rocks of volcanic origin). They immediately suggest seeking a compact description of the spatio-temporal dynamics based on the relationships *among* macroscopic elements rather than lingering on their inner structure. In a word, it is useful and possible to condense information. This is not a mere technical stratagem to cope with a plethora of distinct unrelated patterns. On the contrary, similar structures evidently arise in different contexts, which indicates that universal rules are possibly hidden behind the evolution of the diverse systems that one tries to comprehend. Hexagonal patterns are found in fluid dynamics, as well as in the spatial profile of the electric field of laser sources. Vortices naturally arise in turbulent fluids, chemically interacting systems, and toy models such as cellular automata.

Systems with a few levels of coherent formation, while interesting and worth studying, are incomparably simpler than systems characterized by a *hierarchy* of structures over a *range* of scales. The most striking evidence of this phenomenon comes from the ubiquity of *fractals* (Mandelbrot, 1982), objects exhibiting nearly scale-invariant geometrical features which may be nowhere differentiable. A pictorial representation of this can be obtained by zooming in on, e.g., a piece of a rugged coastline: tinier and tinier bays and peninsulae are revealed when the resolution is increased, while gross features are progressively smeared out. Nonetheless, one perceives an almost invariant structure during the magnification process. Among the several examples of fractals discussed in Mandelbrot (1982), we recall cauliflowers, clouds, foams, galaxies, lungs, pumice-stone, sponges, trees. If the coarse-graining is interpreted as a change in the level of description, an exact scale-invariance, whenever it is observed, testifies to the simplicity of the system. Although such *self-similar* objects can be assimilated to translationally invariant patterns, by interchanging the operations of shift and magnification, there is a fundamental difference. The dynamical rules that support a given pattern are, in general, invariant under some symmetry group (e.g., translation). Hence, it is not surprising that the pattern exhibits the same symmetry (e.g., periodicity, as in a crystal). This is not the case of fractals, the "symmetry" of which is not built in the generation mechanism. The puzzling issue is that the same physical laws account for both types of behaviour as well as for the astounding variety of forms that we habitually experience.

A hierarchical organization in nested subdomains is particularly evident in the vicinity of a (continuous) phase transition, as occurring, e.g., in magnetic materials or superconductors. The coarse-graining procedure (Kadanoff, 1966) has led to the formulation of the renormalization-group theory (Wilson, 1971) which, in spite of its conceptual simplicity, has explained the observed phenomenology with high precision. Phase transitions, however, occur at special parameter values (e.g., melting points) whereas hierarchical structures appear to be a much more general characteristic of nature. As an example, we cite $1/f$ noise, which is the result of signals showing self-similar properties upon rescaling of the time axis. This phenomenon, although one of the commonest in nature, has so far withstood any global theoretical approach. Many other systems exhibit various levels of organization which are neither too strict, as in a crystal, nor too loose, as in a gas, nor are they amenable to any known theoretical modelling. The difficulty of obtaining a concise description may arise from "fuzziness" of the subsystems, which prevents a univocal separation of scales, or from substantial differences in the interactions at different levels of modelling.

Summarizing this introductory section, we remark that the concept of complexity is closely related to that of *understanding*, in so far as the latter is based upon the accuracy of *model* descriptions of the system obtained using a condensed information about it. Hence, a "theory of complexity" could be viewed as a theory of modelling, encompassing various reduction schemes (elimination or aggregation of variables, separation of weak from strong couplings, averaging over subsystems), evaluating their efficiency and, possibly, suggesting novel representations of natural phenomena. It must provide, at the same time, a definition of complexity and a set of tools for analysing it: that is, a system is not complex by some abstract criterion but because it is intrinsically hard to model, no matter which mathematical means are used. When defining complexity, three fundamental points ought to be considered (Badii, 1992):

1. Understanding implies the presence of a *subject* having the task of describing the *object*, usually by means of model predictions. Hence, complexity is a "function" of both the subject and the object.

2. The object, or a suitable representation of it, must be conveniently

divided into *parts* which, in turn, may be further split into subelements, thus yielding a *hierarchy*. Notice that the hierarchy need not be manifest in the object but may arise in the construction of a model. Hence, the presence of an actual hierarchical structure is not an infallible indicator of complexity.

3. Having individuated a hierarchical encoding of the object, the subject is faced with the problem of studying the *interactions* among the subsystems and of incorporating them into a model. Consideration of the interactions at different levels of resolution brings in the concept of *scaling*. Does the increased resolution eventually lead to a stable picture of the interactions or do they escape any recognizable plan? And if so, can a different model reveal a simpler underlying scheme?

1.2   Historical perspective

Although the inference of concise models is the primary aim of all science, the first formalization of this problem is found in discrete mathematics. The object is represented as a sequence of integers which the investigator tries to reproduce exactly by detecting its internal rules and incorporating them into the model, also a sequence of integers. The procedure is successful if a size reduction is obtained. For example, a periodic sequence, such as 011011011..., is readily specified by the "unit cell" (011 in this case) and by the number of repetitions.

This approach has given rise to two disciplines: computer science and mathematical logic. In the former, the model is a computer program and the object sequence is its output. In the latter, the model consists of the set of rules of a formal system (e.g., the procedure to extract square roots) and the object is any valid statement within that system (as, e.g., $\sqrt{4} = 2$). Compression means that knowledge of the whole formal system permits the deduction of all theorems automatically, without any external information. In this view, the complexity of symbol strings is called *algorithmic* and is defined as the size of the minimal program which is able to reproduce the input string (Solomonoff, 1964; Kolmogorov 1965; Chaitin 1966). As a consequence, completely random objects have maximal complexity because no compression is possible for them. This characterization of clearly structureless patterns makes the definition unsuitable in a physical context. Actually, algorithmic complexity coincides, in most cases, with entropy and is therefore a measure of disorder.

A detailed analysis allowed the models to be regrouped in computational classes with qualitatively and quantitatively different ability in the manipulation of symbolic objects. The main four classes form a hierarchy, named after Chomsky (1956, 1959), which culminates in the Turing machine (Turing, 1936), the prototype of a universal computer: that is, of an automaton which is able to simulate any other one if suitably programmed. Chomsky's hierarchy classifies sequences of integers according to the minimal computational capabilities necessary to reproduce them. Thus, the complexity of the machine (automaton) is accounted for, rather than that of the input string.

In practice, however, finding the minimal program or, equivalently, the computational class is possible only in special cases. In fact, it has been proved that no general algorithm exists for the solution of such problems with any input string. In the same way, the problem of assessing whether a theorem (akin to an integer sequence) is a member of a certain formal system (akin to a Turing machine) cannot be resolved in general. This impossibility (technically called "undecidability") is related to the existence of endless computations which prevent a Turing machine from halting. In a mathematical context, this is tantamount to saying that "given any reasonable (consistent and effective) theory of arithmetic, there are true assertions about the natural numbers that are not theorems in that theory" (Gödel, 1931). For any formal logic $\mathscr{L}$ that contains arithmetics and is consistent, there is no Turing machine that can decide of an arbitrary formula of $\mathscr{L}$ whether it is a theorem or not.

Under the strong assumption that relevant physical and biological processes are, in all their aspects, nothing but examples of universal computation, automata have been constructed for specific tasks in many fields, such as control and communication theory (Wiener, 1948), theoretical biology (von Neumann, 1966), and artificial intelligence (Minsky, 1988). Among the many objectives of the investigators, we recall movements of robots in simple landscapes, pattern recognition, trial and error learning, solution of puzzles, playing games (chess), modelling of biological evolution (cellular automata), imitation of mental processes (artificial neural networks). Of course, only approximate solutions are searched for in practical implementations in order to avoid wasting space and time or being stuck by undecidable situations.

This mechanistic hypothesis is implicitly accepted by the vast majority of researchers, at least as a tool to be profitably used until contrary evidence is found. When, instead, this view is pushed to the limit case of the functioning of animal brain, there is much less agreement in the scientific community (Penrose, 1989).

More recently, complexity has been associated with the presence of a large number of elements in a system and, for example, has been attributed to ecology, economy, and immune systems. The first evident common characteristic of these fields is the apparent difficulty in achieving good agreement between model and observation (not necessarily arising from chaotic behaviour). Clearly, this similarity is much too vague to justify claims that a unified theory of multicomponent systems can be constructed. Similarity of behaviour does not necessarily imply the existence of general principles governing all these systems. Rather than looking for an omnicomprehensive theory, one should first agree on a set of typical characteristics of complex behaviour, thus restricting the attention to a well defined class of systems which actually share these features. Among these, certainly, are the simultaneous presence of elements of order and disorder, some degree of unpredictability, the indication of interactions between subsystems which, however, change in dependence on how the system is subdivided (i.e., different interactions appear at different resolution scales).

The first chapters of this book illustrate the above mentioned aspects of complex behaviour before introducing a formalization of the object of study, which will mainly be a one-dimensional, stationary symbol sequence. In fact, not only is DNA, the prototype of the complexity of biological systems, a concatenation of four basic units, but general physical systems with continuous spatio-temporal structure are amenable to symbolic encoding. This restriction, which is only apparently severe, has the merit of clearly defining the mathematical context in which a sound discussion about complexity can be undertaken. In the absence of clear boundaries, in fact, one may be easily diverted from a scientific track and stray, at best, towards philosophical considerations.

Despite the unification of heterogeneous systems which is obtained by adopting a common language, it is still necessary to clarify the goals of the investigation and to distinguish among different sources of complexity. This further technical specification is the subject of the next section.

1.3   Self-generated complexity

The intricacy of a symbolic pattern of unknown origin is often attributed to either of two quite different mechanisms. On the one hand, it may be envisaged as the result of many nearly independent stimuli (either internal to the system or not). This is, for instance, the usual framework when the interaction with a heat bath must be taken into account. On the other hand, the pattern may be postulated to arise from a generic, structureless initial condition, under the action of a simple dynamical rule. It is clearly this second type of complexity, called *self-generated*, which we intend to characterize. More precisely, we speak of self-generated complexity whenever the (infinite) iteration of a few basic rules causes the emergence of structures having features which are not shared by the rules themselves. Simple examples of this scenario are represented by various types of symmetry breaking (superconductors, heat convection) and long-ranged correlations (phase transitions, cellular automata). The relevance of these phenomena and their universality properties discovered by statistical mechanical methods indicate self-generation as the most promising and meaningful paradigm for the study of complexity. It must be remarked, however, that the concept of self-generation is both too ample and too loosely defined to be transformed into a practical study or classification tool. In fact, it embraces fields as diverse as chaotic dynamics, cellular automata, and optimization algorithms for combinatorial problems.

Whatever the unknown underlying dynamics is, the observer is first confronted with the problem of reproducing the pattern with a model to be selected in a suitable class. The choice depends on a few fundamental questions. In fact, the pattern may be either interpreted as a single item to be reproduced exactly or as one of many possible outcomes of an experiment. In the latter case, the model should rather apply to the source itself and describe the set of rules common to all patterns produced by it. The "single-item" approach is commonly adopted in computer science and in coding theory, while the "ensemble" approach has a physical, statistical motivation.

In any case, two extreme situations may occur: the model consists either of a "large" automaton, specified by many parameters, which achieves the goal in a short time, or of a "small" one which must operate for a long time. Consequently, the complexity of the pattern is often visualized either through the size of the former automaton or through the computing time needed by the latter. These two processes are frequently postulated as the basis of biological evolution. Present-day DNA molecules are thought to be the result of the repeated application of elementary assembly operations (self-generated complexity) and of random mutations followed by selection, that is, by the verification of the effective fitness of the product to the environment. This step is equivalent to testing whether a symbolic pattern (DNA) is recognized by some extremely complicated automaton (the environment). It must be noticed, however, that an elaborate machine need not always produce a complex output. A randomly designed Turing machine, indeed, may yield rather dull patterns.

In general, one should avoid a careless identification of the pattern's and the model's complexities since the two do not generally coincide, especially as long as the optimal model has not been yet found. Therefore, another aspect of the problem emerges: namely, that complexity is associated with the disagreement between model and object rather than with the size of the former. This, in turn, calls for the role of a subject (the observer) in determining the complexity itself, through his/her ability to infer an appropriate model. Of course, a system looks complex as long as no accurate description has been found for it. For example, the modelling of deterministic chaos, with the latter's unpredictability and exotic geometric structures, is completely unsuccessful if carried out within the framework of linear stochastic processes.

These considerations and the limited domain of applicability of all existing complexity measures strongly suggest that there cannot be a unique indicator of complexity, in the same way as entropy characterizes disorder, but that one needs a set of tools from various disciplines (e.g., probability and information theory, computer science, statistical mechanics). As a result, complexity is seen through an open-ended sequence of models and may be expressed by numbers or, possibly, by functions. Indeed, it would be contradictory if the "complexity function", which must be able to appraise so many diverse objects, were not itself complex!

Further reading

Anderson (1972), Anderson & Stein (1987), Atlan (1979), Caianiello (1987), Casti (1986), Davies (1993), Goodwin (1990), Kadanoff (1991), Klir (1985), Landauer (1987), Löfgren (1977), Ruelle (1991), Simon (1962), Weaver (1968), Zurek (1990).

Chapter 2

Examples of complex behaviour

In this chapter, we present some of the most frequently quoted examples of "complex" behaviour observed in nature. Far from proposing a global explanation of such disparate systems within a unique theoretical framework, we select those common properties that do cast light on the ways in which complexity exhibits itself.

Natural macroscopic systems are usually characterized by intensive parameters (e.g., temperature T or pressure P) and extensive ones (volume V, number of particles N) which are taken into account by suitable thermodynamic functions, such as the energy E or the entropy S. When the only interaction of a system with its surroundings consists of a heat exchange with a thermal bath, an *equilibrium state* eventually results: the macroscopic variables become essentially time independent, since fluctuations undergo exponential relaxation. The equilibrium state corresponds to the minimum of the free energy $F = E - TS$ and is determined by the interplay between the order induced by the interactions, described by E, and the disorder arising from the multiplicity of different macroscopic states with the same energy, accounted for by the entropy S.

The commonest case is, however, represented by systems that are *open* to interactions with the environment, which usually takes the form of a *source* of energy and a *sink* where this is dissipated. Under these nonequilibrium conditions, the system either is able to recover a stationary configuration or, depending on the strength of the external forces, it evolves towards a spatio-temporally periodic or aperiodic behaviour.

The transition from the ordered, uniform, phase to more structured patterns is usually driven by gradients of heat, pressure, or concentration of components, and may pass through a sequence of different states which lose their stability upon changes of some control parameter. The main characteristic is the non-linearity of the response of the system for sufficiently strong amplitude of the perturbations. Various stages of this process are often regarded as complex, usually because of their aperiodicity or because of the high number of degrees of freedom involved.

2.1 Instabilities in fluids

Fluid systems are often cited as a paradigm of the transition from order to disorder (turbulence). One of the best understood phenomena in this context is *Rayleigh–Bénard*'s thermal convection (Chandrasekar, 1961; Busse, 1978) in a cell with small *aspect ratio* $\gamma = w/h$, where w is the width and h the height of the container. This condition ensures that spatial inhomogeneities are negligible in comparison with the irregularity of the time behaviour.

The fluid flow is usually modelled by the following set of partial differential equations for the velocity $\mathbf{v}(\mathbf{x}, t)$ and the temperature $T(\mathbf{x}, t)$, at the position $\mathbf{x}$ and time t:

$$\frac{\partial \mathbf{v}}{\partial t} + \mathbf{v} \cdot \nabla \mathbf{v} = -\frac{\nabla p}{\rho_0} + v \nabla^2 \mathbf{v} + \mathbf{q}$$

$$\nabla \cdot \mathbf{v} = 0 \qquad\qquad (2.1)$$

$$\frac{\partial T}{\partial t} + \mathbf{v} \cdot \nabla T = \kappa \nabla^2 T \,,$$

where the first is the *Navier–Stokes* equation in the Boussinesq approximation (Chandrasekar, 1961; Behringer, 1985), the second expresses the incompressibility condition and the third is the heat equation. The symbols p, ρ_0, v, $\mathbf{q}$, and κ represent the pressure, the constant density, the kinematic viscosity, the force per unit mass, and the heat diffusivity, respectively.

2.1.1 Temporally "complex" dynamics

In the Rayleigh–Bénard (RB) experiment, a positive temperature gradient $\beta = (T_0 - T_1)/h$ is kept between the bottom (0) and the top (1) plate in the cell. The temperature can be expressed as $T(\mathbf{x}) = T_0 - \beta \hat{\mathbf{z}} \cdot \mathbf{x} + \theta(\mathbf{x})$, where $\hat{\mathbf{z}}$ is the z-axis unit vector and $\theta(\mathbf{x})$ describes the fluctuations around the unperturbed temperature field in the *pure conduction* regime. This is a potentially unstable situation since a hot droplet ($\theta > 0$) has lower density than the surrounding fluid and is, therefore, pushed upwards by buoyancy. In this motion, the droplet enters colder regions where the density is even higher, so that it keeps rising. The initial fluctuation is thus amplified. This destabilizing effect is counterbalanced by viscous dissipation and heat diffusion from the droplet to the environment.

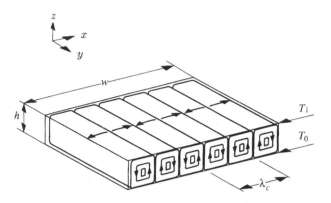

Figure 2.1 Schematic illustration of Rayleigh-Bénard convection rolls. The arrows indicate the direction of the fluid motion. Symbols T_0 and T_1 denote the plates' temperatures.

It turns out that the gradient β must be larger than a threshold value β_c in order for conduction to be taken over by *convection*. In this new regime, the flow develops a pattern of parallel rolls, whereby warm fluid elements reach the upper plate, lose heat and descend again towards the lower plate (see Fig. 2.1). The rolls have a fixed spatial wavelength $\lambda_c \approx 2h$ (i.e., only one Fourier mode is "active") so that the amplitude $A(t)$ of the temperature field at $h/2$ is an appropriate variable to describe the motion:

$$\theta(x, t; z = h/2) \approx A(t) \cos(x/\lambda_c) .$$

The amplitude A grows from zero when β increases beyond β_c and is gradually constrained by nonlinear mechanisms. For small aspect ratio (i.e., small number of rolls), the convective motion keeps spatial coherence while its time evolution ranges, depending on β, from periodicity or quasiperiodicity (with two or more incommensurate frequencies) to an aperiodic behaviour whose irregular features justify the appellation of "chaotic" (Bergé *et al.*, 1986; Libchaber *et al.*, 1983). In this regime, a (usually small) number of mutually interacting Fourier modes with different spatial frequencies may arise. The commonest route to *chaos* (formally defined in Chapter 3) results from an accumulation of successive doublings of the oscillation period upon increase of the control parameter β (*period-doubling*, Feigenbaum, 1978). Analogously with the transition from quasiperiodicity to chaos (Ruelle & Takens, 1971; Feigenbaum *et al.*, 1982), it possesses universal properties (Feigenbaum, 1979a): they are quantitatively the same for systems as diverse as lasers, electronic circuits, fluids, string vibrations, and many others.

Analogous considerations apply to the Couette–Taylor instability, in which a fluid is contained between two coaxial cyclinders rotating at different speeds (Brandstater & Swinney, 1987), to the Bénard–Marangoni convection, where large-scale surface tension inhomogeneities are induced by temperature fluctuations or by chemical variations, and to other related systems such as RB convection in a rotating cylinder (Steinberg *et al.*, 1985; Zhong *et al.*, 1991) or nematic liquid crystals, where electro-hydrodynamic instabilities occur (Manneville, 1990; Cross & Hohenberg, 1993).

When the spatial structures are simple, an adequate model is provided by a set of ordinary differential equations describing the time evolution of the amplitudes of the main Fourier modes for a few relevant variables (e.g., temperature, velocity). Complexity is, in this context, often associated with the existence of different possible asymptotic "states" (*multistability*) or with the unpredictability of the trajectories (see Chapter 3).

2.1.2 Spatio-temporal "complexity"

Systems characterized by sufficiently large aspect ratios ($\gamma > 10$), for which a straightforward modelling through ordinary differential equations fails, are usually called *weakly confined* or *extended*. They have the tendency to develop spatio-temporally irregular motion for rather low values of the control parameters, for which a confined system would still be stable. The dynamics, in fact, may be the result of a competition between different basic patterns which coexist in space. Their shape depends on control parameters, such as Rayleigh, Prandtl, and Nusselt numbers, which are related to the fluid's and the container's characteristics (Manneville, 1990). Accordingly, transitions to various regimes are possible: for example, stationary convection with hexagonal cells (Busse, 1967), crossing of perpendicular roll-patterns, compression–dilatation of wavelengths (*Eckhaus instability*), drifts, onset of zig-zag or skewed-varicose patterns. Moreover, different patterns may compete (Ciliberto *et al.*, 1991b) giving rise to nucleation (creation) of "domain walls" and "defects". The former are boundaries of different ordered patches, the latter are singular points of the wave pattern with particle-like properties. Indeed, they may diffuse and annihilate (Goren *et al.*, 1989; Rehberg *et al.*, 1989; Rubio *et al.*, 1991; Joets & Ribotta, 1991).

The description of these phenomena can still be reduced by focusing on the motion of domain walls and defects. When the external force is further increased, however, the system exhibits fully spatio-temporal chaotic behaviour and this reduction is no longer possible. An order–disorder transition of this type in a fluid subject to periodic forcing has been reported in Tufillaro *et al.* (1989). By increasing the driving amplitude ϵ beyond the onset of a stable capillary wave pattern, long-wavelength modulations are generated. At their intersection, the wave amplitude becomes small and structural defects form. When their density is high (see Fig. 2.2), no preferred pattern orientation is recognizable any longer. At even higher ϵ, complete disorder dominates. The onset of highly irregular bursts alternating with smooth (*laminar*) phases, both in space and time, signals the progressive development of spatio-temporal chaos. This transition scenario, called spatio-temporal intermittency (Manneville, 1990), has

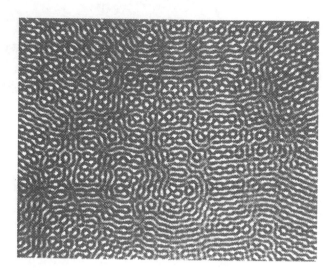

Figure 2.2 Capillary wave pattern on a fluid surface subject to a vertical modulation (from Gollub, 1991).

been experimentally observed in nearly one-dimensional geometries with large aspect ratios (Ciliberto & Bigazzi, 1988; Rabaud *et al.*, 1990). Although there is at present no universally accepted definition of spatio-temporal chaos or "weak turbulence", its quantitative description clearly requires a statistical analysis. In the example of Fig. 2.2, the disorder can be quantified as a lack of *correlation* (see Chapter 5) between many finite regions which are not individually chaotic. Other typical signatures of this phenomenon are temporal evolution with a continuum of low-frequency components (Ciliberto *et al.*, 1991a) and fluctuations of spatial Fourier amplitudes with approximately Gaussian statistics (Gollub, 1991). The observations can be roughly classified as follows (Hohenberg & Shraiman, 1988): temporal chaos occurs when the spatial correlation length λ (e.g., the roll-size or the wavelength; see Chapter 5 for a precise definition) is larger than the system size L; spatio-temporal chaos (weak turbulence) sets in with the appearance of spatially incoherent fluctuations when λ is much smaller than L. In this situation, the amplitudes of the Fourier modes decay exponentially with the wavenumber k.

The complexity of hydrodynamic instabilities is not only represented by the difficulty of finding appropriate dynamical equations (explaining the "cause" of the phenomenon) but it is of a more fundamental nature: providing even just a statistical description is an arduous task. The pattern is neither periodic nor quasiperiodic and, although the correlations decay, it does not appear completely disordered either. The "correct" equations of motion do not constitute a simple explanation of the process and a concise description still must be found. Notice that the necessity of a statistical analysis does not imply by itself the complexity of the phenomenon, as the examples of a Poisson process or a random walk (both rather simple) show.

2.2   **Turbulence**

When a fluid is supplied with an increasing flux of energy, it undergoes a
sequence of transitions from an ordered state (either quiescent or laminar),
characterized by a uniform velocity field, to a disordered one (both in space
and time), called *turbulent*. In the intermediate, weakly turbulent, regime
the fluctuations do not extend over a wide range of frequencies and length
scales. Fully *developed turbulence*, instead, involves motion with conspicuous
energy transfer down to tiny length scales, so that the fluctuations acquire a
fundamental relevance. Hydrodynamic "modes" in a wide range of wavelengths
contribute, with substantial mutual interactions, to the formation of coherent
vortices in the midst of a disordered background (Monin & Yaglom, 1971 and
1975). The interplay between order and disorder is especially intricate because
of the lack of a clear scale separation. The onset of coherent macroscopic
structures heavily depends on the dynamics near the boundaries of the fluid,
where small length-scale processes pump energy into the large scales. Energy is
then passed back into the small length scales in the *interior* of the fluid, until it is
thermally dissipated by the viscous drag. The resulting *energy cascade*, with the
generation of vortices of many different sizes, exhibits "universal" properties,
independent of the specific fluid.

The degree of irregularity of the motion depends on the relative weight of
the (nonlinear) advection term $\mathbf{v} \cdot \nabla \mathbf{v}$ compared with the dissipation term $\nu \nabla^2 \mathbf{v}$
(see Eq. (2.1)). The higher the velocity, the less effective the viscous drag is in
driving the system towards a quiescent state. The ratio of the two terms yields
the *Reynolds number* $R(l)$, which is usually expressed as $R(l) = v(l)l/\nu$, where
$v(l)$ is the characteristic velocity difference observed across the length-scale l.

Although distinct scales cannot be precisely discriminated, the fluid motion
can be seen as the simultaneous excitation of many vorticous structures (eddies)
having a minimum size $l_d = (\nu^3/\bar{\varepsilon})^{1/4}$, called *Kolmogorov length*, where $\bar{\varepsilon}$ is the
mean energy-dissipation rate. At $l = l_d$ the Reynolds number is so small that
turbulence is suppressed. A typical turbulent pattern can be observed in Fig. 2.3
which represents a water jet emerging from a nozzle with Reynolds number
$R \approx 4000$ (Sreenivasan, 1991). The number N_d of active degrees of freedom
can be roughly identified with that of the smallest eddies present in the fluid,
which is estimated to be of the order of $R^{9/4}$ per unit volume. Since R easily
exceeds the value 10^{18} in the atmosphere, N_d can be such an enormous number
that extremely unpredictable, wild behaviour is the rule. Despite this apparent
intractability, the turbulent flow can be recognized as involving a *hierarchy* of
vortices, the largest being more predictable than the smaller ones. In fact,
eddies of size l become unstable and decay into sub-eddies of size rl which,
in turn, give rise to sub-eddies of size $rr'l$, and so on where $0 < r, r' \ldots < 1$.
This process repeats itself in statistically *self-similar* manner until the dissipation

Figure 2.3
Two-dimensional image
of a water jet obtained by
laser-induced fluorescence
(from Sreenivasan, 1991).
The jet was injected into
a water tank from a
well-contoured nozzle
at a Reynolds number
$R \approx 4000$.

takes over: a *homogeneous self-similar object* displays the same structure at all powers of some fundamental spatial (or temporal) scale factor r (Mandelbrot, 1982). A simple, although partial, explanation of this scenario is provided by the invariance of the zero-viscosity Navier–Stokes equations (2.1) under the simultaneous rescaling of distance and time by the rates r and $r^{1-\alpha}$, respectively, where r is arbitrary and the exponent α is to be fixed with the help of physical considerations. Assuming a constant flux of energy ($\propto v^3(l)/l$) from large to small scales (energy balance), the exponent α is found to be $1/3$ (Kolmogorov, 1941).

The scale dependence of the velocity field is usually studied through the *structure function*

$$\left\langle |\mathbf{v}(\mathbf{x} + \mathbf{l}) - \mathbf{v}(\mathbf{x})|^m \right\rangle \propto |\mathbf{l}|^{\zeta_m}, \tag{2.2}$$

where a time average (denoted by the angular brackets) has been taken to eliminate irrelevant fluctuations. The scale-invariance conjecture implies $\zeta_m = m/3$, a relation which is approximately verified in the so-called *inertial range* $l_d \ll l \ll l_s$ (where l_s is the size of the system). Noticeable deviations of ζ_m from the predicted linear behaviour have been, however, detected also in that range (Anselmet *et al.*, 1984). These results have led to the formulation of several phenomenological models in which the scaling factor α is explicitly position-dependent (Meneveau & Sreenivasan, 1991). The above mentioned invariance hypothesis is thereby assumed to hold only in a statistical sense. More precisely, the velocity field is considered as the result of a *random multiplicative process*: velocity differences at the scale r^i are of the order of $\prod_{j=1}^{i} r^{\alpha_j}$, where α_j is a random variable (see Chapter 5) equal, "on average", to $1/3$ (Kolmogorov, 1962; Novikov, 1971; Mandelbrot, 1974; Frisch *et al.*, 1978; Benzi *et al.*, 1984). It must be stressed, however, that no definite conclusion has yet been reached about the scaling behaviour of the structure function from experiments (Frisch & Orszag, 1990) since a reliable estimate of ζ_m requires an enormous amount of independent data. Theoretical and numerical investigations (Constantin & Procaccia, 1991), on the other hand, indicate that isothermal or isoconcentration surfaces of a turbulent fluid develop a *fractal* shape (see Chapter 5) above some Reynolds-number dependent scale l^*. For the boundary of the jet displayed in Fig. 2.3, $l^* \approx 10 l_d$.

Naively, turbulence can be considered complex because of the high dimensionality of the dynamics. A more careful examination of the problem, however, shows that complexity is to be ascribed to the existence of manifest structures at all scales. The basic question is whether "a closed representation exists that is simple enough to be tractable and insightful, but powerful enough to be faithful to the essential dynamics" (Kraichnan & Chen, 1989).

2.3 Biological and chemical reactions

Impressive examples of formation, propagation and interaction of spatially coherent structures are provided by biological and chemical reactions taking place in liquid solutions (Ross *et al.*, 1988). The patterns emerging in these systems consist of variations in the concentration of chemical or biological species which can be visualized by means of spectroscopic and microphotographic techniques. After a suitable data-processing, they appear as *fronts* or *waves* travelling in space and evolving in time.

One of the most studied cases is the *Belousov–Zhabotinsky* reaction, the oxidation of malonic acid (an organic substrate) by acidic bromate catalyzed by cerium or ferroin, which can be schematically represented by

$$2BrO_3 + 3CH_2(COOH)_2 + 2H^+ \rightarrow 2BrCH(COOH)_2 + 3CO_2 + 4H_2O .$$

Various types of waves may occur, depending on the experimental conditions, with geometry varying from plain or circular shapes to spirals, multiarmed vortices, and scrolls (see Fig. 2.4).

The study (mostly concentrated on two-dimensional wave patterns) has focused on dispersion relations, time evolution (velocity, amplitude, overall shape) of the waves, and interactions (e.g., repulsion or annihilation) among them. The concentrations may vary periodically in time, by several orders of magnitude (about five). The cores of the spirals are sometimes stable in time (with the tips tracing circular paths). For other conditions, however, an instability occurs and the tips follow complicated trajectories, which may be quasiperiodic (Jahnke *et al.*, 1989; Skinner & Swinney, 1991) or even unpredictable and, possibly, chaotic (Tam & Swinney, 1990; Vastano *et al.*, 1990; Kreisberg *et al.*, 1991). As already seen for hydrodynamic systems, the perception of complexity is induced by the

Figure 2.4
Two-dimensional wave
pattern in the
Belousov–Zhabotinsky
reaction in a thin solution
layer (from Müller &
Hess, 1989).

puzzling coexistence of order (spirals, circles) and disorder (motion of centres, fronts).

Stationary configurations occurring in reactors with concentration gradients (called *Turing patterns*, from Turing (1952)) present analogies with coherent hydrodynamic structures (White, 1988). It must be stressed, however, that no hydrodynamic effects are involved here: they are usually suppressed by a gel (Skinner & Swinney, 1991). A pure *reaction–diffusion* mechanism rules these phenomena (although convection is sometimes used to accelerate the dynamics): the macroscopic patterns arise from the coupling between chemical reactions and molecular diffusion (transport). The usual model is a partial differential equation of the type

$$\frac{\partial \boldsymbol{\Psi}}{\partial t} = \mathbf{D} \cdot \nabla^2 \boldsymbol{\Psi} + \boldsymbol{F}(\boldsymbol{\Psi}) , \qquad (2.3)$$

where $\boldsymbol{\Psi}(\mathbf{x}, t)$ is a vector of state variables, such as concentrations of species and temperature, $\mathbf{D}$ is the diffusion matrix and $\boldsymbol{F}$ is a nonlinear function representing the effect of the reactions (Kuramoto, 1984).

There is a functional analogy between reaction–diffusion chemical processes and autocatalytic reactions in biological systems (Martiel & Goldbeter, 1987). Similar dynamics is also observed in the development of embryos, in phage-bacterium systems, in the physiology of the heart tissue, to name a few, all of these being examples of *excitable systems* (Goldbeter, 1989), capable of reaction to an external stimulus. The problem of pattern initiation, nucleation and evolution is largely unsolved. Moreover, these processes in biology are controlled by yet unknown functions within the genome of some cellular system: the typical size of the transported volume elements is of the order of a few microlitres in chemical systems, whereas it is not determined in biological systems (it might be that of a single cell or of a group of cells forming a cooperative unit).

In spite of the intricacy of these microscopic interactions, however, a common description of the macroscopic features seems to be achievable. Moreover, the similarities with hydrodynamics suggest the possibility of a universal (i.e., system-independent) understanding, although the observations are strongly influenced by the boundaries and the available amount of data is, for certain purposes, sometimes insufficient.

2.4   Optical instabilities

A wide range of complex phenomena can be observed in nonlinear optics, including various types of temporal instabilities and disordered patterns, as well as the spontaneous formation of cellular structures and vortices (Harrison & Uppal, 1993). The basic mechanism underlying such behaviour is the nonlinear interaction between two linear systems: electromagnetic waves (described by the Maxwell equations) and an atomic medium (microscopically described by the Schrödinger equation).

The prominent optical instability occurs in *lasers*: the atomic medium is confined inside a resonant cavity and supplied with a continuous flow J of energy which pumps the atoms to some excited level. If J is small enough, the energy is dissipated through incoherent emission of light, atomic collisions, and dispersion in the optical cavity. At larger J-values, a more effective process of energy removal sets in: the well known lasing action with emission of *coherent light*.

The laser instability is conceptually equivalent to the onset of the Rayleigh–Bénard convection: a time-periodic behaviour spontaneously emerges when the incoming flux of energy is strong enough compared with intrinsic losses. This analogy has led some researchers to speculate that the mechanisms underlying the onset of this instability could represent a sort of paradigm for the emergence of increasingly complex, if not even disordered, structures. Nowadays, this phenomenon has been recognized as a *Hopf bifurcation*, i.e., just one of the many qualitative changes that may occur in systems far from equilibrium.

The optical cavity plays the same role as the fluid container does in hydrodynamics. A "short" cavity is equivalent to a cell with a small aspect ratio, since only one cavity mode (one Fourier mode in the case of plane mirrors) can become excited and contribute to the dynamics substantially, all others being quickly damped. Although strongly constraining for possible spatial structures, this geometry allows for a rich time evolution as shown, for instance, by one of the simplest experimental setups: namely, the CO_2 laser with periodic modulation of the cavity losses (Arecchi *et al.*, 1982).

Lasers are not only interesting examples of nonlinear optical devices but, because of the strong fields they generate, they also serve as primary energy sources in several experimental configurations. This is of great importance,

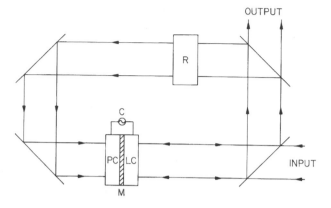

OUTPUT

INPUT

Figure 2.5 Schematic illustration of an experimental setup involving a liquid crystal (LC) light valve. The beam is rotated (R) and fed back to the valve. Symbols M, PC, and C denote a dielectric mirror, a photoconductor, and a current generator, respectively.

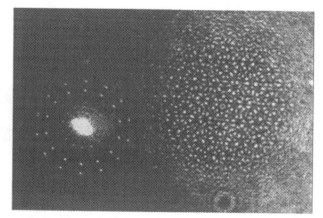

Figure 2.6 Light pattern observed in the liquid crystal valve for a rotation angle $2\pi/7$. A quasiperiodic lattice can be recognized in the right image (near-field), while the left image (far-field) corresponds to its Fourier transform which presents two sequences of 7 peaks.

e.g., for the simultaneous excitation of many modes, a necessary ingredient for the onset of spatial structures. Spontaneous formation of patterns can be observed whenever the propagation of a light beam modifies the optical properties of an atomic medium and is, at the same time, affected by the medium itself. D'Alessandro and Firth (1991) have theoretically predicted that hexagonal patterns may appear even in a simple geometry in which a mirror provides the feedback mechanism. The formation of spatial structures requires the presence of a strong nonlinearity. Akhmanov *et al.* (1988) have shown that a device consisting of a liquid-crystal light valve (LCLV) can be profitably used for this purpose. The control light beam induces a spatial modulation of the refractive index of the liquid crystal which, in turn, modulates the phase of the input light. Fascinating experiments can be carried out by feeding the input light back to the opposite side of the LCLV. Pampaloni *et al.* (1995) have added a transversal rotation of the light beam along the path (see Fig. 2.5) which breaks the original rotational symmetry and induces quasiperiodic structures of the type displayed in Fig. 2.6. On the one hand, the pattern exhibits some degree of disorder since it is not a mere periodic arrangement of identical cells; on the other, its Fourier transform presents clear peaks, which reveal the presence of long-range order. This is analogous to the three-dimensional quasicrystals (see Section 7.3) observed at the atomic scale (Shechtman *et al.*, 1984), the quasiperiodicity of which is demonstrated by peaks in their X ray diffraction spectra which cannot be produced by any ordinary lattice with allowed crystallographic symmetry. This point can be illustrated with a simple example: consider a two-dimensional pattern formed by the superposition of N waves, all with the same spatial period L but different directions $\mathbf{k}_1$, $\mathbf{k}_2$, ..., $\mathbf{k}_N$. The configuration in the vicinity of the origin $\mathbf{x} = \mathbf{0}$ is reproduced exactly around all points $\mathbf{x}$ such that the scalar products $\mathbf{k}_i \cdot \mathbf{x}$ are *all* simultaneously integer multiples of 2π, for $i = 1, 2, \ldots, N$. For this to occur, the number of unknowns (the components of $\mathbf{x}$) must be equal to the number N of constraints. Then, a periodic structure results. For $N = 2$, this is always possible (the unit cell being identified by the unit vectors $\mathbf{k}_1$ and $\mathbf{k}_2$). For $N > 2$, a solution exists only if $(N - 2)$ vectors can be expressed as rational combinations of the remaining two as, for instance, in a hexagonal pattern obtained from 3 vectors forming the angle $2\pi/3$ with one other. The condition is not fulfilled in the quasiperiodic pattern depicted in Fig. 2.6, which results from the superposition of vectors rotated by the angles $2k\pi/7$ (with $k = 0, 1, \ldots, 6$).

2.5 Growth phenomena

Patterns with quite differently shaped components emerging from a dull background are frequently associated with complexity. Typical examples of this behaviour are aggregates of elementary objects, usually spherical particles or liquid droplets, which form under the effect of spatially isotropic forces (Family & Landau, 1984). Notwithstanding the simplicity of the constituents and of the dynamical rules, such structures develop a ramified aspect with approximate symmetries which do not reflect those of the interparticle forces (Arneodo *et al.*, 1992a and b). Moreover, they do not possess a characteristic length scale, but rather display self-similar features. Examples include metal colloids, coagulated aerosols, viscous fingering (obtained, e.g., by injecting water into layers of plaster of Paris), flows through porous media, liquid-gas phase separation (spinodal decomposition), dielectric breakdown, and electrodeposition (Stanley & Ostrowsky, 1986; Jullien & Botet, 1987; Aharony & Feder, 1989; Meakin, 1991). In all these systems, the *growth* of the pattern starts with an initial nucleus (particle or droplet) to which other particles eventually stick. The accretion of the developing cluster may take place at any point of its boundary, although with different probability. The most frequently observed shapes closely resemble coral formations, trees, or thunderbolts. A typical gold colloid aggregate (Weitz & Oliveira, 1984) is shown in Fig. 2.7. The picture, a transmission electron-

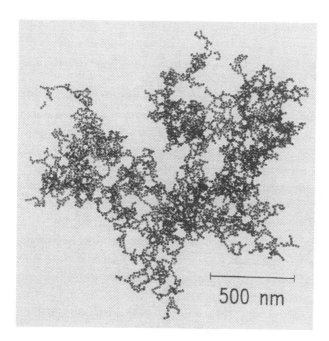

Figure 2.7 An electron micrograph of a cluster of 4739 approximately spherical gold particles in an aqueous suspension, from Weitz & Oliveira (1984).

microscope image, represents a two-dimensional projection of the real cluster. In general, the number N of the particles in the cluster scales as R^D, where R is either the end-to-end distance or the gyration radius and D is a dimension-like quantity which depends on the geometrical constraints and, in particular, on the dimension d of the space in which the dynamics takes place. The value estimated for the aggregate of Fig. 2.7 is $D \approx 1.7$.

The peculiar shape of these patterns and the absence of an absolute length scale indicate that they are fractal objects. Nevertheless, self-similarity does not strictly hold (e.g., the configuration around the origin differs from that at the tips). Hence, the exponent D alone cannot account for the multitude of shapes observed in different systems. In order to discriminate them more precisely and to identify possible universal features, the kinetics of growth phenomena must be considered as well.

Aggregation is an irreversible process in which particles perform a random walk in a fluid and stick together under the action of attractive forces. The clusters formed in this way, in turn, may join to form still larger clusters. Two characteristic times are thus involved: the diffusion time t_d, needed by the particles to come into contact, and the reaction time t_r, needed to form a bond. The description is simplified when the time scales are very different. When $t_d \gg t_r$, one speaks of *diffusion-limited* aggregation (DLA); when $t_r \gg t_d$, of *reaction-limited* aggregation (RLA). In both cases, the cluster formation at the time scale of the slow process can be described by the *Smoluchowski equation* (Friedlander, 1977)

$$\dot{c}_k = \frac{1}{2} \sum_{i+j=k} K_{ij} c_i c_j - c_k \sum_{j=1}^{\infty} K_{kj} c_j \, ,$$

where c_k is the concentration of k-particle clusters, K_{ij} is the rate of the reaction $[i] + [j] \rightarrow [i+j]$ between two clusters of sizes i and j, and the first sum runs over all pairs of clusters with total number of particles k. In the derivation, it is assumed that the probability of finding an i-particle and a j-particle cluster is the product of the single probabilities (statistical independence, see Chapter 5). Moreover, the equation is tantamount to a mean-field approximation, in that the positions of the particles are ignored. The physical properties of the system are to be accounted for by the rates K_{ij}. For a review of the best known choices, see Ernst (1986). The initial condition is usually $c_k(0) = \delta_{k1}$, which describes isolated particles (monodispersion case).

The solution, by yielding the cluster distribution at all times, permits one to establish whether gelation (the formation of an infinite cluster) occurs or not. Unfortunately, exact or even approximate results are known for a few families of rates K_{ij} only. Therefore, special models have been designed to simulate more general systems. A particularly simple one which lends itself to various extensions was proposed by Witten and Sander (1981). A particle is fixed at the centre of a circle of radius R_0 from which another particle is released from a random position. The latter particle performs a *random walk* (Brownian motion) until either it comes back to the border of the circle and is thereby discarded, or it sticks to the former particle as soon as it touches it (i.e., the reaction probability p is 1). In either case, a new particle is subsequently released from the circle and the process continues. After some time, the cluster has such an intricate ramified shape that the newly arriving particles hardly succeed in visiting the "fiords" in depth without touching their border first. In fact, accretion is most likely to occur at the tips. The dimension D of these clusters is estimated to be 1.7 in a planar experiment, 2.5 in three dimensions, and 3.3 in four (Jullien & Botet, 1987). These exponents are different if the simulation is made on a lattice and depend on the lattice type. Hence, the Witten–Sander model is not universal.

In order to compare these findings with real experiments, one is forced to allow the clusters to perform a random walk as well. Such cluster–cluster aggregation models (Kolb *et al.*, 1983; Meakin, 1983) yield $D \approx 1.75$ in three dimensions, a result which is in close agreement with the experiment of Fig. 2.7 (Weitz & Oliveira, 1984).

In the DLA model, the ensemble of all particles performing the random walk simulates a scalar field ϕ that obeys the Laplace equation

$$\nabla^2 \phi = 0 .$$ (2.4)

In fact, the probability $P(i, j)$ for a walker on a square lattice to be at position (i, j) satisfies

$$P(i, j) = \frac{1}{4} \left[P(i-1, j) + P(i+1, j) + P(i, j-1) + P(i, j+1) \right] ,$$

which is the discretized version of Eq. (2.4). Because the particles stick irreversibly on the cluster, this acts as a sink and $\phi = 0$ at its boundary. At a distance $R_0 \gg 1$, vice versa, the particles are supplied isotropically and $\phi = \phi_0$ is a constant. An important extension of the DLA model is obtained by selecting unoccupied perimeter sites according to the probability $P \propto \phi^\eta$ or $P \propto (\nabla \phi - a)^\eta$, where ϕ is the field at the site, a a constant, and η a control parameter (Meakin, 1990 and 1991). Various physical phenomena can be simulated with similar algorithms. Accordingly, a multitude of exponents, distributions, and scaling relations is observed. Complexity, therefore, cannot be accounted for by the overall fractality of the pattern only, but requires one to estimate, e.g., how much this is an agglomerate with "holes" rather than a tree-shaped object, what is its degree of self-similarity, its apparent symmetry, its average branching rate, or how it grows in time. Some of these specific answers will be addressed in Chapter 9.

2.6   DNA

Life, with its astonishing myriad of diverse expressions, constitutes the highest example of a complex architecture. Each single living organism is, in fact, very complex itself, being composed of a huge number of parts in mutual interaction. Even more striking is the capability of living beings to reproduce themselves, a feature that is unique to them, in a way that still allows for small modifications to occur from generation to generation, so that *biological evolution* arises. The overwhelming complexity of biological systems has so far baffled all scientist's attempts at finding a comprehensive mathematical model. Not even specific problems such as the evolution of a single species have received a satisfactory answer (Eigen, 1986).

In the replication process, an organism does not build a copy of itself but rather provides its progeny with *genetic material* containing the information needed to construct a new organism. This inheritance consists of DNA (*deoxyribonucleic acid*) molecules, linear polymers composed of *nucleotides* (or *bases*) of four different types: adenine (A), cytosine (C), guanine (G) and thymine (T).

They are linked together by a sugar-phosphate backbone. Adenine and guanine are *purines*, cytosine and thymine are *pyrimidines*: the two types of molecules differ by their chemical structure (Lewin, 1994). The genetic material of all known organisms and many viruses is DNA. Some viruses use another nucleic acid: the RNA (ribonucleic acid). In its molecule, thymine is replaced by uracil (U, also a pyrimidine). Furthermore, various types of RNA are involved in the process of extracting (transcribing) and transmitting the information contained in DNA for subsequent use (e.g., construction of the organism). Both DNA and RNA can be considered as messages written in a four-letter alphabet and they are equally important for the study of the complexity of biological species.

The spatial structure of DNA is generally a double helix, composed of two complementary strands: each A in one strand is paired with a T in the other and similarly for C and G. Therefore, the two halves are equivalent and it is sufficient to analyse a single sequence. The order of the bases determines the function of DNA (a specific reading direction is naturally selected as a consequence of asymmetries in the chemical bonds along the chain). The amount of DNA is generally measured in units of base pairs (*bp*) and it is referred to as the *C* parameter. A human cell contains approximately 6×10^9 bp, organized in 46 chromosomes. Only about $1/1000$ of the human DNA has been sequenced so far (Bell, 1990). The total amount of DNA in the genome gives only a qualitative indication of the complexity of the species. In fact, although bacteria normally contain less DNA than lower eukaryotes (e.g., protists and fungi) and much less DNA than higher eukaryotes (e.g., mammals and plants), there are many examples which contradict common sense: amoeba dubia, onion, and chicken contain 6×10^{11}, 2×10^{10}, and 10^9 bp, respectively! This is the so-called *C-value paradox*. A more satisfactory ordering is obtained by looking at the minimum genome size found in each phylum (an evolutionary group composed of similar species) (Lewin, 1994) which seems to grow with the apparent complexity of the corresponding organisms.

2.6.1   The genetic code

The *C*-value paradox is resolved once it is realized that not all the information stored in DNA sequences is equally relevant. It has been discovered, in fact, that only some segments of DNA, called *genes* (the inheritance units), have the capability of synthetizing *proteins*, the "building blocks" of the organism. Proteins are linear polymers formed by sequences of 20 different types of monomers: the *amino acids*. The *genetic code* assigns a single amino acid to each triplet of bases (a *codon*) in DNA. In the table, we report the 64 codons together with their occurrence frequency (see later) in a sample of 1952 human genes containing 1 794 792 bases. The symbol to the left of each codon represents the corresponding amino acid (e.g., *F* stands for phenylalanine, *L* for

Table **2.1**. *The genetic code. Of the 64 codons, 61 represent amino acids; the other three (marked with a star) are terminators. All of the amino acids except tryptophan (W) and methionine (M) correspond to more than one codon. The percentage occurrence frequency of each codon, as found in a sample of 1952 human genes, is also reported.*

codon table with percentage occurrence										
F	TTT	1.5	S	TCT	1.4	Y	TAT	1.2	C	TGT	1.0
F	TTC	2.1	S	TCC	1.7	Y	TAC	1.6	C	TGC	1.4
L	TTA	0.6	S	TCA	1.1	*	TAA	0.1	*	TGA	0.3
L	TTG	1.1	S	TCG	0.4	*	TAG	0.1	W	TGG	1.5
L	CTT	1.1	P	CCT	1.8	H	CAT	0.9	R	CGT	0.5
L	CTC	1.9	P	CCC	2.1	H	CAC	1.4	R	CGC	1.1
L	CTA	0.6	P	CCA	1.7	Q	CAA	1.2	R	CGA	0.6
L	CTG	4.0	P	CCG	0.7	Q	CAG	3.3	R	CGG	1.0
I	ATT	1.5	T	ACT	1.3	N	AAT	1.6	S	AGT	1.0
I	ATC	2.2	T	ACC	2.2	N	AAC	2.1	S	AGC	1.9
I	ATA	0.6	T	ACA	1.5	K	AAA	2.2	R	AGA	1.2
M	ATG	2.2	T	ACG	0.7	K	AAG	3.4	R	ACG	1.2
V	GTT	1.0	A	GCT	2.0	D	GAT	2.1	G	GGT	1.4
V	GTC	1.5	A	GCC	2.8	D	GAC	2.7	G	GGC	2.5
V	GTA	0.6	A	GCA	1.6	E	GAA	2.8	G	GGA	1.9
V	GTG	2.9	A	GCG	0.7	E	GAG	3.9	G	GGG	1.7

leucine and so on). The three codons labelled with a star are terminators: they mark the end of protein synthesis along the strand. Notice the redundancy of the code: a single amino acid may correspond to a few codons. This feature minimizes the effects of *mutations*, which are changes in the structure of the strand caused by the environment. Mutations can be of many types: substitution, deletion, insertion, duplication, and transposition of bases, the former being the most frequent one (Fitch, 1986). According to a generally accepted theory of molecular evolution (Kimura, 1983), the majority of mutations are neutral. Some of them are deleterious in that they reduce the *fitness*, i.e., the capability of the individual to perpetuate the species, and may lead to the disappearance of its progeny: they are removed from the population by natural selection. Deadly mutations, instead, are quite rare. Neutral and beneficial mutations have contributed to the diversification among the species through a slow accumulation of changes. The reconstruction of the evolutionary history (*phylogeny*), by means of a descendance tree from a common ancestor DNA sequence, is one of the fundamental problems in biology. This requires the comparison of similar sequences from different species and classification of the mutations occurred in the course of evolution (Tavaré, 1986).

Biological information is processed in two steps: *transcription*, with the generation of a single-stranded RNA (*messenger* RNA or mRNA) identical with one of the two DNA strands (except for the T→U substitution), and *translation*, in which the RNA sequence is turned into a sequence of amino acids comprising a protein. The mRNA template is scanned sequentially and the codons are "read" as nonoverlapping words (Lewin, 1994). In principle, there are three possible reading frames, starting at any of the first three positions of the strand. A statistical analysis shows that the first two frames usually yield meaningful (i.e., functional) amino acidic sequences, whereas the third one does not (Staden, 1990). The simplest quantity to evaluate is the occurrence frequency $f(c)$ of each codon c (see the table): it is defined as the ratio $n(c)/n$ between the number $n(c)$ of times the codon c is observed during a nonoverlapping scan of the sequence and the total number n of codons in the sequence. Frequencies obviously depend on the frame. A strong nonuniformity in the values suggests that the frame is indeed encoding a protein, whereas little variation (and consequent high percentage of terminators) implies disorder. The occurrence frequency of the bases A, C, G, and T in each of the three positions in a codon is also a useful indicator (Staden, 1990). It is easily seen that the third base is often irrelevant for the determination of the amino acid (i.e., of the biological function). Its choice, however, is not entirely neutral, since it influences the transcription process in the local biochemical context (*codon usage*). The first and second bases, instead, contribute nearly to the same extent: for example, an initial C leaves five possibilities open, whereas a C in the second position is compatible with four different amino acids.

2.6.2   Structure and function

The researcher is not only required to understand the *structure* of the message, i.e., the rules governing the appearance of the "words", but also its "meaning", that is, the *function* that each part of the DNA chain is able to perform when interacting with the environment. Hence, the study of the sequences must proceed in concurrence with a biochemical analysis (Doolittle, 1990; Miura, 1986). One of the difficulties in the study of DNA is that the structure and function of sequences, although related, are not completely equivalent (different structures may perform the same function). Hence, the sole analysis of sequences as messages conveying information may be insufficient. Further evidence of this is provided by DNA segments without any known function: indeed, the genes constitute only a small fraction of DNA. Even more surprising was the discovery that the genes themselves are interrupted by noncoding regions, the *introns*. In the process of formation of mRNA, they are removed by a complicated splicing mechanism, so that mRNA is colinear with the translated proteins (Lewin, 1994). The coding sections in DNA, called *exons*, are joined together in a continuous stretch in mRNA. The number of introns grows along the evolutionary tree.

Introns are absent in prokaryotes (unicellular organisms that lack a distinct membrane-bound nucleus) whereas, in eukaryotic genes, they are about 90% of the total DNA (although there are uninterrupted genes in eukaryotes as well). The fraction of noncoding DNA in the human genome is estimated to be about 97%.

The initiation of the transcription by RNA has been found to be marked by a sequence, the *promoter*, located "upstream" of the gene (with respect to the reading direction) in a region of length up to about hundred bases. A common feature of all promoters is the occurrence of TATA and CAAT "boxes" about 10 and 80 bases before the first base of the coding region (Lewin, 1994). The TATA box is recognized by the RNA-polymerase, a protein which starts the transcription from a position whose dependence on the context is not yet fully understood. A further hindrance to the recognition of a promoter is the infrequent observation of the entire TATA sequence: indeed, base T is just the most frequently observed one at the first and third of the four positions in the box, and base A at the second and fourth. The mechanism by which the RNA-polymerase identifies (quite rapidly, indeed) a promoter in a genome with a length of over 10^6 bases is unknown. The process appears to involve a few types of reaction in combination with diffusion along the chain, until tight binding occurs at the right sites.

The transcription ceases when a *terminator* is encountered, either immediately after the coding region or within it. A terminator may present itself as a palindrome sequence[1]. The presence of such a sequence favours a local folding of the single strand of DNA, since it leads to a perfect base pairing. The folding of DNA represents, in turn, an obstacle to the RNA-polymerase which thus stops the transcription. A last important process in the transcription of DNA is the recognition of introns. Although their extremities are often signalled by the pairs GU and AG, this is not sufficient in general. Other "marks" have been found well inside the introns.

It is thus clear that recogniton methods based on properties of the sequences would bring a major advancement in the research, making an automatic search possible and providing, at the same time, a clue to the understanding of DNA. Since coding regions carry information necessary to perpetuate life, they are expected to be less random than noncoding ones and presumably more "complex". A quantitative analysis is, however, hindered by many factors, the salient one being the nonuniformity of the genome: experiments indicate that different regions are involved in different biological processes. This produces *nonstationary* patterns (see Chapter 5) and may prevent the application of a standard statistical analysis. While only a little fraction of DNA is actually translated, the rest is by no means all "junk": besides promoters and terminators,

---

1. A DNA sequence is palindrome if it is invariant simultaneously under the inversion of the reading direction and the exchange $(A \leftrightarrow T, C \leftrightarrow G)$ as, e.g., GACCTAGGTC.

there are target sites, which can be bound to proteins, and repeated sequences, which may act as regulators of biochemical processes (Zuckerkandl, 1992).

Finally, the hierarchical nature of biological organization must be pointed out. In the assembly of each new organism, low-order complexes (amino acids) form higher-order ones (proteins) which, in turn, interact to give rise to even more diversified aggregates. The rather low number of stable molecules contributes to the acceleration of the evolutionary process: most of the concatenations of bases are discarded by selection, which hence acts as a feedback of information from the environment (Simon, 1962). As a consequence of these mechanisms, one is led to associate DNA with texts written in natural languages, with a hierarchy of complexes: words, sentences, paragraphs, and chapters. In spite of several attempts at the construction of a formal theory in specific cases (e.g., description of functional signals, prediction of splicing sites), a general theory is, however, yet to come (Gelfand, 1993).

Further reading

Arecchi (1994), Batchelor (1982 and 1991), Cross and Hohenberg (1993), Falconer (1989), Field and Burger (1985), Müller and Plesser (1992), Stryer (1988), Tennekes and Lumley (1990).

Chapter 3

Mathematical models

Most of the physical processes illustrated in the previous chapter are conveniently described by a set of partial differential equations (PDEs) for a vector field $\Psi(\mathbf{x}, t)$ which represents the state of the system in phase space X. The coordinates of Ψ are the values of observables measured at position $\mathbf{x}$ and time t: the corresponding field theory involves an infinity of degrees of freedom and is, in general, nonlinear.

A fundamental distinction must be made between *conservative* and *dissipative* systems: in the former, volumes in phase space are left invariant by the flow; in the latter, they contract to lower dimensional sets, thus suggesting that fewer variables may be sufficient to describe the asymptotic dynamics. Although this is often the case, it is by no means true that a dissipative model can be reduced to a conservative one acting in a lower-dimensional space, since the asymptotic trajectories may wander in the whole phase space without filling it (see, e.g., the definition of a fractal measure in Chapter 5).

A system is conceptually simple if its evolution can be reduced to the superposition of independent oscillations. This is the *integrable* case, in which a suitable nonlinear coordinate change permits expression of the equations of motion as a system of oscillators each having its own frequency. The propagation of solitons without dispersion falls in this class: the peculiarity of this phenomenon must be entirely attributed to the coordinate change in an infinite-dimensional phase space.

Sometimes, nonintegrable conservative systems can also be described in a compact way. This occurs whenever statistical mechanical methods can be applied. The success of that approach relies on a direct estimate of the probability of individual states from the equations of motion (Hamiltonian function) without explicit integration (for a more accurate discussion see Section 3.2, 3.5, and Chapter 5).

Of course, a comprehensive study of the immense variety of dynamical structures that nonlinear systems may generate (solitons, vortices, spirals, defects, interfaces, etc.) lies beyond the present capabilities of research. Indeed, only specialized, *ad hoc* methods are available (Cross & Hohenberg, 1993), often without mutual relationships. Whether the construction of a general theory is effectively achievable or not, it must be remarked that "interesting" behaviour does not necessarily follow from the infinite dimensionality of phase space. For this reason, in Section 3.1, we briefly review some of the methods for reducing a PDE to a finite set of ordinary differential equations (ODEs), without altering the essential features of the phenomenology. In the following section, we recall basic notions of the theory of ordinary differential equations. A deeper study of nonlinear systems is undertaken in Section 3.3 for the equivalent class of discrete-time transformations (maps). The original spatio-temporal nature of the problem is then reexamined by considering cellular automata, in which the main simplification is given by the discreteness of the observable Ψ (Section 3.4). Finally, classical thermodynamic spin systems are discussed in Section 3.5 to illustrate order–disorder transitions in a stochastic context.

3.1   Reduction methods for partial differential equations

Notwithstanding the broad range of possible dynamical behaviours in PDEs, universal features are observed at certain transition points. Simplified models, accounting for deviations from a spatially homogeneous, time invariant structure, turn out to have a form which depends only on the symmetries of the problem. This is the case of the Ginzburg–Landau equation (Newell & Moloney, 1992), which describes the evolution of a spatially extended system in the vicinity of a Hopf bifurcation: namely, when the field Ψ acquires an oscillating component. Accordingly, Ψ is written as $\Psi(\mathbf{x}, t) \approx \mathbf{A}(\mathbf{X}, T)e^{i\mathbf{k}_c \cdot \mathbf{x} - \omega t} + c.c. + \Psi_0$, where $\mathbf{X}$ and T are suitably rescaled space and time coordinates, $\mathbf{k}_c$ and ω are, respectively, the wave vector and the frequency of the (destabilized) uniform pattern, $c.c.$ denotes the complex conjugate, and Ψ_0 is the homogeneous solution. The amplitude $\mathbf{A}(\mathbf{X}, T)$ is a slowly-varying envelope containing the relevant information on the motion about the "carrier" $e^{i\mathbf{k}_c \cdot \mathbf{x} - \omega t}$. The complex Ginzburg–Landau equation (CGL) then reads

$$\frac{\partial A}{\partial t} = A + (1 + i\alpha)\nabla^2 A - (1 + i\beta)A|A|^2 , \qquad (3.1)$$

where the real numbers α an β are the only free parameters. This equation, considered as the prototype of spatio-temporal chaos, applies to various experimental conditions and can therefore be seen as a "normal form" for this type of bifurcation. In fact, it holds for systems having more than one field, any order of spatial derivatives, and any type of nonlinearity (all compatible with a Hopf bifurcation).

An effective reduction of a PDE to a set of ODEs is usually achievable in a confined geometry. The commonest method, named after Galerkin, consists of the expansion of the fields over suitable basis functions and of a subsequent truncation at some order (Fletcher, 1984). A classical example of this procedure is the derivation of the Lorenz model (Lorenz, 1963) from the equations describing the Rayleigh–Bénard convection. A crude approximation, which retains the least number of terms necessary to keep a coupling between temperature and velocity (see Appendix 1), yields

$$
\begin{aligned}
\dot{x} &= \sigma(y - x) \\
\dot{y} &= -y + rx - xz \\
\dot{z} &= -bz + xy \,,
\end{aligned}
\tag{3.2}
$$

where the dot indicates the time derivative and the three variables are proportional to velocity and temperature Fourier modes. System (3.2) is considered as a milestone in the theory of dynamical systems. It is the first example of a deterministic flow which was found to exhibit aperiodic (chaotic) behaviour. Interestingly, this model is much closer to physical reality in optics, where it turns out to describe the evolution of a wide class of lasers. As shown by Haken (1975), the Maxwell–Bloch equations for a single-mode laser can be transformed into the Lorenz equations with a change of coordinates. Even more surprisingly, an experiment performed by Weiss and Brock (1986) reveals that the dynamics of far-infrared lasers displays many of the characteristic features of the Lorenz model. Conversely, in hydrodynamics, more realistic models must be devised, with the inclusion of a larger number of modes. Even in such a case, however, satisfactory convergence is observed well before the shortest wavelength of the model becomes comparable with the Kolmogorov length l_d.

In general, it is plausible to assume that the asymptotic dynamics takes place on a manifold $\mathcal{M}$ of relatively low dimension D. When this holds, any choice of D coordinates that parametrize $\mathcal{M}$ provides an exact description of the motion. If, however, the dynamics is projected onto an arbitrary set of basis functions (e.g., the Fourier modes), one has to retain a number N_b of terms determined by the relevant components of $\mathcal{M}$ in the expansion. It is clear that inappropriate bases may lead to unnecessarily large N_b. A considerable reduction of the number of variables can still be obtained with a linear method, called *singular value decomposition* or SVD (Lumley, 1970), which tailors the choice of the base to the dynamics of interest. The directions are defined recursively: the first is the one along which the projection of the asymptotic trajectory has maximal extension. The successive directions enjoy the same property in the residual subspaces obtained by eliminating the previously resolved directions. This procedure is automatically carried out with a simple evaluation of an autocorrelation matrix.

The substantial dimension reduction achievable by this method is indicated by simulations of the Rayleigh–Bénard instability. The 70 principal components are sufficient to reproduce the energy spectrum to 90% accuracy in a case in which a direct simulation requires including up to 10^4 variables (Sirovich, 1989). A further improvement has been obtained for a weakly turbulent regime by treating separately the contributions of laminar and "stochastic" domains: the definition of two different bases represents a step towards a fully nonlinear treatment of the problem. The implementation of the SVD method, however, requires the prior integration of the complete model, so that its practical utility is questionable.

An extension that is amenable to analytical treatment is based on the identification of a set of "independent" variables such that all others (the "slaved" variables) can be expressed through them. The resulting nonlinear expression defines the *inertial manifold* $\mathcal{M}$ (Temam, 1988) and constitutes an essential improvement over the plain Galerkin method, since the slaved variables are not neglected altogether, but adiabatically eliminated.

In this context, the concept of complexity may be associated not only with the unpredictability of the trajectories (present also in low-dimensional systems), but also with the minimum dimension D that an accurate model must have in order to reproduce the dynamics. In general, however, no systematic procedure is available for the identification of the proper variables.

The above described methods are obviously no more applicable or even meaningful when the dimension D is very large (even for numerical simulations). One can then devise simplified models which retain the main characteristics of the system, at least in a statistical sense. We illustrate this point with the Navier–Stokes (NS) equations (2.1) in the turbulent regime. As already commented in the previous chapter, fluid turbulence increases with the Reynolds number. Accordingly, the effects of viscosity become negligible over an increasing range of spatial scales and this regime can be approximated by the inviscid NS equations (called Euler equations)

$$\dot{y}_i + \sum_{jk} A_{ijk} y_j y_k = 0 , \qquad (3.3)$$

where the variables y_i denote the Fourier components of the velocity field, after a suitable reordering (Kraichnan & Chen, 1989). The parameters A_{ijk}, arising from advection and pressure terms, are identically zero if any two indices coincide, and satisfy $A_{ijk} + A_{jki} + A_{kij} = 0$. This condition expresses the conservation of the energy in all triadic interactions. As a consequence of these relations, both the kinetic energy $K(t) = \sum_i y_i^2(t)/2$ and volumes in phase space are conserved. In fact, the divergence $\sum_i \partial \dot{y}_i / \partial y_i$ is identically zero (see next section). It is interesting to notice that such conservation laws survive a Galerkin truncation, in which all modes that lie above a cutoff wavenumber k_c are discarded. One can therefore construct a formal statistical mechanics and invoke the equipartition principle: namely, the uniform distribution of energy over all degrees of freedom. This assumption, however, implies the concentration of the energy in the high wavenumbers, as confirmed by numerical experiments, and eventually (for $k_c \to \infty$) leads to an ultraviolet catastrophe. Hence, there is a profound difference between the equilibrium dynamics of the truncated Euler equations and the actual NS turbulence. In the latter, the energy is distributed over relatively low wavenumbers below the dissipation threshold: its "spectrum" decays as $k^{-5/3}$. More important, the transfer of energy from large to small scales appears to approach a constant rate independently of the viscosity v, for $v \to 0$.

This example shows that viscosity, no matter how small, cannot be neglected altogether. Its effect is retained in an extreme simplification of the NS equations, the *cascade model* (Gledzer, 1973), obtained through a nonuniform coarse-graining of the wave-vector space. All Fourier modes with spatial frequencies $|\mathbf{k}|$ in the "shell" $[k_{n-1}, k_n]$, with $k_n = 2^n$, are represented by a single complex variable u_n. By neglecting interactions between distant shells, while preserving the energy balance, the dynamical equations reduce to

$$\dot{u}_n = -v k_n^2 u_n + i(2k_n u_{n+1}^* u_{n+2}^* - k_{n-1} u_{n-1}^* u_{n+1}^* - k_{n-2} u_{n-1}^* u_{n-2}^*) + q\delta_{n,n_0},$$

for all $n \geq 1$. For $v = 0$, both energy $E = \sum |u_n|^2$ and phase-space volumes are conserved as in the case of the Galerkin truncation of the Euler equations (3.3). Obviously, recourse to the equipartition principle leads to the same inconsistency as for the Euler equations. In spite of its simplicity, this model preserves certain features of the NS equations as, for example, Kolmogorov's scaling law, which is reproduced by the fixed-point solution $u_n = k_n^{-1/3}$ for $v = 0$.

Numerical analyses (Jensen *et al.*, 1991) exhibit deviations of ζ_m (see Eq. (2.2)) from linear behaviour, in fair agreement with the experimental results, when the small-viscosity limit is taken. This finding is, however, in conflict with the results of simulations performed with a more realistic model, recently introduced by Eggers and Großmann (1991). At variance with the cascade model, they retain several wavevectors in each shell, so that the three-dimensionality of the dynamics is maintained and the advection process is reproduced more accurately. Since the numerical integration of these equations at high Reynolds numbers ($\sim 10^7$) does not show any evident deviation from Kolmogorov's scaling (Großmann & Lohse, 1994), the presence of fluctuations in the moment scaling exponent ζ_m must still be considered controversial.

A relevant question about the energy flow in Fourier space concerns the universality of the underlying mechanism and, in turn, its observability in areas other than fluid dynamics. A positive answer comes from the propagation of light in nonlinear dielectrics which, in a first approximation, is described by the nonlinear Schrödinger equation, a conservative model obtained from the CGL equation in the limit α, $\beta \to \infty$. In one spatial dimension the equation is integrable and, depending on the sign of the nonlinearity, gives rise to solitons. In two and more dimensions, the equation is no longer integrable, it possesses three constants of motion, and develops discontinuities in a finite time. For large but finite α and β, the energy flow towards small scales is eventually blocked by the dissipative forces. Thus, one can say that CGL and nonlinear Schrödinger equations play a similar role to Navier–Stokes and Euler equations, respectively, in fluid dynamics. In fact, approaches similar to those developed for high-Reynolds-number turbulence have been introduced to describe what may be called optical turbulence (Dyachenko *et al.*, 1992). The scenario is also confirmed by analytical estimates of the structure function for the field A of Eq. (3.1) (Bartuccelli *et al.*, 1990).

3.2   Ordinary differential equations

Apart from their mathematical legitimacy, the above discussed approximations are physically justified by the agreement between the time evolution of many real systems and that obtained from systems of ordinary differential equations of the form

$$\dot{\mathbf{x}} = \mathbf{f}(\mathbf{x}) \quad , \quad \mathbf{x} \in \mathbb{R}^d . \tag{3.4}$$

The symbol $\mathbf{x}$ replaces now the observable Ψ and represents the state of the system in the *phase space* $\mathbb{R}^d$. Usually, $\mathbf{x}$ denotes the spatially averaged values of a set $\mathbf{Q} = \{Q_1, \ldots, Q_d\}$ of observables. The "force" $\mathbf{f}$, describing the reduced dynamics, typically depends on various control parameters $\mathbf{a} = (a_1, a_2, \ldots)$ (e.g., Reynolds number, viscosity, dissipation constants).

The volume $V(t)$ of a domain $D(t) \subset \mathbb{R}^d$ evolves, under the dynamical law (3.4), according to

$$\dot{V}(t) = \int_{D(t)} \nabla \cdot \mathbf{f}(\mathbf{x}) \, d^d x . \tag{3.5}$$

The system is said to be *conservative* if the divergence

$$\nabla \cdot \mathbf{f} = \sum_i \frac{\partial \dot{x}_i}{\partial x_i} \tag{3.6}$$

of the flow is everywhere zero. An exponential decrease of phase-space volumes at a rate $\gamma = -\nabla \cdot \mathbf{f} > 0$ implies the *dissipativity* of the system. When γ is a function of the position $\mathbf{x}$ in phase space, these criteria must be referred to an average along the trajectory. In this case, there are "mixed" systems with either conservative or dissipative behaviour, depending on the initial condition (Politi *et al.*, 1986; Roberts & Quispel, 1992).

In the construction of the model, the motion is usually required to unfold within a compact subset X of phase space $\mathbb{R}^d$. The (smooth) *vector field* $\mathbf{f}$ generates a *flow* $\boldsymbol{\Phi}_t(\mathbf{x}) : X \to X$, which represents the image $\mathbf{x}(t)$ of the initial condition $\mathbf{x} = \mathbf{x}(0) \in X$ at time t. A set X is *invariant* for a flow $\boldsymbol{\Phi}_t$ if $\boldsymbol{\Phi}_t(\mathbf{x}) \in X$, for any $\mathbf{x} \in X$ and for all $t \in \mathbb{R}$. The pair $(\mathbf{f}, X)$ constitutes a *dynamical system*. Whenever $\mathbf{f}$ does not depend on time explicitly, the system is called *autonomous*.

In the linear case (i.e. when $\mathbf{f}(\mathbf{x}) = \mathbf{A} \cdot \mathbf{x}$, with $\mathbf{A}$ a constant matrix), the dynamics can be globally decomposed into the superposition of independent motions in $\mathbf{f}$-invariant subspaces (spanned by the eigenvectors of $\mathbf{A}$). Bounded solutions are characterized by a finite number of (incommensurate) frequencies. For a nonlinear field $\mathbf{f}$, the invariant subspaces are curved manifolds. The *local stable manifold* $W_\lambda^s(\mathbf{x}; \varepsilon)$ at a point $\mathbf{x} \in X$ is defined as

$$W_\lambda^s(\mathbf{x}; \varepsilon) = \left\{ \mathbf{y} \in X : \| \boldsymbol{\Phi}_t(\mathbf{y}) - \boldsymbol{\Phi}_t(\mathbf{x}) \| \leq \varepsilon e^{\lambda t} \text{ for all } t \geq 0 \right\}, \qquad (3.7)$$

where $\lambda < 0$ and $\varepsilon > 0$. The *local unstable manifold* $W_\lambda^u(\mathbf{x}; \varepsilon)$ is defined analogously, with $\lambda > 0$, by replacing t with $-t$. The exponent λ characterizes the motion of points close to $\mathbf{x}$. A negative (positive) value implies exponential attraction (repulsion). A fixed point $\mathbf{x}^*$ of (3.4) is called a *saddle* if it has mutually transversal stable and unstable manifolds with λ's strictly different from zero. When $\lambda = 0$, definition (3.7) and its counterpart for W^u can be broadened to include algebraic attraction (i.e., $\| \boldsymbol{\Phi}_t(\mathbf{y}) - \boldsymbol{\Phi}_t(\mathbf{x}) \| \leq \varepsilon t^{-\alpha}$, with $\alpha > 0$) or even weaker laws.

The direction of the flow itself is *marginal*: in fact, all pairs $(\mathbf{x}, \mathbf{y})$, with $\mathbf{y} = \boldsymbol{\Phi}_{t_0}(\mathbf{x})$ for some t_0 and $|\mathbf{x} - \mathbf{y}|$ sufficiently small, satisfy condition (3.7) (and the one for W^u) with λ identically zero ($\mathbf{x}$ and $\mathbf{y}$ belong to the same "branch" of the orbit). Each local manifold can be globally extended by taking the union of all its preimages (for W^s) or images (for W^u) under $\boldsymbol{\Phi}$. Although locally smooth, the global manifolds may bend in a very complicated manner.

Sometimes it is possible to construct a global coordinate transformation which "straightens" the manifolds and, correspondingly, linearizes the dynamics. In this case, $\mathbf{x}(t)$ is characterized by a finite number of frequencies, together with their harmonics which arise from the change of coordinates. When the nonlinearity is sufficiently strong, however, the decomposition of the phase space in independent subspaces cannot be pursued to such an extent that periodic oscillations are eventually isolated. Indeed, the manifolds are so sharply bent as to intersect each other, making the linearization intrinsically impossible. The

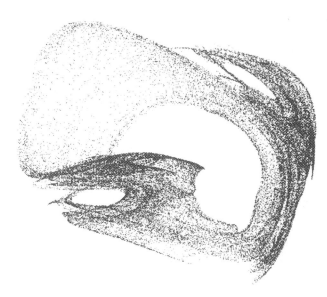

Figure 3.1 Intersection points of a chaotic orbit with a two-dimensional plane, computed from a model of a nuclear-magnetic-resonance laser with delayed feedback (Simonet *et al.*, 1995). Notice the higher-dimensional topology of the set, clearly visible also in this two-dimensional projection.

resulting aperiodic behaviour has been recognized to be ubiquitous in nature and is termed *deterministic chaos*. The name is further justified by the exponential sensitivity of the trajectories to small perturbations of the initial conditions, which renders long-term predictions virtually impossible. The intricacy of the orbits typical of chaotic motion (see Fig. 3.1) is often associated with the concept of complexity. As will become clear later, however, the stretching and folding of phase space follow rules which, in many cases, may be described with a simpler model than the differential equation (3.4).

3.3   Mappings

The analysis of a flow $\boldsymbol{\Phi}_t$ in $\mathbb{R}^d$ is greatly simplified by a discretization of time, which can be introduced without losing relevant information about the structure of the motion. This is obtained by choosing a $(d-1)$-dimensional (*Poincaré*) surface Ξ in $\mathbb{R}^d$ such that $\boldsymbol{\Phi}_t(\mathbf{x}) \in \Xi$ for some $t > 0$ and for any initial condition $\mathbf{x}$ in the invariant set X: that is, every trajectory in X must intersect Ξ. It is further required that, for any $\mathbf{x} \in \Xi$, $\mathbf{f}(\mathbf{x})$ is not tangent to Ξ. Whenever the motion takes place in a compact set X, as assumed, it is possible to find a Poincaré surface. For example, since any component $x_i(t)$ of $\mathbf{x}(t)$ is a bounded function, the condition that $\dot{x}_i(t)$ be zero is certainly satisfied for some t: rewriting it as $f_i(\mathbf{x}) = 0$ defines a possible surface of section Ξ.

The successive returns of the system onto the surface Ξ are indicated by $\mathbf{x}_n$, where $n \in \mathbb{Z} = \{\ldots,-2,-1,0,1,2,\ldots\}$ is the new discrete time; they can be interpreted as the outcome of a *Poincaré map*

$$\mathbf{x}_{n+1} = \mathbf{F}(\mathbf{x}_n)\,, \tag{3.8}$$

where $\mathbf{F}(\mathbf{x}_n) = \boldsymbol{\Phi}_{t_n}(\mathbf{x}_n)$ and t_n is the time elapsed between the events $\mathbf{x}_n$ and $\mathbf{x}_{n+1}$. If $\boldsymbol{\Phi}_t$ is a smooth function, $\mathbf{F}$ is a *diffeomorphism* (i.e., it is one-to-one and both $\mathbf{F}$ and its inverse $\mathbf{F}^{-1}$ are differentiable). In general, the symbol $\mathbf{F}^n$ denotes the n-th composition of $\mathbf{F}$ with itself, with $n \in \mathbb{Z}$: negative n-values represent iterates of the inverse map.

The main obvious advantage of this procedure is the reduction of the dimensionality from d to $d-1$ which, however, still preserves all relevant features of the flow. In fact, the fixed points of $\mathbf{f}$ and $\mathbf{F}$ coincide since the surface Ξ passes through them. Moreover, the periodic orbits of $\mathbf{f}$ are transformed into a finite number of points which are mapped cyclically by $\mathbf{F}$. The invariant manifolds present the same degree of intricacy. Therefore, most of the definitions given for a flow can easily be translated into the language of mappings by changing the continuous time index t to n. The study of bounded dynamics for $n \to \infty$ requires concentration on invariant sets that satisfy certain recurrence properties.

Definition: A point $\mathbf{x}$ is *nonwandering* for the diffeomorphism $\mathbf{F}$ if, for every neighborhood U of $\mathbf{x}$ and for some $n_0 > 0$, there exists a time $n > n_0$ such that $\mathbf{F}^n(U) \cap U \neq \emptyset$.

The set $\Omega \subset X$ of all nonwandering points of $\mathbf{F}$ (the *nonwandering set*) is closed. Typical examples of nonwandering sets are fixed or periodic points and the unions of these with the trajectories connecting them (when they exist). Such orbits are called *homoclinic* if they connect a point to itself and *heteroclinic* if they connect different ones. They are branches of invariant manifolds.

For certain nonlinear systems, it may happen that the nonwandering set Ω is separable into distinct domains (closed and invariant themselves) which are not dynamically connected. Any closed, invariant set $\Lambda \subset \Omega$ is *indecomposable* if, for every pair of points $\mathbf{x}$ and $\mathbf{y}$ in Λ and $\varepsilon > 0$, one can find a sequence $(\mathbf{x}_1,\ldots,\mathbf{x}_{n-1})$ such that $\|\mathbf{x}_i - \mathbf{F}(\mathbf{x}_{i-1})\| < \varepsilon$, for $1 \leq i \leq n$, where $\mathbf{x}_0 = \mathbf{x}$ and $\mathbf{x}_n = \mathbf{y}$. Moreover, Λ is *maximal* if there is no $\Lambda' \supset \Lambda$ which is itself indecomposable. Maximal, indecomposable, invariant sets are the fundamental objects of investigation in nonlinear dynamics. Intricate or "complex" dynamics, as illustrated in Fig. 3.1, is generated when the stable and unstable manifolds $W^s(\mathbf{x}^*)$ and $W^u(\mathbf{x}^*)$ of a saddle point $\mathbf{x}^*$ intersect transversally at a so-called *homoclinic point*. The presence of one homoclinic point $\mathbf{p}$ implies the existence of an infinity of them, obtainable as images $\mathbf{F}^n(\mathbf{p})$ and preimages $\mathbf{F}^{-n}(\mathbf{p})$ of $\mathbf{p}$, for $n \in \mathbb{N}$. As a consequence, a point $\mathbf{x}$ in the neighbourhood $U(\mathbf{x}^*)$ of $\mathbf{x}^*$ is initially repelled from $\mathbf{x}^*$ "along" $W^u(\mathbf{x}^*)$ and is successively reinjected into

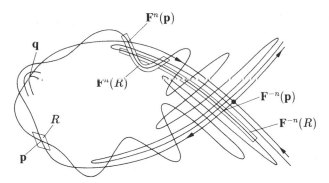

Figure 3.2 Stable and unstable manifolds of a fixed point (full circle). A homoclinic point (**p**) and a homoclinic tangency (**q**) are indicated. The n-th image and preimage of a "rectangle" R centred at **p** are shown to form a horseshoe at their intersection.

$U(\mathbf{x}^*)$ when it approaches $W^s(\mathbf{x}^*)$. This process repeats indefinitely, thus yielding erratic behaviour. Choosing $U(\mathbf{x}^*)$ as a "rectangle" R and considering the image $\mathbf{F}^{n_0}(R)$ that first intersects R itself (Fig. 3.2), one realizes that the restriction of the dynamics to R is approximated by Smale's *horseshoe map* $\mathbf{G}$, illustrated in Appendix 2. Although $\mathbf{G}$ admits only a subset of the orbits of the complete map $\mathbf{F}$, it exhibits a rather interesting dynamics. Its invariant set Λ contains

1. a countable set of periodic orbits of all periods,
2. an uncountable set of aperiodic orbits,
3. a dense orbit.

Moreover, all periodic orbits are of the saddle type, and they are dense in Λ. Their closure is a Cantor set. Finally, the restriction $\mathbf{G}|\Lambda$ of map $\mathbf{G}$ to the invariant set Λ is *structurally stable*: i.e., any small smooth perturbation of $\mathbf{G}$ does not alter the above properties (for a rigorous definition, see Guckenheimer & Holmes (1986)). The dynamics on Λ is characterized by a strong sensitivity to the initial conditions: nearby orbits separate exponentially in time. All these fundamental features constitute a definition of (hyperbolic) chaos and are found both in conservative and dissipative systems. Notice that, for a map, the divergence (3.6) is replaced by $\gamma(\mathbf{x}) = \ln|J|$, where $J = \det(\partial\mathbf{F}/\partial\mathbf{x})$ is the Jacobian determinant.

3.3.1 Strange attractors

In dissipative systems any d-dimensional set in phase space shrinks continuously to a lower-dimensional, zero-volume object Λ, the attractor, which is actually reached only after an infinitely long transient. Among the several definitions of attractor, we recall Milnor's (1985), which requires the concept of ω-limit set $\omega(\mathbf{x})$ of a point $\mathbf{x}$. A point $\mathbf{y} \in \mathbb{R}^d$ belongs to $\omega(\mathbf{x})$ if and only if, for any $n_0 > 0$ and $\varepsilon > 0$, there exists an $n > n_0$ such that $\|\mathbf{y} - \mathbf{F}^n(\mathbf{x})\| < \varepsilon$. That is, $\omega(\mathbf{x})$ is the set of all points that can be asymptotically reached from the initial condition $\mathbf{x}$.

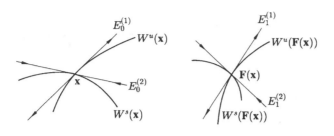

Figure 3.3 Sketch of invariant manifolds W^i at a point $\mathbf{x}$ and at its image $\mathbf{F}(\mathbf{x})$ and corresponding linear subspaces.

Definition: An indecomposable closed set Λ, invariant under a diffeomorphism $\mathbf{F}$, is an *attractor* for $\mathbf{F}$ if the *basin of attraction* $\mathscr{B}(\Lambda)$, composed of all points $\mathbf{x} \in \mathbb{R}^d$ such that $\omega(\mathbf{x}) \in \Lambda$, has positive volume (i.e., Lebesgue measure: see Appendix 3 for a definition).

The advantage of this formulation is that not all points in a neighbourhood of Λ are required to be mapped to Λ itself. This remark is not a mere mathematical oddity: in fact, there is a broad class of systems for which, no matter how close the initial condition $\mathbf{x}_0$ is to the attractor, it is impossible to foresee whether the trajectory will eventually approach the attractor or not (Sommerer & Ott, 1993b).

An attractor Λ is *chaotic* or *strange* if it exhibits exponential separation of nearby orbits (Ruelle, 1981). This is quantified by the *Lyapunov exponents* λ_i $(i = 1, \ldots, d)$, the definition of which involves the matrix $\mathbf{T}^n(\mathbf{x}) = \left(\partial F_i^n / \partial x_j \right)$ of partial derivatives of the n-th iterate $\mathbf{F}^n$ at $\mathbf{x}$ and its adjoint $\mathbf{T}^{n\dagger}(\mathbf{x})$ (Eckmann & Ruelle, 1985). Let us indicate with $E_n^{(1)} \supset E_n^{(2)} \supset \ldots \supset E_n^{(d)}$ a sequence of subspaces in the tangent space at $\mathbf{F}^n(\mathbf{x})$ and with $\lambda_1 \geq \lambda_2 \geq \ldots \geq \lambda_d$ numbers such that

1. $\mathbf{T}\left(E_n^{(i)} \right) = E_{n+1}^{(i)}$,
2. $\dim \left(E_n^{(i)} \right) = d + 1 - i$,
3. $\lim_{n \to \infty} (1/n) \ln \| \mathbf{T}^n(\mathbf{x}) \cdot \mathbf{v} \| = \lambda_i$ for all $\mathbf{v} \in E_0^{(i)} \setminus E_0^{(i+1)}$.

The numbers $\lambda_i(\mathbf{x})$ are the Lyapunov exponents of $\mathbf{F}$ at $\mathbf{x}$. In particular, if $\mathbf{x}$ is a saddle fixed point, the subspaces $E^{(i)}$ are simply the eigenspaces associated with the eigenvalues μ_i of the matrix $\mathbf{T}(\mathbf{x})$ and $\lambda_i = \ln |\mu_i|$ (see Fig. 3.3). Thus, the concepts of stability and eigenvalues can be extended, via the Lyapunov exponents, to a generic point $\mathbf{x}$ (not necessarily belonging to a periodic orbit). All vectors in the subspace $E^{(1)} \setminus E^{(2)}$ are stretched at the fastest possible rate, those in $E^{(2)} \setminus E^{(3)}$ at the next fastest one, and so forth. Lyapunov exponents have been proved to exist under very general conditions (Oseledec, 1968; Pesin, 1977).

When passing from a flow to a map, the Lyapunov exponents are all rescaled by the same constant (the average return time on the Poincaré surface). The only exception is the exponent (equal to zero) associated with the marginally stable flow direction, which is eliminated by taking the Poincaré section. A strange attractor is a union of unstable manifolds. If these intersect the local stable manifolds transversally at every point $\mathbf{x}$, the attractor is *hyperbolic* and possesses the properties (1)–(3) listed above. The most important consequence of hyperbolicity is the structural stability of the dynamical system (Guckenheimer & Holmes, 1986).

Although the presence of a horseshoe represents the commonest indication of an underlying chaotic dynamics, other mechanisms exist which can give rise to strange attractors. A famous example is the solenoid map (Smale, 1967), in which a solid torus in $\mathbb{R}^3$ is stretched (with a reduction of its cross-sectional area), twisted, and folded to fit inside itself, thus ensuring global confinement.

In the presence of a smooth folding mechanism, physical systems are not expected to be everywhere hyperbolic because the invariant manifolds form families of curves which meet tangentially at certain points called *homoclinic tangencies* (see Fig. 3.2, point $\mathbf{q}$). Because of the local coincidence between stable and unstable directions, an incipient instability may eventually turn, upon iteration of the map, into a contraction, thus giving rise to stable periodic solutions. Since finite regions of phase space are densely filled by the folds of the stable manifold, tangencies occur at "almost all"[1] parameter values. The variation of a parameter, in fact, smoothly displaces the invariant curves. The set of tangency points and, consequently, the attractor Λ are, hence, deeply altered by arbitrarily small perturbations of the dynamical equations. Therefore, nonhyperbolic systems are not structurally stable. In particular, arbitrarily close to a nonhyperbolic attractor one finds an infinity of stable periodic orbits (Newhouse, 1974, 1980). Hence, it cannot be excluded that numerically computed trajectories are just portions of long, stable periodic orbits. In spite of this disconcerting result, nonhyperbolic chaotic attractors do exist, as proved by Benedicks & Carleson (1991).

This phenomenology is suggestive of complexity not just because the motion is unpredictable, but because the asymptotic behaviour of the map itself is extremely sensitive to the parameter values. It must be remarked, however, that hyperbolic behaviour is still predominant, since the set of transverse homoclinic points has a higher (fractal) dimension than that of the homoclinic tangencies, as will be discussed in Chapter 6. Therefore, mathematical procedures developed for the analysis of hyperbolic structures can be profitably carried over to generic systems.

We introduce below a few basic examples of nonlinear maps which will be recalled throughout the book. For an exhaustive presentation of the phenomenology of low-dimensional chaos we refer the reader to the cited literature.

---

1. See Chapter 5 for a precise definition of this expression.

Examples:

[3.1]   The Bernoulli map $B(p_0, p_1)$ is defined as

$$y_{n+1} = f(y_n) = \begin{cases} y_n/p_0 & \text{if } 0 < y_n \le p_0, \\ (y_n - p_0)/p_1 & \text{if } p_0 < y_n \le 1, \end{cases} \tag{3.9}$$

with $p_0 + p_1 = 1$. The motion is chaotic for any choice of the parameters $0 < p_k < 1$ and for almost all initial conditions. An interval with length $\varepsilon \ll 1$ is mapped in n steps to an interval with length $\varepsilon_n = \varepsilon \prod_{i=1}^{n} f'(y_i) = \varepsilon e^{n\lambda_n}$, where λ_n converges to the Lyapunov exponent $\lambda = -p_0 \ln p_0 - p_1 \ln p_1$, for $n \to \infty$. A transformation consisting of N monotonically increasing linear parts which map N disjoint intervals with lengths p_1, p_2, ..., p_N *onto* $[0, 1]$ is a Bernoulli map, usually indicated with $B(p_1, \ldots, p_N)$, where $\sum_i p_i = 1$.   □

[3.2]   The (generalized) baker transformation (Farmer *et al.*, 1983)

$$(x_{n+1}, y_{n+1}) = \begin{cases} (r_0 x_n, y_n/p_0) & \text{if } 0 < y_n \le p_0, \\ (r_1 x_n + 1 - r_1, (y_n - p_0)/p_1) & \text{if } p_0 < y_n \le 1, \end{cases} \tag{3.10}$$

with $r_0 + r_1 < 1$ and both r_0 and $r_1 > 0$, is the hyperbolic analogue of the horseshoe map (see Appendix 2). It is obtained by coupling the Bernoulli map with a y-dependent contraction along the x-direction (this construction is called a *skew product*). The part of the unit square R lying above (below) $y = p_0$ is stretched by a factor $1/p_1$ ($1/p_0$) vertically and contracted by a factor r_1 (r_0) horizontally. The two rectangles (V_1, V_0) obtained in this way are then placed within R on opposite sides. Since no points are mapped outside R, the invariant set is an attractor (at variance with the horseshoe map which supports a repeller). It is easily checked that the transformation is invertible, although the Bernoulli map is not.   □

[3.3]   The Bernoulli map is chaotic for all parameter values. This is not the case for nonhyperbolic transformations, such as the logistic map

$$x_{n+1} = 1 - a x_n^2. \tag{3.11}$$

For small positive a, the attractor is the fixed point $x^* = (\sqrt{1 + 4a} - 1)/(2a)$. At the value $a = a_1 = 3/4$, the fixed point becomes unstable and a period-two attractor sets in for $a > a_1$ until, at some value $a = a_2 > a_1$, it also loses stability and gives place to a period-four attractor. This process, called *subharmonic* or *period-doubling cascade*, continues indefinitely and yields all periods of order 2^n until, at $a = a_\infty = 1.401155189\ldots$, chaos occurs. The accumulation point a_∞ is named after Feigenbaum who studied this route to chaos, showing its universality (Feigenbaum, 1978 and 1979a): all unimodal (one-hump) maps F with quadratic maximum (or minimum) exhibit the same behaviour during period-doubling (PD), up to a smooth change of variables. The rate $\delta_n = (a_{n+1} - a_n)/(a_n - a_{n-1})$

converges to the universal value $\delta_\infty = 4.669201609\ldots$. Universality has been established by observing that the iterate of order 2^n of F can be rescaled in a region around $x = 0$ so as to resemble the iterate of order 2^{n-1}. Feigenbaum showed that, in the limit $n \to \infty$, the rescaling of any map $f(x)$ with quadratic maximum converges to a unique, even, real, analytic mapping $g(x)$ such that $g(0) = 1$, $g''(0) < 0$ and $g(x) = \alpha g^2(x/\alpha)$, where $\alpha = -2.502907875$, (see Collet & Eckmann (1980) for a detailed study).   □

[3.4]   The two-dimensional "extension" of the logistic map is Hénon's transformation (Hénon, 1976)

$$\begin{cases} x_{n+1} = 1 - ax_n^2 + by_n \\ y_{n+1} = x_n \, , \end{cases} \tag{3.12}$$

where $|b| \leq 1$ is the modulus of the Jacobian. The map is invertible as long as $b \neq 0$. For $|b| < 1$, the map is dissipative. In the limit $b \to 0$, the logistic map is recovered. Period-doubling again unfolds according to Feigenbaum's scenario (the only difference occurring in the area-preserving case $|b| = 1$, when the constants δ_∞ and α have different values). In fact, this route to chaos is so universally valid that it can be observed in many different experimental setups (see, e.g., Libchaber et al., 1983).

In the chaotic regime, the iterates of the map (3.12) appear to lie on a set which is indistinguishable from the unstable manifold of the saddle point $x^* = y^* = (b - 1 + \sqrt{(b-1)^2 + 4a})/(2a)$ and is locally the product of a curve and a Cantor set. Beyond the numerical evidence, the existence of a strange attractor has been recently proved by Benedicks & Carleson (1991).   □

[3.5]   Nonlinear phenomena associated with oscillatory dynamics (coupled pendula, periodically driven systems, biological rhythms) are often modelled by circle maps. They act on one or more angular variables (i.e, on the unit interval modulo 1 or on the d-torus). The simplest case is represented by the bare circle map

$$x_{n+1} = x_n + \alpha \bmod 1 \, , \tag{3.13}$$

which models a rigid rotation by an angle $\alpha \in (0, 1)$. If $\alpha = p/q$ is a rational number (p and q being integers), every initial condition lies on a (marginally stable) period-q orbit of the map F: the Lyapunov exponent $\lambda = \ln|F'(x)|$ is identically zero. The q points are visited in p revolutions around the circle (unit interval): α is called the bare rotation number. When α is irrational, the motion is quasiperiodic: the orbit fills the whole circle. When a nonlinearity is included, as in the sine circle map

$$x_{n+1} = x_n + \alpha + K \sin(2\pi x_n) \bmod 1 \, , \tag{3.14}$$

the rotation number ω_r is defined as the average number of revolutions in n

steps in the limit $n \to \infty$ and depends on α and K^2. The first consequence of nonlinearity is the occurrence of *mode-locking*: for $K > 0$, stable cycles with $\omega_r = p/q$ are found in an interval of α-values (and not just at $\alpha = p/q$) which broadens for increasing K. The regions of stability for all p/q-cycles in the (α, K)-plane are called *Arnold tongues* (Arnold, 1983) and keep widening until, at $K = 1$, their total length is 1. The α-values that yield a quasiperiodic motion form a Cantor set on the line $K = 1$. Beyond $K = 1$, the map becomes noninvertible, the Arnold tongues start overlapping and chaos may occur. The transition from quasiperiodicity to chaos presents universal features (Feigenbaum *et al.*, 1982; Shenker, 1982; Ostlund *et al.*, 1983). This scenario has been observed experimentally (Stavans *et al.*, 1985). $\qquad\qquad\square$

[3.6]   In two-dimensional area-preserving maps, such as the standard map

$$\begin{cases} x_{n+1} = x_n + K \sin(2\pi y_n) \\ y_{n+1} = y_n + x_{n+1} \qquad \text{mod } 1, \end{cases} \tag{3.15}$$

chaos coexists with "islands" of quasiperiodic motion, consisting of families of closed invariant curves. Along each of them, the map is conjugate to an irrational rotation. The celebrated theorem of Kolmogorov (1954), Arnold (1963) and Moser (1962) (KAM) states that, for small nonlinearity, such curves exist, provided that their rotation number α is sufficiently irrational: i.e., it satisfies the infinite set of relations

$$\left| \alpha - \frac{p}{q} \right| \geq cq^{-\gamma}$$

for some $c, \gamma > 0$ and all integers $p, q > 0$. In general, for any irrational α, there exist infinitely many rationals p/q such that $|\alpha - p/q| < q^{-2}/\sqrt{5}$ (Baker, 1990; Lorentzen & Waadeland, 1992). The number α can be approximated by the finite continued fraction

$$\alpha_n \equiv \frac{p_n}{q_n} = a_0 + \cfrac{1}{a_1 + \cfrac{1}{a_2 + \cfrac{\ddots}{\quad \frac{1}{a_n}}}} \equiv [a_0, a_1, \ldots, a_n] \tag{3.16}$$

where the *diophantine approximants* p_n, q_n are relatively prime numbers. The ratios p_n/q_n converge to α for $n \to \infty$. They provide the "best" approximation to α in so far as $|q\alpha - p| \geq |q_n\alpha - p_n|$ for all integers p, q with $0 < q \leq q_n$. The *golden mean* $\omega_{gm} = (\sqrt{5} - 1)/2 = [0, 1, 1, 1, \ldots]$ exhibits the slowest convergence among the irrationals (in the unit interval) and is thus viewed as the "most irrational" number: it is the limit of ratios F_n/F_{n+1} of *Fibonacci numbers*, defined by $F_{n+1} = F_n + F_{n-1}$, with $F_0 = F_1 = 1$. In the standard map (3.15), the close invariant curve with rotation number ω_{gm} is the last to disappear when

---

2. The quantity ω_r is also called *winding number*.

K is increased from 0: this happens at $K \approx 0.9716$ (Greene, 1979). In general, however, the last KAM curve to be destroyed need not have such a rotation number. Moreover, other islands with regular (quasiperiodic) motion appear for higher values of the nonlinearity (Lichtenberg & Lieberman, 1992). ⊓

[3.7] The logistic map over the complex field

$$z_{n+1} = c + z_n^2 \qquad (3.17)$$

is the prototype of a complex analytic mapping $F : \mathbb{C} \to \mathbb{C}$. Transformations in this class have applications in two-dimensional electrostatics (Peitgen & Richter, 1986), diffusion problems (Procaccia & Zeitak, 1988), methods for the determination of the zeros of the partition function in statistical mechanics (Derrida & Flyvbjerg, 1985). The images of the origin $z = (0,0)$ of the complex plane give rise to a bounded or unbounded trajectory, depending on the value of the parameter $c \in \mathbb{C}$. The set $J = J(c)$ of points that, for a fixed c, neither escape to infinity nor converge to a stable periodic orbit is called the *Julia set*: more precisely, J is the closure of the set of the repelling periodic points of F, where a periodic point $z_0 = F^n(z_0)$ is repelling if $|dF^n/dz|_{z=z_0} > 1$. Julia sets are a particular example of the class of *strange repellers*: invariant unstable sets characterized by a chaotic dynamics. For $c = 0$, J is the unit circle and the dynamics is a Bernoulli $B(1/2, 1/2)$ process for the phase of z. In general, Julia sets may have a very complicated shape, consisting of a finite or infinite number of connected components. Even when they are simple closed curves, as for $|c| < 1/4$, they need not contain smooth arcs: their structure is fractal. The locus of all c-values for which the orbit of the origin $(0,0)$ is bounded is called the *Mandelbrot set* (Mandelbrot, 1982), again a fractal object with peculiar properties, and is the subject of current mathematical research (Devaney, 1989; Peitgen & Richter, 1986). □

[3.8] Besides the exponential separation of nearby orbits, deterministic systems may exhibit another source of uncertainty which is related to the unpredictability of the asymptotic attractor. Consider, in fact, the forced oscillator (Sommerer & Ott, 1993b)

$$\ddot{x} - \gamma\dot{x} + \nabla V(x) = A \sin \omega t , \qquad (3.18)$$

where $x = (x, y)$, $A = (A, 0)$, and the potential is

$$V(x, y) = (1 - x^2)^2 + (x - x_0)y^2 .$$

Because of the symmetry $V(x, y) = V(x, -y)$, the plane $\mathcal{M} = \{(x, \dot{x})\}$, defined by $y = \dot{y} = 0$, is invariant under Eq. (3.18). Moreover, the dynamics in $\mathcal{M}$ is the same as for the "ordinary" forced Duffing oscillator (Guckenheimer & Holmes, 1986), one of the prototypical systems that become chaotic for certain parameter values. When this is the case, all orbits unfolding in $\mathcal{M}$ are unstable (irrespective of whether they are periodic or not). The plane $\mathcal{M}$, instead, may be either attracting or repelling. Upon changing the control parameters, the infinitely many periodic orbits which correspond to those of the Duffing system within $\mathcal{M}$ do not change simultaneously their stability properties in the $(y, \dot{y})$ subspace. Thus, transversally to $\mathcal{M}$, stable and unstable orbits may coexist. Since there are trajectories on $\mathcal{M}$ that connect the former to the latter (the attractor is indecomposable), attracting and repelling regions must be intimately intertwined (see also Pikovsky & Grassberger, 1991). In the global system (3.18), the ultimate fate of any trajectory is to be captured by one of the two limit sets $\mathcal{M}$ and $\{\mathbf{x} : y = \pm\infty\}$, each having its own basin of attraction. By the peculiarity of the periodic-orbit structure of the system, points of each basin may be situated arbitrarily close to points of the other. The term *riddled basins* has been coined to denote such a high level of intricacy (Alexander *et al.*, 1992).   □

3.4   Cellular automata

Cellular automata, introduced with the purpose of modelling some of the features of reproduction in biological systems (Von Neumann, 1966), have been later employed in a variety of physical contexts (Wolfram, 1986). A cellular automaton (CA) consists of a dynamical rule which updates synchronously a discrete variable, defined on the sites of an infinite d-dimensional lattice. The values s_i of the observable are taken in the finite set $A = \{0, \ldots, b-1\}$. In the one-dimensional case, the automaton is a function $f : A^{\mathbb{Z}} \to A^{\mathbb{Z}}$, where $A^{\mathbb{Z}}$ is the set of all two-sided infinite sequences of symbols in A. The image s_i' of the ith variable s_i on the lattice depends on the values of the observable in a fixed symmetric neighbourhood $U_r(i) = [i-r, i+r]$ of i, with r being the interaction range. Defining the rule is tantamount to assigning one of the possible b values of s_i' for each of the b^{2r+1} different configurations that can occur in $U_r(i)$. Hence, there are $\mathcal{N}_R(b, r) = b^{b^{2r+1}}$ different rules. This number increases very rapidly with both the size $2r + 1$ of the neighbourhood and the number b of symbols: $\mathcal{N}_R(b, r)$ is larger than 10^{115} for $b = 3$ and $r = 2$. Restrictions on the admissible rules, suggested by physical constraints such as left–right symmetry, do not alter the order of magnitude of $\mathcal{N}_R(b, r)$ significantly.

A compact notation has been introduced to enumerate the various rules. The so-called *elementary* cellular automata (Wolfram, 1986), in which $s_i \in \{0, 1\}$ is a boolean variable and the neighbourhood $U_1(i) = [i-1, i+1]$ contains just three sites, are usually classified by suitably ordering the different $2^{2^3} = 256$ rules. By writing the admissible configurations in the inverse *lexicographic order* $\{111, 110, 101, \ldots, 000\}$, the sequence $\{f_7 = f(111), f_6 = f(110), \ldots\}$ of their images is interpreted as the binary expansion of an integer number

$$I = \sum_{k=0}^{7} f_k 2^k , \tag{3.19}$$

which is used to label the rule. For example, *Rule 22* is the short-hand notation for the block-renaming table

111	110	101	100	011	010	001	000
0	0	0	1	0	1	1	0

Sometimes a subclass of all rules can be considered in order to restrict the domain of the investigation, especially for large b and r. In the so-called *totalistic* automata, the ith symbol is uniquely determined by the value of the sum $\sigma_U(i) = \sum_{j \in U_r(i)} s_j$ of the variables in its neighbourhood. Totalistic rules can also be encoded through a relation of the form (3.19), where f_k is now the value assumed by the image symbol s' when $\sigma_U = k$, for all $k = 0, 1, \ldots, (b-1)(2r+1)$.

Useful information about CAs is provided by a formal analogy with discrete-time mappings which is established by interpreting the symbols in the infinite configuration $\mathscr{S}$ as the expansion coefficients of two real numbers x and y. By setting

$$x_n \equiv \sum_{i=0}^{-\infty} s_{i+n} b^{i-1} , \quad y_n \equiv \sum_{i=1}^{\infty} s_{i+n} b^{-i} , \tag{3.20}$$

the CA rule $f : A^{\mathbb{Z}} \to A^{\mathbb{Z}}$ can be represented as a two-dimensional map $\mathbf{F}_{CA} : R^2 \to R^2$. For $r = 1$ and $b = 2$, x_{n+1} is a function $F^{(x)}(x_n, y_n) = F^{(x)}(x_n, \lfloor 2y_n \rfloor)$, where $\lfloor 2y_n \rfloor$ is the integer part of $2y_n$. Analogously, $y_{n+1} = F^{(y)}(y_n, \lfloor 2x_n \rfloor)$. The usual requirement that the rule be left–right symmetric (Wolfram, 1986) implies $F^{(x)}(x, \lfloor 2y \rfloor) = F^{(y)}(y, \lfloor 2x \rfloor)$. Hence, the CA is specified by the pair $(F_0, F_1) = (F(x, 0), F(x, 1))$. The existence of the analogy between CAs and planar maps might suggest that CAs do not exhibit new phenomena with respect to smooth dynamical systems. This is, however, not true since $F_0(x)$ and $F_1(x)$ are highly discontinuous functions which present some degree of self-similarity, as shown in Fig. 3.4 for the case of rule 22.

The existence of discontinuities in arbitrarily small intervals prevents any linearization of the dynamics and, in turn, a simple investigation of the stability properties of the trajectories. Nonetheless, some of the tools used to describe and classify discrete-time maps can be extended to CAs. Particularly relevant are the recurrent configurations, because they characterize the asymptotic time evo-

Mathematical models

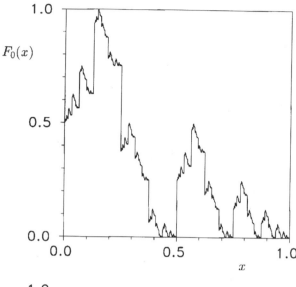

$F_0(x)$

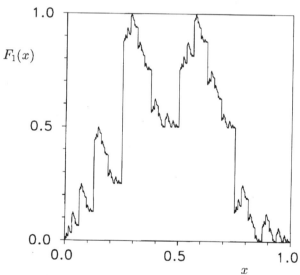

$F_1(x)$

Figure 3.4 Map representation of the elementary CA rule 22. Notice the irregularity of the functions $F_0(x)$ (top) and $F_1(x)$ (bottom).

lution. Let $\Omega^{(n)}$ denote the set of all configurations surviving after n applications of rule f:

$$\Omega^{(n)} = \bigcup_{\mathscr{S}_0 \in A^{\mathbb{Z}}} f^n(\mathscr{S}_0), \qquad (3.21)$$

where $\mathscr{S}_0 = \{\ldots, s_{-1}, s_0, s_1, \ldots\}$ is the initial condition. The maximal invariant set (called the *limit set*) $\Omega^{(\infty)}$, is then defined as

$$\Omega^{(\infty)} = \bigcap_{n=0}^{\infty} \Omega^{(n)}. \qquad (3.22)$$

A CA can be viewed both as a *source* and as a *selector* of patterns. Indeed, when acting on a seed consisting of a uniform background of equal symbols except a few (possibly one), some CAs are able to yield interesting limit patterns, neither periodic nor completely disordered. On the other hand, the effect of a CA on a random configuration that contains all possible concatenations of symbols is the progressive exclusion of some of these, until a subset survives, according to Eq. (3.22). This is of course only true for noninvertible rules: a given finite sequence S may be generated from different initial conditions. Sequences with no ancestors are obviously forbidden. The limit set consists of all infinite patterns that have an infinite sequence of preimages.

Several schemes have been proposed for a systematic classification of cellular automata. In the following, we illustrate a modification of Wolfram's proposal (Wolfram, 1986), which is based on the type of asymptotic dynamical behaviour: CAs are divided into the two broad groups of *regular* and *chaotic* rules, plus a possible third class containing "complex" rules. The few representative examples shown in Fig. 3.5 clearly reveal qualitatively different patterns. The classification can be made more rigorous by defining a suitable metric in $A^{\mathbb{Z}}$ and formalizing the notions of "sensitivity to initial conditions" (Gilman, 1987) or "attractor" (Hurley, 1990).

3.4.1   Regular rules

All configurations in the limit set of a regular rule are periodic in time (either in a stationary, or in a moving reference frame). In the simplest case, $\Omega^{(\infty)}$ reduces to a single pattern, homogeneous both in space and time. This example evidently corresponds to a stable fixed point in the language of dynamical systems. A less trivial, although frequently encountered, case is represented by a random spatial arrangement of "braids", each displaying a time-periodic behaviour without spatial shift (see Fig. 3.5(a)). Such patterns correspond to periodic cycles of the equivalent map. However, because of the spatially random structure, changing the origin in Eq. (3.20) yields an infinity of cycles. Thus, the limit set is the union of infinitely many spatially aperiodic components. The observed configuration depends on the initial condition.

The last example of a regular CA is reported in Fig. 3.5(b), where three distinct types of travelling objects are visible. As a result of simple scattering and annihilation processes, a single, seemingly random "species" eventually survives. The propagation of these structures with constant speed along a diagonal path converts spatial into temporal randomness. Indeed, the trajectory of the associated map $\mathbf{F}_{CA}$ closely resembles the evolution of a baker map (although with many "forbidden" orbits): the leading bits of y_n are continuously transferred to x_n. Thus, although the asymptotic patterns in Fig. 3.5(a) and (b) do not differ very much from each another, the trajectories in the associated map $\mathbf{F}_{CA}$ are completely different: periodic in the former and chaotic in the latter. The

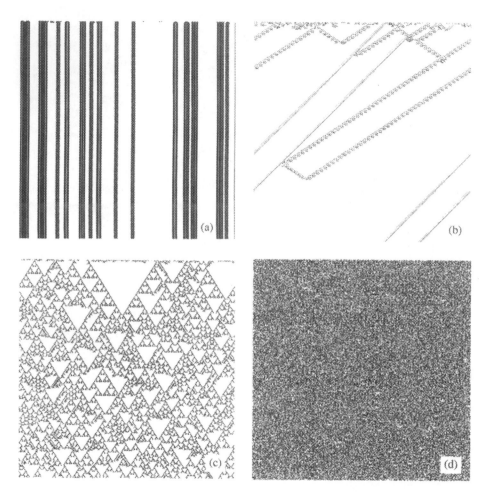

Figure 3.5 Space-time patterns of various totalistic rules with time flowing downwards. In this and in the following four figures, the rules act upon a randomly chosen initial configuration on a lattice with $L = 500$ sites and periodic boundary conditions. In two-symbol automata, black and white correspond to 0 and 1, respectively. When $b = 3$, black, grey, and white denote 0, 1, and 2, respectively, unless otherwise stated. Asterisks indicate that the configuration is shown every second time step. (a) Rule 71*, with $b = 2$ and $r = 3$; (b) rule 134135, with $b = 3$ and $r = 2$; (c) rule 173015, with $b = 3$ and $r = 2$; (d) rule 42 with $b = 2$ and $r = 2$.

origin of the paradox lies in the many-to-one relation between map trajectories and CA patterns as defined by Eq. (3.20). In fact, the translational equivalence of all sites on the lattice $\mathbb{Z}$ under the CA contrasts with the hierarchical order implied by the binary expansion (3.20) which yields the coordinates of a point in the map's phase space.

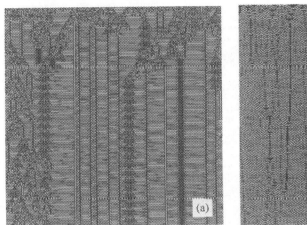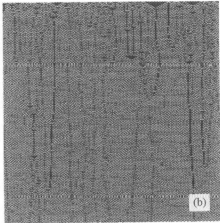

Figure 3.6 Totalistic rules 5 with $b = 2$ and $r = 2$ (a), and 153 with $b = 3$ and $r = 2$ (b).

3.4.2 Chaotic rules

While the spatio-temporal configuration of a regular CA is periodic in at least one direction, chaotic rules yield "disordered" patterns with exponentially decaying correlation functions (see Chapter 5 for the definition) in any direction. Two representative examples are reported in Figs. 3.5(c) and (d): the former exhibits a fairly long-ranged order (as the presence of large triangles indicates), whereas the latter resembles a random pattern, except for tiny details. It is conjectured that in this case there is a single relevant *ergodic* (see Chapter 5) component: a sufficiently long trajectory passes arbitrarily close to any configuration in the limit set. The only exceptions are (irrelevant) configurations that are generated by a negligibly small set of initial conditions. We leave as an exercise to the reader to show that the only preimage of the doubly-infinite sequence $\mathscr{S} = \{\ldots 1010101 \ldots\}$ under the elementary rule 22 is the sequence $\mathscr{S}$ itself. Accordingly, $\mathscr{S}$ is in the limit set but is practically unreachable: in the language of dynamical systems it corresponds to an unstable period-2 orbit.

3.4.3 "Complex" rules

The remarkable variety of patterns produced by cellular automata (see Figs. 3.6–3.9) has suggested that a third, "complex", class of rules might exist between the two previously discussed ones. Intuition, however, may be misleading since it is based on the observation of finite spatio-temporal samples which, moreover, appear after a finite-time evolution. On the one hand, a seemingly ordered, although aperiodic, pattern might exhibit exponentially decaying correlations, even when the numerical results are not precise enough to exclude an algebraic law. On the other hand, it is possible that irregular behaviour is observed only for

Figure 3.7 Totalistic rules 1024* with $b = 3$ and $r = 2$ (a), and 1051* with $b = 3$ and $r = 2$ (b).

Figure 3.8 Totalistic rules 88 with $b = 2$ and $r = 3$ (a), and 130976* with $b = 3$ and $r = 2$ (b).

a long but finite transient time. Hence, a careful quantitative analysis is required in order to distinguish among different kinds of behaviour. While deferring further discussion to Chapters 5 and 7, we outline here two classification schemes based on transient times.

The first approach requires the definition of the *first recurrence time* $t(L;S)$ for a configuration S of length L evolving under a CA f (periodic boundary conditions are also assumed). By the finiteness of the system, all solutions are asymptotically periodic. Indicating with $f^n(S)_i$ the ith symbol in the nth image of S, $t(L;S)$ is the least integer n for which $f^n(S)$ coincides with a configuration $f^m(S)$ which appeared at some previous time $m > 0$, up to a possible shift of the lattice by j sites (to account for rules admitting travelling structures): that is,

$$t(L;S) = \min \left\{ n : f^n(S)_i = f^m(S)_{i+j}, n > m \geq 0, \right.$$
$$\left. \text{for all } i = 1, 2, \ldots, L \text{ and some } j \right\}, \tag{3.23}$$

where the index $i+j$ is to be taken modulo-L. The dependence on S is eliminated by averaging over the b^L different initial conditions S, thus obtaining the mean first recurrence time

$$T(L) = b^{-L} \sum_S t(L;S). \tag{3.24}$$

The time $T(L)$ depends on both the transient time needed to approach a periodic state and the period of that state. In the absence of propagation phenomena, $T(L)$ is determined by local processes and is hence independent of the size L, as in the case of the rule shown in Fig. 3.5(a). In the presence of travelling structures

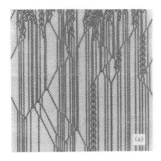 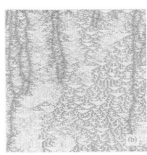

Figure 3.9 Totalistic rules 284 (a) and 129032 (b), both with $b = 3$ and $r = 2$. The colours corresponding to 0 and 1 are interchanged with respect to Fig. 3.5.

in a homogeneous background, as in Fig. 3.5(b), the leading contribution to $T(L)$ is given by the time needed for all collisions to occur, that is, $T(L) \sim L$.

For chaotic rules, instead, $T(L)$ grows exponentially with L, so that, in the large-L limit, the detection of recurrent orbits is deferred to enormous, unobservable times. This is reminiscent of finite precision iteration of chaotic maps: a strange attractor with fractal dimension D displays at most $N(\varepsilon) \sim \varepsilon^{-D}$ distinct points in a simulation performed with an accuracy of L bits, with $\varepsilon = 2^{-L}$ (see Chapter 5 and Appendix 4). As a consequence, any orbit must close onto itself in no more than $T(L) \sim 2^{LD}$ steps, which thus represents an upper bound to the recurrence time. The exponential growth of $T(L)$ can be considered as a signature of chaotic evolution also in CAs. Hence, the first recurrence time permits distinguishing different classes of behaviour.

Cellular automata's relation with computers can also be exploited to clarify some aspects of their behaviour. According to it, each cell is considered to perform logical operations on the input data. The problem is coded into the initial configuration and the end of the computation is signalled by the occurrence of a target state. Although the equivalence between CAs and general purpose computers has been proved only in a few cases as, e.g., the totalistic two-dimensional automaton known as the "Game of Life" (Berlekamp *et al.*, 1982), it is speculated that an entire class of CAs is capable of universal computation (Wolfram, 1986). Rules within this class are often referred to as complex. In fact, in the same way as a formal statement may be undecidable (i.e., unprovable in a finite number of steps), the ascertainment that the final state has been reached may require an infinite time.

If the problem is to determine which component of the invariant set eventually attracts the specific initial condition, the corresponding computation time coincides with the above defined first recurrence time. In this spirit, one is tempted to define all chaotic rules as complex because of their exponentially long computation times. Although plausible, this identification makes sense only whenever it is possible to prove that no short-cut to the direct computation exists.

A comparison between the results of the recurrence time analysis with alternative approaches helps clarifying the subject. Numerical simulations by Gutowitz (1991a) have shown that the growth of $T(L)$ with L is exponential for the elementary rule 22 and sub-exponential for rule 54 ($T(L) \sim \exp[0.6(lnL)^{1.8}]$). Both results confirm the visual impression: chaotic and a mixture of order and disorder, respectively. More instructive is the totalistic rule 20 with $b = 2$ and $r = 2$ which, although usually classified as complex, displays a presumably sublinear growth of $T(L)$. The complexity of a rule (and its ability to simulate a computer) is, in fact, sometimes assessed from the existence of localized travelling objects (gliders) which interact in a complicated way. This is because gliders appear to transfer and process information in analogy with common computer operations. Meaningful configurations for which such processes do occur can, however, be extremely unlikely, so that an average indicator like $T(L)$ may fail to point out possible computational skills of the CA. The large variability of $t(L;S)$ with the initial condition S might provide a more accurate tool for the investigation. On the other hand, it must be noticed that the association between CAs and computers sketched above might be too simple to assess the computing power of strictly chaotic rules as well.

Further information about cellular automaton behaviour is provided by the convergence time of a random initial condition S with uniformly distributed subsequences to a configuration having the asymptotic distribution of the given CA. In this framework, not only regular but also evidently chaotic rules have short transients, in the same way as ordinary strange attractors are rapidly approached from generic initial conditions. "Complex" rules should, instead, be characterized by anomalously long transients. This approach is essentially suggested by the slowing down effects that occur in the vicinity of bifurcations in dynamical systems (e.g., at period doubling) and of phase transitions in statistical mechanics (see Chapter 6). In cellular automata, however, no parametrization is *a priori* available for a quantitative study of transition phenomena. The rules' code number I (see Eq. 3.19) cannot be used for this purpose since its uniform increase yields wild jumps in the type of evolution. A more sensible proposal has been put forward by Langton (1986) who suggested parametrizing CAs by using the fraction v_s of neighbourhoods $U_r(i)$ that yield a symbol s' different from s. He claimed that, upon increasing v_s from 0 to 1, one moves from regular to chaotic rules, passing through a "transition" region where complex evolution is concentrated. Although very appealing, this scenario is, however, weakened by the many-to-one correspondence between rules and v_s, since different automata with the same v_s do not necessarily exhibit the same type of evolution. This ambiguity can be progressively reduced by introducing more parameters: an effective procedure, based on Markov approximations of the configurations, is illustrated in Chapter 5.

The situation is much simpler in probabilistic automata, where the symbol yielded by each neighbourhood is known only in a probabilistic sense. In this class of systems, it is possible to observe transitions between different states by tuning some parameter p (usually a probability). The most widely known of such phenomena is *directed percolation*, a transition from an asymptotically homogeneous (i.e., ordered) state to a fully disordered one. A power-law of the type $Q(p) \sim |p - p_0|^\beta$ is observed for various observables Q (as, e.g., the fraction of disordered subdomains) around the transition point $p = p_0$ (Stauffer & Aharony, 1992).

Notwithstanding their simplicity, CAs have been used as models for various physical processes, among which we recall lattice-gas automata (Frisch *et al.*, 1986), introduced to simulate a turbulent flow at a microscopic level. A nonzero value $s = 1, 2, \ldots, p$ represents a particle moving in one of the p possible directions on a d-dimensional lattice (these automata are defined for $d \geq 2$), while symbol 0 refers to an empty node. The rule, besides allowing each particle to move in the proper direction, describes scattering processes with mass and momentum conservation at each collision. Apart from these constraints, the evolution of a lattice-gas automaton does not substantially differ from that of a generic chaotic CA.

Another demonstration of the usefulness of CA-simulations is found in the study of excitable media. The simplest application (Winfree *et al.*, 1985) requires three symbols to denote the relevant states of a biological cell: quiescient (Q), excited (E), and tired (T). State Q is followed by E if at least one neighbouring cell (in a two-dimensional lattice) is already excited. A cell remains in E for one time step; then it becomes tired and cannot be immediately re-excited. After one time step, it is set back to Q. This essentially regular rule gives rise to propagating spirals qualitatively similar to those observed in the Belousov–Zhabotinsky reaction.

3.5   Statistical mechanical systems

So far, it has been tacitly assumed that only simple spatio-temporal dynamics takes place within the framework of equilibrium statistical mechanics. This is the case of isolated systems, in which the free energy $\mathscr{F}$ gradually decreases to a minimum value and no time dependence of macroscopic variables is eventually observed. In fact, the macroscopic properties of the system can be deduced from the Gibbs–Boltzmann assumption, which attributes the weight $\exp(-\mathscr{H}(Q)/k_B T)$ to each configuration $Q = (q_1, \ldots, q_i, \ldots, p_1, \ldots, p_i, \ldots)$ with energy $\mathscr{H}(Q)$ at the temperature T (k_B being the Boltzmann constant). A thermodynamic description is made possible by the identification of an appropriate macrostate, i.e., an ensemble of microstates[3] which give the leading contribution

---

3. We shall equivalently use the terms "configurations" and "phase-space points".

to the statistical averages when the system size diverges (the so-called thermodynamic limit, in which fluctuations become negligible). Such a correspondence is the result of a delicate balance between the weight of the various configurations (controlled by their energy E) and their number (quantified by the entropy S).

Although straightforward, this construction may lead to surprisingly fascinating phenomena as, for example, when the free-energy landscape presents several minima. The state of the system may then depend on its history (i.e. on the way it has been preparared), so that the probability of each microstate is no longer given by its Gibbs–Boltzmann weight; moreover, transitions among different states may occur over macroscopic time scales. This scenario calls for the concept of ergodicity which deals with the question whether the entire phase-space is accessible from almost any initial condition (see Chapter 5 for a formal discussion of this topic). It turns out that the most interesting phenomena observed at equilibrium, notably phase transitions and glassy behaviour, descend from a breakdown of such global ergodicity.

Classical discrete spin systems with two-body interactions on a d-dimensional lattice provide the simplest framework to elucidate this subject. The spin variable σ_i is a scalar that assumes the two values ± 1 which correspond to the up ($\uparrow$) and down ($\downarrow$) state, respectively, and the Hamiltonian reads

$$\mathscr{H} = -\sum_{i,j} J_{ij}\sigma_i\sigma_j - \sum_i h_i\sigma_i \,, \tag{3.25}$$

where J_{ij} is the coupling constant between the ith and jth spin and h_i the strength of the external magnetic field. When the interaction extends to nearest neighbours only, Eq. (3.25) defines the *Ising–Lenz model* (Huang, 1987). The evolution of a spin configuration $\mathscr{S} = \ldots\sigma_{-1}\sigma_0\sigma_1\ldots$ in contact with a heat bath at temperature T cannot be obtained directly from dynamical equations for the spin variables σ_i because of their discreteness. Two solutions to this inconvenience have been proposed. In the former, the spins are treated as continuous variables and a confining double-well potential $V = \sum_i a(-\sigma_i^2/4 + \sigma_i^4)$ is added to the Hamiltonian: the evolution equations take the form $\dot{\sigma}_i = -(\nabla\mathscr{H})_i + \xi_i(t)$, where $\xi_i(t)$ is a continuous stochastic process (see Chapter 5) with suitable statistical properties[4]. In the latter, known as the Monte Carlo method (Metropolis *et al.*, 1953), a stochastic dynamics is simulated by choosing a site at random and by flipping the associated spin according to a transition probability $W(\mathscr{S} \to \mathscr{S}')$, where $\mathscr{S}$ and $\mathscr{S}'$ are the configurations before and after the move, respectively. Any form of W satisfying the *detailed balance* condition

$$\frac{W(\mathscr{S} \to \mathscr{S}')}{W(\mathscr{S}' \to \mathscr{S})} = \exp\left(-\frac{\Delta\mathscr{H}}{k_B T}\right) \,, \tag{3.26}$$

---

4. Discrete and continuous Ising–Lenz models are equivalent, as shown by the Kac–Hubbard–Stratonovich transformation (Fisher, 1983).

with $\Delta \mathscr{H} = \mathscr{H}(\mathscr{S}') - \mathscr{H}(\mathscr{S})$, guarantees the convergence to the desired equilibrium distribution. The most frequent choice is $W(\mathscr{S} \to \mathscr{S}') = 1$ if $\Delta \mathscr{H} \leq 0$, and $W(\mathscr{S} \to \mathscr{S}') = \exp\{-\Delta \mathscr{H}/k_B T\}$ otherwise. Detailed balance implies that, at equilibrium, there are as many transitions per unit time from $\mathscr{S}$ to $\mathscr{S}'$ as from $\mathscr{S}'$ to $\mathscr{S}$. This condition is satisfied in isolated systems (van Kampen, 1981), while it is typically violated in nonequilibrium situations. In fact, it is precisely in the absence of detailed balance that macroscopic structures may evolve in time (biological systems being the most striking example thereof).

The Monte Carlo method, which can be viewed as a probabilistic cellular automaton with an asynchronous updating of the spins, permits evaluation of thermodynamic averages in systems having a sufficient size for significant predictions of physical observables. The naive sampling method in which configurations are picked up at random (with the same *a priori* probability) is, in comparison, inefficient since the majority of the contributions is collected from high-energy microstates with negligible weight. Monte Carlo dynamics, instead, selects each configuration according to its actual Gibbs–Boltzmann weight.

Phenomena with strong implications in the study of complexity, such as ergodicity breaking and phase transitions, take place already in the simple Hamiltonian (3.25) with nearest-neighbour (nn) couplings. In the absence of an external field and with isotropic ferromagnetic interactions ($J_{ij} = 1$), the Ising model possesses two mutually symmetric ground states (having all spins either up or down). Therefore, statistical averages at low temperature are dominated by configurations with large patches of homogeneous phases. The energy (and hence the probability) of each microstate is essentially determined by the domain walls separating opposite phases. Accordingly, the dimension d of the lattice crucially affects the properties of the system. No interesting behaviour arises for $d = 1$ since, in that case, the energy depends only on the number of walls (which are pointlike). In fact, the following situation may verify: a finite energy fluctuation $\Delta \mathscr{H}$ flips a spin in one of the two ground states and creates a pair of walls which, in turn, start diffusing and reversing all other spins. As a result, the system can drift from one ground state to the other. Moreover, if the system is large enough, the free energy $F = E - TS$ is lowered when a domain wall is created. In fact, the finite increase in the energy E is overridden by the increase of the entropy S which is proportional to $\ln N$, where N is the number of the possible distinct positions of the domain wall. Therefore, random configurations are favoured in the thermodynamic limit. In two- and higher-dimensional lattices, instead, the energy of a domain wall depends on its length (area). Thus, a homogeneous nucleus, which may spontaneously arise within a background of opposite spins, requires further energy to expand and reach the system size. At sufficiently small temperatures, the entropy growth due to all possible realizations of a given domain wall is not able to counterbalance the energy increase, and the ordered configuration is stable against random perturbations[5]. Hence, the system presents two macroscopically distinct, but *a priori* equivalent, ensembles of configurations which contribute to its thermodynamic properties. The representative point in phase space X is effectively confined, on physical time scales, in a subregion (component) of X. This is perhaps the simplest example of ergodicity breaking: the evaluation of statistical averages is possible only when the single components can be identified and properly weighted (Palmer, 1982).

To clarify this point further, consider again the concept of macrostate: namely, an ensemble of microstates sharing some macroscopic property (e.g., a given fraction of "up" spins). This ensemble is meaningful, i.e., *pure* if any statistical indicator (e.g., the probability of finding m consecutive "up" spins) has zero fluctuations in the thermodynamic limit. Whenever this is not the case, the initial ensemble should be suitably split into the largest possible subsets enjoying such a property (more technically, one should identify the maximal ergodic components: see also Chapter 5). Unfortunately, there is no general procedure to perform such decomposition. In the Ising–Lenz model, at sufficiently low temperature, the energy alone does not allow identification of a pure macrostate, since the magnetization $M = \sum_{i=1}^{n} \sigma_i/n$ exhibits a bimodal distribution with two peaks at $M_{\pm} = \pm 1$. Ergodicity breaking is the key feature of phase transitions which indeed occur when, upon variation of a control parameter $\mathscr{P}$ (typically the temperature) across some value $\mathscr{P}_0$, the "generic" configurations (responsible for the thermodynamic properties) are no longer representative of the whole phase space but rather exhibit a distinctive structure.

3.5.1   Spin glasses

The example above, although useful to introduce relevant subjects such as ergodicity and phase transitions, is too elementary to be considered as a prototype of a complex scenario. A more fascinating one is provided by spin glasses (Fischer & Hertz, 1991). Experimental observations indicate that the response of such systems to a time-dependent external magnetic field significantly depends on the time scale, and possibly manifests itself after astronomical times. In fact, at sufficiently low temperature, a spin glass presents small fluctuations around a frozen disordered state accompanied by transitions to similar states occurring over a wide range of time scales. These phenomena descend from the existence of an infinity of metastable configurations organized in a hierarchical way which, in turn, testify to the existence of an infinity of ergodic components (Palmer, 1982). The interesting point is that this structure is not a feature of

5. In general, the breaking of ergodicity depends on the height of the "free-energy barrier" between different configurations. Thus, a transition between two phase-space regions may be inhibited either by the high energy difference $\Delta\mathscr{H}$ or by the small number of "connections" between them (i.e., by a small entropy S).

the microscopic interactions, but rather emerges if the following two general ingredients are present: conflicting interactions, which prevent a straightforward minimization strategy for the free energy, and some degree of disorder in the interactions.

The former mechanism is well illustrated by the so called RKKY interactions (Fischer & Hertz, 1991) which are alternately ferro- and anti-ferromagnetic in space. A simple model of this coupling scheme appears in the axial next-nearest-neighbour Ising (ANNNI) system (Elliott, 1961), in which the first- and second-nn interaction constants (I_1 and I_2, respectively) have opposite character. If $I_1 = -I_2 > 0$, each spin strives to be aligned with its nns and counter-aligned with its next nns at the same time. Because of this competition, the minimum-energy configuration in one dimension is the periodic chain ($\ldots \uparrow\uparrow\downarrow\downarrow\uparrow\uparrow\downarrow\downarrow\uparrow\uparrow \ldots$). Clearly, the energy is minimized for all next-nn pairs but not for counter-aligned nns. Such spins are said to be *frustrated* (Toulouse, 1977) since, upon flipping one of them, the "dissatisfaction" would just be transferred to the next site.

Frustration alone is not sufficient to ensure spin-glass behaviour. On the one hand, in fact, ergodicity breaking may just yield long-range order as in the previous example, in which 4 periodic ground states are found (given by the spatial shifts of the basic configuration). On the other hand, an irregular metastable state may indeed exist and be separated from other patterns of (locally) minimal energy, but just by finite barriers which can be surmounted in the thermodynamic limit.

Disorder is the other important ingredient for the generation of spin-glass behaviour. The model that has attracted most of the interest is discussed in Edwards and Anderson (1975). It is essentially an Ising–Lenz system with a spatially random, Gaussian distribution of the interactions J_{ij}. It is important to distinguish *quenched* disorder, with fixed J_{ij}, from *annealed* disorder, in which the J_{ij}s fluctuate in time. The latter can be treated analytically in a much simpler way since the average over disorder can be exchanged with the statistical average. As a result, however, no ergodicity breaking occurs. Indeed, the couplings J_{ij} can be formally treated as further statistical variables and the average over different realizations of the J_{ij}s be interpreted as a suitable average for a quite standard deterministic Hamiltonian function involving both the σ_is and the J_{ij}s (Fischer & Hertz, 1991). From a physical point of view, the energy landscape may be imagined to undergo fluctuations, so that valleys are exchanged with hills and the system is prevented from being trapped within a local minimum.

Although disorder does appear in nature (e.g., materials containing magnetic impurities randomly placed in an inert substrate), it is not clear to which extent its idealizations, such as the Edwards–Anderson model, constitute realistic models. A particularly relevant question is whether a random distribution of the J_{ij}s can *spontaneously* emerge from an "ordered" Hamiltonian. If this were the case, one could speak of self-generated complexity. Glasses *tout court* provide a significant example of such a problem. They are generated by rapidly cooling (i.e., quenching) a liquid and can therefore be considered as "snapshots" of its time evolution. One then asks whether this procedure leads only to metastable states (i.e., stable over a finite time) which, eventually, leave place to a crystalline configuration, or also disordered configurations which remain stable even in the thermodynamic limit. Although there is no definite answer yet (Stein & Palmer, 1988), convergence to an ordered stable configuration is expected to occur in such a long time that the utility of simple mathematical models of disorder is thereby confirmed.

The origin of the (infinitely) many metastable states can be elucidated with reference to a chain of classical nonlinear oscillators with competing interactions between first and second nearest-neighbours. The potential energy is given by

$$V(Q) = \sum_i V_1(q_{i+1} - q_i) + V_2(q_{i+2} - q_i) \,, \tag{3.27}$$

where q_i is the displacement of the particle at the ith lattice site. The stationary configurations are given by the extrema of $V(Q)$. Setting $\partial V(Q)/\partial q_i = 0$, one obtains an implicit equation in the variables $q_{i-2}, q_{i-1}, \ldots, q_{i+2}$ which can be interpreted as a four-dimensional mapping $q_{i+2} = F(q_{i-2}, q_{i-1}, q_i, q_{i+1})$, with a time-reversal symmetry. In the nonlinear case, chaotic evolution is possible and each trajectory of F corresponds to a stationary configuration Q_k. Reichert and Schilling (1985) proved the existence of infinitely many metastable configurations for certain piecewise-linear potentials with local nn repulsion and next-nn attraction, or vice versa. This competition is analogous to that of the ANNNI system. A similar situation occurs in continuous Ising–Lenz systems. In Eq. (3.27), however, no bistable potential is explicitly present in the Hamiltonian. This example shows that nonlinear interactions, as in Eq. (3.27), may cause a breaking of the translational invariance of the Hamiltonian and give rise to random (chaotic) frozen configurations Q_k. The reader may notice that this phenomenon is not much different from the frozen disordered patterns observed in some cellular automata and reported in the previous section.

Having realized that quenched disorder may autonomously arise also in translationally invariant systems, it is important to stress that a random layout of interaction coefficients J_{ij} (akin to an extremal configuration Q_k of Eq. (3.27)) is neither necessary nor sufficient for the generation of a spin glass phase. On the one hand, there are disordered systems with random ferromagnetic couplings which remain ferromagnets in all respects. On the other, recent theoretical studies have shown that certain *deterministic* Hamiltonians also give rise to glassy behaviour (Bouchaud & Mézard, 1994; Marinari *et al.*, 1994a and 1994b; Chandra *et al.*, 1995). One of these constitutes the so-called *sine model* (Marinari

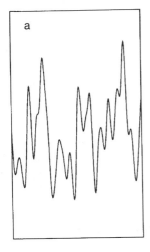

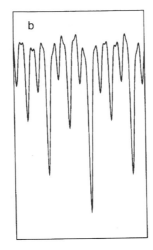

Figure 3.10 A random energy landscape (a) and an approximately hierarchical one (b). Notice the clustering of the minima in the latter.

et al., 1994b), in which all spins interact with one another with the coupling constants

$$J_{ij} = \frac{2}{\sqrt{1 + 2N}} \sin \left(\frac{2\pi ij}{1 + 2N} \right)$$

where N is the number of spins (the anomalous normalization of the J_{ij}s is a consequence of the infinite interaction range: it is introduced in order to mantain the proportionality between the total energy and the system size: see Chapter 6 for a discussion of this point).

Nevertheless, completely random distributions of (positive and negative) couplings provide the simplest context for the understanding of the glassy phase. The first difference with ordinary statistical mechanics is that the Hamiltonian is here known only in a probabilistic sense. Finding a complete exact solution for a specific realization of the disorder is, hence, *a priori* impossible. In particular, detailed features such as the form of the ground state are not accessible to analytical treatment. In spite of this, relevant functions (as, e.g., free energy and entropy) can be determined, since they have the notable property of being self-averaging: i.e., their fluctuations vanish in the thermodynamic limit (Mézard *et al.*, 1986).

Remarkably, most of the progress in this area has been obtained with a simpler system than Edwards and Anderson's, which is due to Sherrington and Kirkpatrick (1975). There, each spin interacts with all others, irrespectively of the distance. This is the standard simplification introduced to construct a self-consistent *mean-field* theory. Accordingly, each spin is assumed to experience a mean field produced by all others. The expectation that all spins evolve in the same way is, however, wrong. It is precisely the failure of this naive picture that led to the discovery of a hierarchical organization of the metastable states (for a simple illustration of this concept, see the two profiles in Fig. (3.10).

3.5.2   Optimization and artificial neural networks

Although the Sherrington–Kirkpatrick model may be too rough an approximation of a real spin glass, its solution must be considered as a true milestone in statistical physics since it revealed profound relationships with many seemingly unrelated fields. Among them, we mention combinatorial optimization and artificial neural networks, both of which can be profitably recast in a spin-glass formalism. In an optimization problem, the suitability of a layout $\{\ldots, q_{i-1}, q_i, q_{i+1}, \ldots\}$ of variables (i.e., a microstate Q) to a given task is quantified by a cost function $F(Q)$ which plays the role of a Hamiltonian. The smaller the value of $F(Q)$, the better the test solution Q: hence, the optimal configuration Q_0 is just the ground state. The tight relationship with spin-glass theory is clearly revealed by the graph (bi)partitioning problem, which has applications in computer manifacturing. Consider an idealized electronic circuit with a (large) set W of wires connecting several nodes. The goal is to divide W into two equal-size subsets W_+ and W_- (to be physically mounted on two different chips) such that the total number of interconnections is minimized. In simple cases, as when only nearby nodes in a plane are connected together, the optimal solution can be easily found (it is sufficient to cut W with a line). In general, however, there are many "long distance" connections which prevent the use of simple topological arguments. In these cases, disorder enters the problem by the assumption that the links w_k are placed at random in W. The ultimate goal is then not solving a single specific instance $W = W_0$ of the problem but finding a general algorithm that is able to cope with all possible Ws.

It is interesting to notice that the hypothesis of quenched disorder, which may be questionable in the context of magnetic systems, is here indisputable, since the weights are fixed at the beginning once for all, as shown in the following. Let us associate the value $\sigma_i = \pm 1$ with node n_i depending on whether it belongs to W_+ or to W_- and set $J_{ij} = 1/2$ if there is a link between n_i and n_j and $J_{ij} = 0$ otherwise. Indicating with $Q = \{W_-, W_+\}$ a generic partition of W, the number $N(Q)$ of links between W_- and W_+ can be written as

$$N(Q) = \sum_{i,j} J_{ij}(1 - \sigma_i \sigma_j) .$$ (3.28)

The desired equality of the number of nodes in W_- and W_+ is expressed by $\sum_i \sigma_i = 0$. The problem then approximately reduces to the minimization of the cost function

$$F(Q) = N(Q) + \mu \left(\sum_i \sigma_i \right)^2$$

where μ weights the importance of the equal-size constraint. Simple algebra shows that

$$F(Q) = \sum_{i,j} J'_{ij} \sigma_i \sigma_j + C$$

where $J'_{ij} = \mu - J_{ij}$ and C is a constant. If the wires are randomly placed and $\mu = 1/4$, $F(Q)$ reduces to a Sherrington–Kirkpatrick Hamiltonian with a two-peak distribution of the J'_{ij}s $(= \pm 1/4)$. The reader should notice that although $N(Q)$ has the same form as $F(Q)$, it is not able to yield a glassy phase since it does not contain antiferromagnetic couplings; frustration appears only after including the constraint on the number of edges (Fu and Anderson, 1986).

Spin glass and optimization problems primarily differ in the final goal. For example, there is no practical interest in developing a statistical mechanics for the graph partitioning problem with an average over all possible realizations $\{J_{ij}\}$ of disorder (which weight should be attributed to each wire configuration W?). Conversely, it is of primary importance to find an approximate ground state, for a generic instance W. The accomplishment of this goal is, however, severely limited by various difficulties. On the one hand, there is presumably no general algorithm that finds the optimal solution for any W (see the discussion of NP-complete problems in Chapter 8); on the other, finding the optimal solution Q_0 by trying all possible partitions Q is already unfeasible for systems with as few as 100 elements, even with the *a priori* exclusion of certainly inappropriate partitions. An alternative strategy is offered by *simulated annealing* (Kirkpatrick *et al.*, 1983), in which Monte Carlo dynamics is combined with a slow cooling process. As long as the temperature is nonzero, the system explores several minima of the free-energy landscape and, as the temperature approaches zero sufficiently slowly, it should find itself in the ground state $\mathscr{F}_0$. Clearly, success depends on the time scales of the process. The system, in fact, may be trapped in a secondary minimum of $\mathscr{F}$ whenever the absolute one is found in too narrow a valley to be visited in a realistic computer time. For systems such as Sherrington–Kirkpatrick's, however, one can prove that several equally good "ground states" exist (Mézard *et al.*, 1986). Then, there is no need for the absolute minimum, since good approximations are easily found. Accordingly, knowledge of the number of metastable configurations with a given energy permits estimation of the average error expected in a simulated annealing process. This is one of the objectives of the theory of combinatorial optimization.

The analogy with spin glasses proved very effective also for artificial *neural networks* (Müller *et al.*, 1995): namely, systems composed of nodes and connections that change configuration under an external influence until they are able to perform a certain task (e.g., pattern recognition or execution of mathematical operations). This process is called automatic learning. Using nodes (neurons) with only two states (firing and nonfiring) and simple interconnections, it is again possible to find a formal equivalence with the evolution of a spin glass. The goal is here approximately the inverse as in optimization. In fact, one seeks the Hamiltonian (set of interactions among neurons) that guarantees convergence of the neuron pattern to a preassigned configuration (corresponding to a metastable state in a spin glass).

From the point of view of complexity, interest primarily lies in the intricate relationship between a given generic realization of the couplings J_{ij} and the structure of the ground state. This is very different from the examples discussed in the previous sections, in which the complex object was the spatio-temporal pattern produced by the system. Here, complexity arises as a strongly relative concept from the comparison between the solution and the formulation of a problem. Nevertheless, this does not prevent the occurrence of the former type of complexity too, as shown by the hierarchical organization of metastable states, which may be present in real spin glasses with nn-interactions (see Chapter 6).

Further reading

Beck & Schlögl (1993), Cvitanović (1984), Farmer *et al.* (1984), Gutowitz (1991b), Ott (1993), Schuster (1988), Toffoli & Margolus (1987).

Part 2

Mathematical tools

Chapter 4

Symbolic representations of physical systems

The setting of a theory of complexity is greatly facilitated if it is carried out within a discrete framework. Most physical and mathematical problems, however, find their natural formulation in the real or complex field. Since the transformation of continuous quantities into a symbolic form is much more straightforward than the converse, it is convenient to adopt a common representation for complex systems based on integer arithmetics. This choice, in fact, does not restrict the generality of the approach, as this chapter will show. Moreover, discrete patterns actually occur in relevant physical systems and in mathematical models: consider, for example, magnets, alloys, crystals, DNA chains, and cellular automata. We recall, however, that a proposal for a theory of computational complexity over the real and complex fields has been recently advanced (Blum, 1990).

The symbolic representation of continuous systems also helps to elucidate the relationship between chaotic phenomena and random processes, although it is by no means restricted to nonlinear dynamics. Indeed, von Neumann's discrete automaton (von Neumann, 1966) was introduced to model natural organisms, which are mixed, "analogue–digital" systems: the genes are discrete information units, whereas the enzymes they control function analogically. Fluid configurations of the kind reproduced in Fig. 2.2 also lend themselves to discretization: owing to the constancy of the wavelength (complexity being associated with the orientation of the subdomains), a one-dimensional cut through the pattern yields a binary signal (high-low). A similar encoding is possible for lasers in a spiking regime (Hennequin & Glorieux, 1991), where the maxima (or minima) of the signal are compared with a reference threshold. A less straightforward labelling can be obtained for ramified systems, such as a DLA aggregate (Meakin, 1990; Argoul *et al.*, 1991). Exploiting the relationship between a DLA pattern and a suitable map over the complex field (Procaccia & Zeitak, 1988; Eckmann *et al.*, 1989), the former can be labelled through the periodic orbits of the latter. Other examples of a hierarchical symbolic encoding can be found in studies of turbulence (Novikov & Sedov, 1979).

4.1 Encoding in nonlinear dynamics

Smooth dynamical systems can be treated in a unique framework with cellular automata and discrete spin systems on a lattice, in spite of the apparent unlikeness of the two worlds. In order to achieve this, time must first be discretized by choosing a suitable Poincaré surface Ξ. The phase space of the map $\mathbf{F}$ associated with the flow $\boldsymbol{\Phi}_t$ through Ξ is then carefully divided (coarse-grained) into cells, so as to obtain a sort of "coordinate grid" for the dynamics.

For simplicity, the compact set $X \in \mathbb{R}^d$ within which the motion is circumscribed will be identified with the phase space itself. The encoding of X is accomplished by introducing a *partition* $\mathcal{B} = \{B_0, \ldots, B_{b-1}\}$ consisting of b disjoint subsets: i.e., $\bigcup_{j=0}^{b-1} B_j = X$ and $B_j \cap B_k = \emptyset$ for $j \neq k$. Finite partitions ($b < \infty$) will always be considered in the following, unless explicitly stated. The set $A = \{0, \ldots, b-1\}$ of the labels of the partition elements is called the *alphabet*.

Under the action of the dynamics, the system describes an orbit $\mathcal{O} = \{\mathbf{x}_0, \mathbf{x}_1, \ldots, \mathbf{x}_n\}$ which touches various elements of $\mathcal{B}$. Denoting with the symbol $s_i \in A$ the index of the domain $B \in \mathcal{B}$ visited at time i, the itinerary $\mathcal{O}$ is associated with the symbolic sequence $S = \{s_0, s_1, \ldots, s_n\}$, where $\mathbf{x}_i \in B_{s_i}$ for all $i = 0, \ldots, n$. Of course, no relevant feature of the dynamics must be lost in this process. In particular, it is desirable that the original trajectory $\mathcal{O}$ can be retraced to some extent from the knowledge of the symbolic sequence, of the partition $\mathcal{B}$ and of the dynamical law $\mathbf{F}$. This is indeed possible with a careful choice of $\mathcal{B}$. In fact, the sequence S, consisting of $n+1$ symbols, can be produced only by the points $\mathbf{x}_0$ that belong to the intersection

$$B_S \equiv B_{s_0} \cap \mathbf{F}^{-1}(B_{s_1}) \cap \ldots \cap \mathbf{F}^{-n}(B_{s_n}). \tag{4.1}$$

The observation of the first two symbols $\{s_0, s_1\}$ in S, for example, implies that the representative point was, at time $i = 0$, not only in B_{s_0} but also in the first preimage $\mathbf{F}^{-1}(B_{s_1})$ of B_{s_1}. Hence, the joint measurement at times $i = 0$ and 1 reduces the uncertainty on the initial condition $\mathbf{x}_0$, as illustrated in Fig. 4.1, provided that $\mathbf{F}^{-1}(B_{s_1}) \cap B_{s_0} \subset B_{s_0}$. Symbolic sequences S of increasing length n are expected to identify smaller and smaller subsets in phase space X. This is formalized through the concept of *refinement*: a partition $\mathcal{C}$, containing c

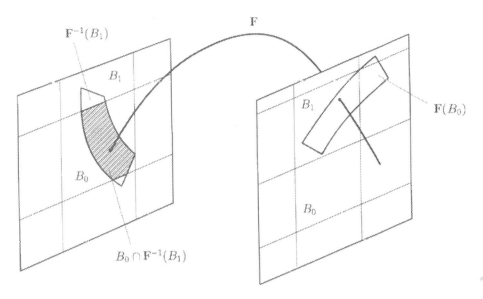

Figure 4.1 Symbolic labelling of orbits: all points in the intersection of B_0 with $\mathbf{F}^{-1}(B_1)$ share the same initial symbols $\{0, 1\}$.

elements, is a refinement of $\mathscr{B}$ if each element of $\mathscr{B}$ is a union of elements of $\mathscr{C}$. The first refinement $\mathscr{B}_1$ of $\mathscr{B}$ under $\mathbf{F}$ consists of the subsets $B_{s_j} \cap \mathbf{F}^{-1}(B_{s_k})$, for all $s_j, s_k \in A$ for which the intersection is not empty; its elements are labelled by pairs $\{s_j, s_k\}$ of symbols. Three-symbol sequences label the second refinement $\mathscr{B}_2$. In general, one has

$$\mathscr{B}_n \equiv \bigvee_{i=0}^{n} \mathbf{F}^{-i}\mathscr{B} = \mathscr{B} \vee \mathbf{F}^{-1}\mathscr{B} \vee \ldots \vee \mathbf{F}^{-n}\mathscr{B}, \tag{4.2}$$

where $\mathbf{F}^{-i}\mathscr{B} = \{\mathbf{F}^{-i}(B_0), \ldots, \mathbf{F}^{-i}(B_{b-1})\}$ and $\mathscr{B} \vee \mathscr{C} = \{B_j \cap C_k : 0 \leq j \leq b-1, 0 \leq k \leq c-1\}$ denotes the *join* of the two partitions $\mathscr{B}$ and $\mathscr{C}$. Notice that the refinements $\mathscr{B}_n$ are produced by the dynamics itself (dynamical refinements).

The study of the symbolic signal $\mathscr{S} = \{s_0, s_1, \ldots\}$ (infinitely long sequences will be distinguished from finite ones by using different typographical fonts, such as $\mathscr{S}$ and S) is called *symbolic dynamics* (SD). It is equivalent to that of the real trajectories $\mathscr{O}$ of the system if every infinitely long symbol sequence corresponds to a single point (initial condition) $\mathbf{x}_0$ in phase space: that is, if there is a map ϕ such that $\mathscr{S} = \phi(\mathbf{x}_0)$. This result is achieved for partitions $\mathscr{B}$ that refine themselves indefinitely under the dynamics, according to Eq. (4.2): they are called *generating*.

If the map $\mathbf{F}$ is invertible, the backward iterates of $\mathbf{x}_0$ are taken into account by a doubly-infinite sequence $\mathscr{S} = \{\ldots, s_{-2}, s_{-1}; s_0, s_1, \ldots\}$, where the semi-colon marks the "origin" of the sequence on the time-lattice. The nth forward (backward) iterate $\mathbf{x}_n = \mathbf{F}^n(\mathbf{x}_0)$ of $\mathbf{x}_0$ yields, through the map ϕ, a sequence

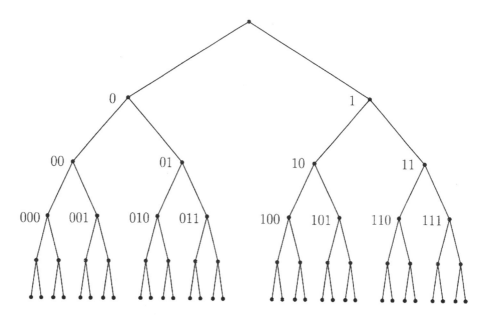

Figure 4.2 Hierarchical tree labelling the refinements of a binary partition.

that is just the nth left- (right-) shifted version of $\mathscr{S}$, with respect to the origin. Noninvertible maps (e.g., the logistic map) naturally define one-sided (forward) sequences because of the nonunicity of the preimages. In order to produce a bi-infinite sequence by iterating such a map, there should be a prescription for the choice of the preimages, such that the properties of the forward and backward symbolic dynamics were guaranteed to be the same. The distinction between one- and two-sided sequences, however, will be neglected in the following, unless explicitly stated.

The encoding is usually represented in a hierarchical way by means of a tree (see Fig. 4.2 for the binary case). The first level below the root contains b *nodes* or *vertices*, corresponding to the elements B_{s_j} of the partition $\mathscr{B}$ and labelled by the symbols in A. Each vertex s_j has b branches pointing to the nodes associated with the subsets $B_{s_j s_k}$ of B_{s_j}, for all k. The vertices at the lth level refer to concatenations of l symbols. All branches leaving a generic vertex $S = s_1 s_2 \ldots s_l$ point to the extensions $S s_m$ of sequence S (refinements of subset B_S), for all m.

Examples:

[4.1] The Bernoulli map $B(p_0, p_1)$ (Eq. 3.9) with the generating partition $\{B_0, B_1\} = \{[0, p_0], (p_0, 1]\}$, produces a symbolic dynamics in which all sequences occur, although with different probabilities (see Chapter 5). The baker map (3.10) yields the same SD, being the partition defined by the line $y = p_0$.   □

[4.2] For a one-dimensional map (Collet & Eckmann, 1980), a generating partition is defined by the coordinates of the critical points (maxima, minima, vertical asymptotes). In this way, distinct preimages of a given point receive different labels (one for each monotonic branch). □

[4.3] A generator of the SD for the *Lozi map*

$$\begin{cases} x_{n+1} = 1 - a|x_n| + by_n \\ y_{n+1} = x_n \end{cases} \tag{4.3}$$

(the hyperbolic analogue of the Hénon map) is defined by the line $y = 0$ (Tél, 1983). □

The problem of constructing generating partitions in multidimensional nonhyperbolic maps has not been entirely solved yet. Grassberger and Kantz (1985) have proposed a practical method for systems ruled by a horseshoe mechanism. They construct the borders of the partition by connecting homoclinic tangencies together. In fact, these are the points around which phase-space volumes are folded and thus represent the natural extension of the extrema of one-dimensional maps. This method has been successfully applied to moderately dissipative systems, including flows (Politi, 1994). Its implementation in a conservative dynamics has been hindered by the difficulty of identifying the proper (*primary*) tangencies to be considered. Very recently, Christiansen and Politi (1995) have been able to overcome the obstacle in the standard map (3.15) by simultaneously using tangencies and finite pieces of invariant manifolds.

Example:

[4.4] Homoclinic tangencies cannot always be readily identified as, e.g., in experimental signals. In such cases, a trial-and-error approach, based on the observation that all periodic points must receive a different encoding, proved to be rather effective. As an example, let us mention the attractor occurring in a chaotic NMR-laser (Flepp *et al.*, 1991; Badii *et al.*, 1994). In Fig. 4.3, a Poincaré section of the system is shown in suitable coordinates, together with the line defining the binary partition and the intersection points of a period-nine unstable orbit. □

If a generating partition $\mathcal{B}$ exists, there is an infinity of them: in particular, every refinement of $\mathcal{B}$ is, *a fortiori*, generating. The difficult task is to find one with the minimum number b of elements.

The analysis of a discrete-time dynamical system $(X, \mathbf{F})$ in terms of symbolic representations is especially useful if the symbolic signals are not only a faithful encoding of $(X, \mathbf{F})$ (ensured by the generating property of $\mathcal{B}$), but also "understandable", as happens with *Markov* partitions.

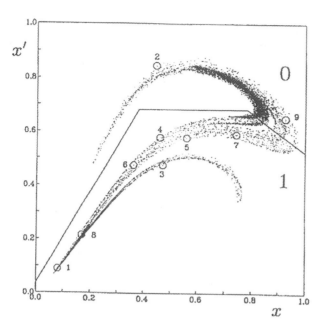

Figure 4.3 Strange attractor of the Poincaré map of an NMR laser (Flepp *et al.*, 1991) reconstructed from experimental data. The curve defining an approximate generating partition is shown: the two domains are labelled by 0 and 1. The intersection points of an unstable period-nine orbit are also displayed.

Definition: (Adler & Weiss, 1967) A finite partition $\mathscr{B} = \{B_0, \ldots, B_{b-1}\}$ has the Markov property for the map $\mathbf{F}$ if $\mathbf{F}(\overline{B}_i)$ is the union of some $\overline{B}_j$'s, for all $i \in \{0, 1, \ldots, b-1\}$ (the overbar indicating the closure operation).

When such a partition can be constructed, the description of the (symbolic) dynamics and the computation of interesting quantities, such as invariant measures and entropies (see Chapter 5), are greatly facilitated.

Example:

   [4.5]   The critical point $x_c = 1/2$ of the *roof map*

$$x_{n+1} = \begin{cases} a + 2(1-a)x_n & \text{if } x_n < 1/2, \\ 2(1-x_n) & \text{if } x_n \geq 1/2 \end{cases} \tag{4.4}$$

belongs to an unstable period-5 orbit for $a = (3 - \sqrt{3})/4$ (see Fig. 4.4). Consider the points $z_1 = F_0^{-1}(z_2)$ and $z_2 = F_1^{-1}(1/2) = 3/4$, where F_0^{-1} and F_1^{-1} denote the left and right inverses of the map F, respectively. It is easily verified that the intervals $B_\alpha = [0, z_1)$, $B_\beta = [z_1, 1/2)$, $B_\gamma = [1/2, z_2)$ and $B_\delta = [z_2, 1]$ constitute a finite Markov partition. For a "typical" value of a (see Chapter 5 for a definition) it is only possible to find a countable Markov partition.   □

Invariant sets Λ of diffeomorphisms admit finite Markov partitions, provided that they are compact, maximal, indecomposable, and hyperbolic (Bowen, 1978). The construction is made by tiling the set Λ with "rectangles" whose edges, formed by pieces of stable and unstable manifolds, are mapped forward (back-

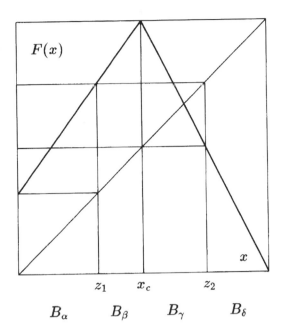

Figure 4.4 Graph of the roof map at $a = (3 - \sqrt{3})/4$. The generating partition $\{[0, x_c), [x_c, 1]\}$ and the Markov partition $\{[0, z_1), [z_1, x_c), [x_c, z_2), [z_2, 1]\}$ are indicated.

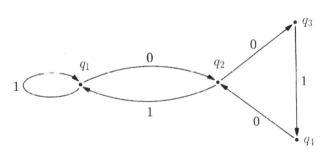

Figure 4.5 Directed graph for the roof map at $a = (3 - \sqrt{3})/4$.

ward) exactly onto those of one or more rectangles in the unstable (stable) direction. As already mentioned, however, hyperbolic systems are not common in nature; in general, one does not expect to have finite Markov partitions.

Of course, a Markov partition is always generating, although the converse is not true: hence, its b elements B_i can be used to define a symbolic labelling of phase space. Since each domain $\overline{B}_i$ is mapped onto the union of other $\overline{B}_j$'s (borders map to borders), the symbolic dynamics can be specified by a $b \times b$ (topological) *transition matrix* $\mathbf{M}$ with entries M_{ij} equal to 0 or 1 according to whether $F(B_i) \cap B_j = \emptyset$ or not, respectively. If points from B_i are mapped by $\mathbf{F}$ to B_j, the transition $i \to j$ is allowed (i.e., the sequence $\ldots ij \ldots$ exists) and $M_{ij} = 1$. Notice that points lying on the boundaries of the elements B_i do not have a unique symbolic representation and must be sometimes treated in a special way (this is reminiscent of numbers with more than one expansion in some base b).

In Fig. 4.5, the legal transitions for the roof map (4.4) at $a = (3 - \sqrt{3})/4$

are illustrated by means of a *directed graph*: any admissible symbolic orbit corresponds to an infinite path on the graph, whereby a symbol is "emitted" each time an arrow is followed. Markov systems yield finite graphs. A graph is called *topologically transitive* or *irreducible aperiodic* if there is a sequence of arrows leading from any node to any other one. In terms of the matrix $\mathbf{M}$, this means that there is an integer n such that no entry of $\mathbf{M}^n$ (the nth power of $\mathbf{M}$) is zero. Topologically transitive transformations are indecomposable and may admit a dense set of periodic points.

4.2   Shifts and invariant sets

The study of symbolic sequences is usually referred to as symbolic dynamics. The reason for this name is clear in the case of nonlinear dynamical systems since a one-step iterate of the map (3.8) corresponds to the reading of the next symbol in the associated sequence $\mathscr{S}$. This mechanism is formalized by introducing a shift transformation $\hat{\sigma}$ as follows. Let $\Sigma_b = A^{\mathbb{Z}}$ indicate the set of all two-sided infinite sequences $\mathscr{S} = \{\ldots, s_{-1}, s_0, s_1, \ldots\}$ over the alphabet A. The index b denotes the number of symbols in A and is usually dropped unless it is necessary. For each $i \in \mathbb{Z}$, s_i can be regarded as the ith coordinate of $\mathscr{S}$ and the mapping $\pi_i : \Sigma \to A$, such that $s_i = \pi_i(\mathscr{S})$, is called the ith projection of the power space Σ onto the base space A. A topology (*product topology*) is induced in the *sequence space* Σ by defining the distance $d(\mathscr{S}, \mathscr{T})$ between sequences $\mathscr{S}$ and $\mathscr{T}$ (with coordinates $s_i, t_i \in A$) as

$$d(\mathscr{S}, \mathscr{T}) = \sum_{i=-\infty}^{\infty} |s_i - t_i| b^{-|i|} . \tag{4.5}$$

Accordingly, two sequences are "close" if their entries "agree" in a neighbourhood of the lattice origin $i = 0$. Other metrics can be adopted, depending on convenience (see Wiggins (1988) for the case $b = \infty$). Neighbourhoods are specified in Σ by *cylinder sets*: the n-cylinder $c_n(S)$, relative to the finite sequence $S = \{s_1, \ldots, s_n\}$, is defined as

$$c_n(s_1, \ldots, s_n) = \left\{ \mathscr{T} \in \Sigma : t_i = s_i \text{ for } - \left\lfloor \frac{n-1}{2} \right\rfloor \leq i \leq \left\lfloor \frac{n}{2} \right\rfloor \right\}, \tag{4.6}$$

where $\lfloor x \rfloor$ denotes the greatest integer in x. Thus, all elements ("points") in $c_n(S)$ are close to S. The cylinder sets form a basis for the topology of Σ. The set Σ, with the metric (4.5), is a compact, totally disconnected, *perfect* space (i.e., it is closed and each of its points is a limit point). Notice that these three properties are usually taken as the definition of a Cantor set.

The (left) shift map $\hat{\sigma} : \Sigma \to \Sigma$ is defined by

$$\pi_i(\hat{\sigma}(\mathscr{S})) = \pi_{i+1}(\mathscr{S}) \tag{4.7}$$

and is a homeomorphism: i.e., a continuous, one-to-one mapping of Σ onto itself, with continuous inverse. In agreement with (4.5), a sequence $\mathscr{S} \in A^{\mathbb{Z}}$ can also be represented as a pair of real numbers x and y expanded to the base b (see section 3.4), or as a concatenation $\ldots s_{-1}.s_0 s_1 \ldots$ of symbols, where the "central" dot separates the coordinates of the two semi-infinite sequences corresponding to x and y (to be read from left to right and vice versa, respectively). With this notation, $\hat{\sigma}(\ldots s_{-1}.s_0 s_1 \ldots) = \ldots s_0.s_1 s_2 \ldots$, and the symbol to the right of the origin may be considered as the currently "observed" outcome in a measurement process described by $\hat{\sigma}$.

4.2.1 Shift dynamical systems

The pair $(\Sigma_b, \hat{\sigma})$ may be regarded as a dynamical system, which is called the *full shift* (or *Bernoulli shift*) on b symbols. For $b = 2$, this transformation has two fixed points ($\ldots 000 \ldots$ and $\ldots 111 \ldots$), a period-two orbit ($\ldots 01.01 \ldots =$ $\hat{\sigma}(\ldots 10.10 \ldots)$), and so on. In general, $(\Sigma_b, \hat{\sigma})$ has a countable infinity of periodic orbits of all possible periods and an uncountable number of aperiodic orbits. This is easily understood by recalling that the irrationals in the unit interval $[0, 1]$ constitute an uncountable set and that every number in $[0, 1]$ can be expanded in any base $b \geq 2$, with the irrationals corresponding to nonrepeating sequences of coefficients (Σ is perfect and has, therefore, at least the cardinality of the continuum). Finally, Σ contains an element $\mathscr{S}$ which is dense in Σ: i.e., for any $\mathscr{S}' \in \Sigma$ and $\varepsilon > 0$, there exists some integer n such that $d(\hat{\sigma}^n(\mathscr{S}), \mathscr{S}') < \varepsilon$ (for a proof, see Devaney (1989) or Wiggins (1988)). That is, by suitably shifting $\mathscr{S}$, it is possible to approach any other sequence $\mathscr{S}'$ with arbitrary precision. This completes the list of the main properties of the full shift.

In general, the domain of $\hat{\sigma}$ may be a subset Σ' of Σ, as a result of incompletely folding dynamics (which prevents the occurrence of certain sequences). A set Σ' is called $\hat{\sigma}$-*invariant* if $\hat{\sigma}(\Sigma') = \Sigma'$. The restriction of $\hat{\sigma}$ to a nonempty, closed, $\hat{\sigma}$-invariant set $\Sigma' \subset \Sigma$ yields a *symbolic dynamical system* or *subshift* (Hedlund, 1969), indicated by the pair $(\Sigma', \hat{\sigma})$. Given a bisequence $\mathscr{S}$, its "orbit" is defined as the set $\mathcal{O}(\mathscr{S}) = \{\hat{\sigma}^n(\mathscr{S}) : n \in \mathbb{Z}\}$ of all its images under the shift. Clearly, $\mathcal{O}(\mathscr{S})$ corresponds exactly to the orbit $\mathcal{O}(\mathbf{x}) = \{\mathbf{F}^n(\mathbf{x}) : n \in \mathbb{Z}\}$ of a point under a map $\mathbf{F}$, if there is an application ϕ from the phase space X of $\mathbf{F}$ to sequences in Σ: i.e., if $\phi^{-1} \circ \hat{\sigma} \circ \phi(\mathbf{x}) = \mathbf{F}(\mathbf{x})$, for all $\mathbf{x}$ in the invariant set Λ of $\mathbf{F}$. The invariant space Σ' is conveniently chosen as the *orbit closure* $\overline{\mathcal{O}(\mathscr{S})}$.

Not all sequences give rise to interesting systems. An important class consists of *topologically transitive* systems, which are characterized by the existence of at least one point $\mathscr{S}_0$ (respectively, $\mathbf{x}_0$) with a dense orbit in Σ' (respectively, Λ). This means that the dynamics visits the neighbourhood of every point in Σ'. The full shift is topologically transitive. If the orbit is dense in Σ' for *all* $\mathscr{S} \in \Sigma'$, the system $(\Sigma', \hat{\sigma})$ is *minimal*. Then, Σ' admits no other nonempty, closed, $\hat{\sigma}$-invariant subset than itself. Minimality implies topological transitivity. Every point $\hat{\sigma}^n(\mathcal{S})$ of a minimal system $(\Sigma', \hat{\sigma})$ returns to each neighbourhood of $\mathcal{S}$ for a *relatively dense* set J of indices n: i.e., any interval of length N contains at least one index $n \in J$, for some N. This is tantamount to saying that every point of a minimal set is *recurrent* under $\hat{\sigma}$. Notwithstanding this, a minimal transformation cannot have finite invariant sets, that is, periodic points, unless it acts just on a finite space: the simplest minimal sets are, in fact, a fixed point, a limit cycle, the surface of a torus where a quasiperiodic motion (without any unstable fixed point) takes place. As a consequence, finite graphs admitting aperiodic infinite paths cannot correspond to minimal subshifts since they necessarily contain cycles. More general examples of minimal systems will be given in the next section.

The previous discussion has illustrated how a sequence $\mathcal{S} \in \Sigma$, irrespective of its origin (DNA or cellular-automaton configuration, nonequilibrium state generated in the course of a Monte Carlo simulation, etc.), can be formally associated with a dynamical process, the shift map, for which time is the lattice coordinate. In the case of a symbolic signal $\mathcal{S}$ arising from a dynamical system such as (3.8) through a generating partition, shifting the sequence by one place to the left with $\hat{\sigma}$ is equivalent to iterating the map $\mathbf{F}$ once. This suggests that a good understanding of the structure of $(\overline{\mathcal{O}(\mathcal{S})}, \hat{\sigma})$ may lead to the construction of a model for the observed signal $\mathcal{S}$ consisting of a nonlinear map $\mathbf{F}$. When the correspondence between points in Λ and sequences in Σ' is established by a homeomorphism ϕ, the dynamical system $(\Lambda, \mathbf{F})$ and the subshift $(\Sigma', \hat{\sigma})$ are *topologically conjugate*. The horseshoe map $\mathbf{G}$, for example, is conjugate to the full shift and has, therefore, a countable infinity of periodic orbits of arbitrarily high periods (all of saddle type), an uncountable infinity of aperiodic orbits and a dense orbit. When using homeomorphisms to study the relations between maps acting in different spaces, it is particularly important to identify properties that are left unchanged by these transformations. Examples of these so-called *topological invariants* are compactness, connectedness, perfectness and the set of the lengths of all periodic orbits. A metric (i.e., probabilistic) characterization of dynamical systems will be introduced in Chapter 5.

4.3   Languages

From a topological point of view, shift dynamical systems can be classified according to the properties of the sequences that belong to their invariant set, i.e., to the (closure of the) orbit $\mathcal{O}(\mathcal{S}) = \{\hat{\sigma}^n(\mathcal{S}) : n \in \mathbb{Z}\}$, produced by iterating a "point" $\mathcal{S} = \ldots s_1 s_2 \ldots$ in Σ'. This can be done in a systematic way by considering all finite subsequences occurring in $\mathcal{S}$. An *n-block* $T = t_1 t_2 \ldots t_n$ (also called *word* or *string*) appears in $\mathcal{S}$ if $T = s_{i+1} s_{i+2} \ldots s_{i+n}$ for some i.

A formal *language* $\mathscr{L}$ is a set of words formed from an alphabet A. The *complement* of $\mathscr{L}$ is the set of words not in $\mathscr{L}$. The length of a word w, denoted by $|w|$, is the number of symbols in w; the empty string ϵ contains no symbols. The *concatenation* of two words v and w is indicated as vw; the symbol v^n indicates the nfold consecutive repetition $vv\ldots v$ of v; vv is a "square" and v^n the nth power of v. A string v is a proper *subword* (or *factor*) of a sequence S if $S = uvw$ (with $|uw| \geq 1$). A number of leading (trailing) symbols in a string constitutes its *prefix* (*suffix*): the word $w = abcd$ has prefixes ϵ, a, ab and abc. The "largest" language over an alphabet A, indicated by A^*, consists of all finite strings with letters from A and contains all other languages: if $A = \{0,1\}$, $A^* = \{\epsilon, 0, 1, 00, 01, 10, 11, 000, \ldots\}$ in the so-called *lexicographic order* (i.e., with the words listed with increasing length and alphabetically ordered within each constant-length group).

In general, a physical system or a mathematical model do not "allow" all concatenations of symbols to occur: there are forbidden sequences. Analogously, in most natural languages one never encounters more than four consecutive consonants and the letter "q" is almost exclusively followed by the "u". In DNA molecules, although all codons are found, some of their concatenations may be forbidden. In CAs there may be no initial condition leading to a given configuration $\mathscr{S}$ after even a single iteration; essentially the same happens in nonlinear dynamical systems when phase space is incompletely folded over itself. Of course, if a word v is forbidden, all its extensions uvw are forbidden as well. A language is *transitive* if, for all u and w in $\mathscr{L}$, there exists a v such that $uvw \in \mathscr{L}$.

The classification of languages according to their structure is part of the task of characterizing complexity. It is the existence of "rules" that makes it difficult to describe a language in a condensed way. Notice, first of all, that a language $\mathscr{L}$ may be specified without any reference to a shift dynamics. It may happen, however, that an arbitrarily defined language finds no counterpart in any aperiodic transitive shift dynamical system. The language $\mathscr{L}' = \{0, 1, 10, 11, 101, 110, 111, \ldots\}$ is an example thereof, since it does not contain sequence 01 explicitly (it only appears as a subword). A language $\mathscr{L}$ is *factorial* if every subword of elements of $\mathscr{L}$ is an element of $\mathscr{L}$; moreover, $\mathscr{L}$ is *extendible* if every element $v \in \mathscr{L}$ is extendible as uvw, with $uvw \in \mathscr{L}$ and $u, w \in A$. Symbolic dynamical systems and factorial extendible languages are related through a bijection (Blanchard & Hansel, 1986). In the following, languages of this kind will be mainly considered. If $\mathscr{L}(\mathscr{S}) = \mathscr{L}(\mathcal{O}(\mathscr{S}))$ arises from an autonomous symbolic dynamical system with a dense orbit $\mathscr{S}$, it is uniquely specified by the set of all finite subwords of $\mathscr{S}$ and is shift-invariant: i.e., $\mathscr{L}(\mathscr{S}) = \mathscr{L}(\hat{\sigma}(\mathscr{S}))$ or, in other words, the set of constraints on $\mathscr{L}$ is invariant under translations.

A convenient specification of a language is provided by the set $\mathscr{F}$ of all *irreducible* forbidden words (IFWs): a forbidden string w is irreducible if it does not contain any forbidden proper subword. The set $\mathscr{F} = \{00\}$ corresponds to a language in which all concatenations of the words $w_1 = 1$ and $w_2 = 01$ are allowed: $\mathscr{L} = \{w_1, w_2, w_1w_1, w_1w_2, w_2w_1, w_2w_2, \ldots\}$. All strings of the form $u00v$, with $|uv| \geq 1$, are forbidden but not irreducible: they can be decomposed into words belonging to $\mathscr{L}$ and $\mathscr{F}$. The "forbidden" set $\mathscr{F} = \{00, 010, 111\}$ corresponds to the periodic signal $\mathscr{S} = (\ldots 011011 \ldots)$, while $\mathscr{F}_1 = \{10^{2n+1}1, n \in \mathbb{N}\}$ defines the *even system*, in which only an even number of 0s may appear between 1s.

Every formal language $\mathscr{L}$ can be associated with a discrete automaton that is able to "recognize" all legal words of $\mathscr{L}$ when acting on a signal $\mathscr{S}$ (a detailed discussion is deferred to Chapter 7). Formal languages can then be classified according to the size of the memory required by the corresponding automata. Intuitively, this depends on the "richness" of the set $\mathscr{L}$ (or $\mathscr{F}$) and on the lengths of the words involved. The lowest-order class consists of the *subshifts of finite type* (SFT) which are specified by a finite list $\mathscr{F}$ of finite forbidden words. Any discrete Markov process gives rise to such a language. Therefore, SFTs are also called topological Markov shifts.

An extension of the subshifts of finite type is represented by the *sofic systems* (SS). A precise definition (Adler, 1991) requires the notion of *follower set*: given a left infinite sequence $S_- = \ldots s_{-2}s_{-1}$, its follower set consists of all semi-infinite sequences $S_+ = s_0s_1 \ldots$ such that S_-S_+ is allowed. A system is sofic if the number of different follower sets associated with all possible sequences S_- is finite (each set, of course, may contain an infinity of sequences). If there exists an n_0 such that the follower set of any S_- is determined by the last n_0 symbols $\{s_{-n_0}, \ldots, s_{-1}\}$ of S_-, the system is an SFT; otherwise, it is called *strictly sofic*. In fact, an SFT is specified by a finite set $\mathscr{F}$ of IFWs: if the maximum of their lengths is set to $n_0 + 1$, it is sufficient to examine the last n_0 symbols in the trajectory in order to predict the set of all possible continuations (the follower set). Although SSs can exhibit an infinity of prohibitions, they can still be represented as finite graphs and for this reason they coincide with the class of *regular languages* (Weiss, 1973; Coven & Paul, 1975), as will be explained in Chapter 7.

Two simple examples of SSs are provided by the above defined even system and by the language having the set $\mathscr{F}_2 = \{01^{2n}0, n \in \mathbb{N}\}$ of IFWs (as an exercise, the reader may construct the graph corresponding to $\mathscr{F}_2$: the solution can be found in Chapter 7). The latter describes the symbolic dynamics of the roof map (4.4) at $a = 2/3$, with the generating partition $\mathscr{B} = \{[0, 1/2), [1/2, 1]\}$. Notice that the refined partition $\mathscr{B}' = \{[0, 1/2), [1/2, x^*), [x^*, 1]\} = \{I_0, I_1, I_2\}$, obtained by splitting the interval $[1/2, 1]$ at the fixed point $x^* = 2/3$, is Markovian. In fact, both I_0 and I_1 are mapped onto I_2 and I_2 onto the union $I_0 \bigcup I_1$. In this new alphabet, the dynamics is an SFT with transition matrix

$$\mathbf{M} = \begin{pmatrix} 0 & 0 & 1 \\ 0 & 0 & 1 \\ 1 & 1 & 0 \end{pmatrix}. \tag{4.8}$$

The intervals I_i can be relabelled by the words $w_0 = 02$, $w_1 = 12$ and $w_2 = 2$, from left to right, thus taking into account the compulsory transitions in the dynamics. With this position, one recovers the full shift over the two words w_0 and w_1: w_2 becomes superfluous. In general, merging elements of a Markov partition yields a "sofic partition" (independently of whether the resulting shift still properly encodes the map's dynamics or not). Vice versa, refinements of the latter may yield Markov partitions (Adler, 1991).

Besides SFTs, a further relevant subclass of sofic systems is represented by *renewal systems* (RS), which consist of all infinite concatenations of a finite set of words. The even system is an RS, with $w_1 = 1$ and $w_2 = 00$.

For a meaningful classification of symbolic sequences, one needs to compare the shifts that arise from them through orbit closure. Two shifts are *conjugate* if their spaces are connected via a homeomorphism which commutes with the respective shifts. The two sofic systems discussed above are not conjugate to any subshift of finite type. This confirms that sofic systems form a broad class which strictly contains not only SFTs but also RSs: in fact, there are examples of SSs that cannot be conjugate to any RS (Keane, 1991). Clearly, the system defined by the set $\mathscr{F}_2$ above is strictly sofic. Finite-time spatial configurations of elementary cellular automata give rise to regular languages if the initial condition also does. This is a consequence of the finiteness of the range of CA-rules. In the infinite-time limit, however, other kinds of languages may be generated (see Chapter 7).

4.3.1 Topological entropy

The "richness" of a language arising from a shift dynamical system $(\Sigma', \hat{\sigma})$ can be roughly characterized by means of the *topological entropy* K_0 which yields the exponential growth rate of the number $N(n)$ of legal sequences as a function of their length n:

$$N(n) \sim e^{nK_0} \quad \text{for} \quad n \to \infty, \tag{4.9}$$

where the symbol $\sim$ denotes the leading asymptotic behaviour in n. The Bernoulli shift admits 2^n different words of length n, so that its topological entropy is $K_0 = \ln 2$: i.e., it saturates to the limit value $\ln b$, where b is the cardinality of the alphabet A. If there are prohibitions, $K_0 \leq \ln b$.

Given an SFT $(\Sigma', \hat{\sigma})$ with transition matrix $\mathbf{M}$, there are two equivalent ways of determining K_0 besides the direct application of its definition (4.9):

1. If $N_p(n)$ denotes the number of fixed points of $\hat{\sigma}^n$ in the invariant set Σ' (i.e., of n-periodic points of $\hat{\sigma}$), then

$$K_0 = \lim_{n\to\infty} \frac{\ln N_p(n)}{n}. \tag{4.10}$$

2. $K_0 = \ln \mu_1$, where μ_1 is the eigenvalue of $\mathbf{M}$ with maximal modulus (if $\mathbf{M}$ is irreducible, μ_1 is unique and positive).

Moreover, $N_p(n) = Tr(\mathbf{M}^n)$. Clearly, the former relation is useless for a minimal system which, by definition, does not admit periodic orbits (unless it consists of a single periodic orbit itself). Interestingly, there are minimal systems with positive topological entropy (Furstenberg, 1967; Grillenberger, 1973; Denker *et al.*, 1976).

Although systems with larger K_0 possess a richer language, in terms of the number of allowed words, they may be simpler to describe from a combinatorial point of view than others in which many more prohibitions occur. This is the case of strictly sofic systems, compared with SFTs, for which $\mathscr{F}$ is an infinite set. In this sense, sofic systems do belong to a higher "complexity class" than SFTs. One can construct languages that are even more difficult to "understand" than SSs: for example, the set of all binary sequences in which 1s are separated by a prime number of 0s (Adler, 1991). Even more elaborate examples are often constructed both in number theory and in language theory. Remaining in a dynamical context, it is worth mentioning that geodesic flows (Bedford *et al.*, 1991) do not produce SSs but rather languages obtained from sofic ones by removing a dense set of points (sequences in $\mathscr{F}$) (Adler, 1991). This set involves an infinite number of rules but can still be specified by a finite number of statements.

4.3.2 Substitutions

Regular languages are generated *sequentially* by following the allowed paths on a finite graph. Many physical, mathematical, or biological systems, however, have an intrinsically *parallel* dynamics. For example, the eddies in a turbulent fluid evolve simultaneously in time, according to the local forces; cellular automata and Monte Carlo processes update all spins on a lattice at each time step; the cells in many living organisms express their function nearly independently of one another. A simple class of mathematical devices has been proposed to model cellular development in elementary systems (Lindenmayer, 1968) and the formation of *quasicrystals* (Steinhardt & Ostlund, 1987; see also the focus issue of the Journal de Physique, Coll. C3, Suppl. 7, **47**, 1986). It is based on *substitutions*: namely, maps $\psi : A^* \to A^*$ which transform each symbol $s_i \in A$ into a fixed word $\psi(s_i) = w_i$ over A. Upon iteration, the substitution is performed simultaneously on all symbols in the currently obtained word. Furthermore, the empty word ϵ obeys $\psi(\epsilon) = \epsilon$ and $\psi(s_j s_k)$ yields the concatenation $\psi(s_j)\psi(s_k)$ (these properties define ψ as a *morphism*). The substitution is *nondestructive* if none of the images $\psi(s_i) = \epsilon$. It is further assumed that, for some $k \geq 1$ and some $s \in A$, $\psi^k(s)$ contains all letters of A and that $|\psi^k(s)| \to \infty$ for $k \to \infty$. This ensures the existence of an infinite limit-word ω. More concisely, ω is always defined if $\psi(s) = sw$ for some $s \in A$ and $w \neq \epsilon$ (i.e., if ψ is a *prefix-preserving* morphism).

Examples:

[4.6] The simplest scheme yielding an aperiodic limit-word is the *Fibonacci* substitution $\psi_F(\{0,1\}) = \{1,10\}$. Starting with $S_0 = 0$, one obtains $S_1 = \psi(S_0) = 1$, $S_2 = \psi^2(S_0) = 10$, ..., $S_{n+1} = S_n S_{n-1}$. Hence, the length $|S_{n+1}| = |S_n| + |S_{n-1}|$ coincides with the Fibonacci number F_{n+1}. The fixed-point sequence $\psi_F^\infty(0)$ is also generated by the bare circle map (3.13), with symbols $s_n = \lfloor x_{n-1} + \alpha \rfloor$ and $x_0 = \alpha$, when α equals the golden mean ω_{gm}. The inverse map yields the equivalent substitution $\psi_{F'}(\{0,1\}) = \{1,01\}$ (since $\psi_{F'}^2$ is prefix-preserving but $\psi_{F'}$ is not, one obtains two limit-words when reading from left to right, both specular to $\psi_F^\infty(0)$). The number of allowed sequences is $N(n) = n + 1$: hence, $K_0 = 0$. $\qquad\square$

[4.7] The symbolic dynamics at the period-doubling accumulation point (Ex. 3.3) is most concisely described by $\psi_{pd}(\{0,1\}) = \{11,10\}$. This rule is closely related to the *Morse–Thue* (MT) substitution $\psi_{MT} : \{0,1\} \rightarrow \{01,10\}$, which is obtained from a period-doubling map by turning one of the two monotonic branches upside-down (Procaccia *et al.*, 1987). In both cases, $|\psi^n(0)| = 2^n$. The MT limit-sequence can also be written sequentially, since $S_n = \psi^n(0)$ satisfies $S_n = \overline{S}_{n-1} S_{n-1}$, where the overbar denotes the complement (i.e., $\overline{0} = 1$, $\overline{1} = 0$). The number $N(n)$ for the MT rule is upper-bounded by $10n/3$ (De Luca & Varricchio, 1988). $\qquad\square$

[4.8] A language is called *kth-power-free* if $w^k \notin \mathscr{L}$, $\forall w \neq \epsilon$, for some integer $k > 1$. The infinite Fibonacci word $\psi_F^\infty(0)$ is 4th-power-free (the sequence 101 occurs three times consecutively) and $\psi_{MT}^\infty(0)$ is cube-free. An even stronger lack of periodic patterns is exhibited by the MT morphism on three symbols $\psi_{MT_3}(\{0,1,2\}) = \{012,02,1\}$ (Lothaire, 1983): when applied to $S_0 = 0$, it is *square-free*. The number $N(n)$ for this language is also linearly bounded. $\qquad\square$

In general, given a language $\mathscr{L}$ generated by a morphism, if all symbols of A appear in every word w with length $|w| > n_0 > 0$, the number $N(n)$ is linearly upper-bounded (Ehrenfeucht & Rozenberg, 1981; Coven & Hedlund, 1973). Hence, the topological entropy K_0 is zero. The set $\mathscr{F}$ of forbidden words is even more rarefied. For the Fibonacci morphism, $\mathscr{F} = \{v_m, m \geq 0\}$, where $v_0 = 00$, $v_{2m+1} = \psi_F(v_{2m})1 = 1w_{2m+1}$, and $v_{2m+2} = 0\psi_F(w_{2m+1})$. It consists of palindrome (i.e., left-right symmetric) words which are obtained from the "images" of the forbidden word 00, according to the rules above. The number $N_f(n)$ of IFWs of length n is 1 if n is a Fibonacci number F_k, for some $k \geq 2$, and 0 otherwise.

These prohibitions are rare but "strong": in fact, $N(n) = n + 1$ grows only linearly. Similar considerations hold for the MT sequences. The systems of Ex. 4.7 and 4.8 are minimal. Their directed graphs are infinite and contain no cycles. Minimality essentially means that the system consists of a single, infinite, aperiodic sequence (together with all its shifted images).

Further reading

Alekseev & Yakobson (1981), Hao (1989), Lind & Marcus (1995), Ott (1993), Schuster (1988).

Chapter 5

Probability, ergodic theory, and information

As discussed in the previous chapter, the action of a shift map $\hat{\sigma}$ associated with a symbolic signal $\mathscr{S}$ can be partly described in terms of a formal language. So far, however, only topological aspects of shift transformations have been mentioned: in particular, all possible words over an alphabet have just been catalogued as either allowed (existent) or forbidden (nonexistent) by the dynamics, without considering their relative weight. "Metric" features such as the frequency of occurrence of a given string S in the signal $\mathscr{S}$ have been ignored.

The aim of the present chapter is to furnish a more complete characterization of complex patterns, by taking into account their statistical properties as well. This is necessary because most interesting signals display definite analogies with stochastic processes, the understanding of which requires evaluating average values and fluctuations of suitable observables.

As already mentioned, the analysis of complex behaviour entails a classification problem. For example, subshifts of finite type are, from a purely topological point of view (recall the finiteness of the set $\mathscr{F}$ of forbidden sequences), simpler than sofic systems which, in turn, belong to a lower class than substitutions (since the corresponding graphs would be infinite). This approach has been followed especially in computer science (see Chapter 7). On the other hand, a statistical study allows discriminating systems with nearly the same language but different degree of complexity (in its acceptation of unpredictability or stochasticity).

Various aspects of the complexity of aperiodic symbolic patterns or time series have been studied in the context of different scientific disciplines. A first, rudimentary classification could be based on the structure of density

functions satisfying certain invariance properties (Section 5.1). In the framework
of stochastic processes, signals are usually studied by evaluating conditional
probabilities (Section 5.2, 5.3) and correlation functions (Section 5.4). Ergodic
theory classifies dynamical systems according to their long-term behaviour, by
means of suitable time-averages (Section 5.5). The degree of unpredictability
in a signal is quantified by its information content or entropy (Section 5.6).
Finally, the study of the correlations between different regions in a pattern
leads to a thermodynamic formulation of the theory of shift dynamical systems
in which the interactions between words play a fundamental role (Chapter 6).
It may be recalled, for example, that the combinatorial problem of counting
periodic points in a shift space can be solved by methods originally introduced
in statistical mechanics.

   Therefore, a wealth of possible complexity measures is available to undertake
a systematic study of spatial or temporal structures with seemingly complex
properties. It must be stressed again that the investigation is here restricted to
one-dimensional configurations.

5.1   Measure-preserving transformations

Physical systems with time-independent statistical properties are described by
measure-preserving transformations (MPTs) acting in a measure space $(X, \tilde{\mathcal{B}}, m)$,
where $\tilde{\mathcal{B}}$ is a σ-algebra of subsets of X and m a probability measure (see
Appendix 3 for a summary of the relevant mathematical notions).

Definition: A mapping $\mathbf{F} : X \to X$ is a measure-preserving transformation if

1. $\mathbf{F}$ is measurable $\left(\text{i.e., } \forall\, B \in \tilde{\mathcal{B}},\ \mathbf{F}^{-1}(B) \in \tilde{\mathcal{B}} \right)$;
2. $m\left(\mathbf{F}^{-1}(B)\right) = m(B),\ \forall\, B \in \tilde{\mathcal{B}}$.

The measure-preservation property clearly depends on both the map $\mathbf{F}$ and the
measure m. The σ-algebra $\tilde{\mathcal{B}}$ may be interpreted as the family of observable
events (outcomes of an experiment). A measurable function $f : X \to \mathbb{R}$
characterizes the measurement process: $f(\mathbf{x})$ represents the value assumed by
some physically relevant observable when the system is in the state $\mathbf{x}$. Its
expectation value $\langle f \rangle$ with respect to the *invariant* measure m is given by the
integral $\langle f \rangle = \int_X f(\mathbf{x})dm$.

   For a map $\mathbf{F} : X \to X$, let $\tilde{\mathcal{B}}$ and $\tilde{\mathcal{B}}_n$ be the σ-algebras generated by the
elements of a partition $\mathcal{B}$ and of its nth refinement $\mathcal{B}_n$ (Eq. (4.2)), respectively.
Consider a trajectory $\{\mathbf{x}_i\}$ of $\mathbf{F}$. The *elementary* events are of the type $\mathbf{x}_i \in B_{s_i}$,
for $n \in \mathbb{Z}$ and $B_{s_i} \in \mathcal{B}$. The *compound* events (of order n) take the form
$(\mathbf{x}_{i+1} \in B_{s_{i+1}}, \ldots, \mathbf{x}_{i+n} \in B_{s_{i+n}})$, which is equivalent to $\mathbf{x}_{i+1} \in B_{s_{i+1}} \cap \ldots \cap \mathbf{F}^{-n+1}(B_{s_{i+n}})$.
Hence, the localization of a phase-space point in the nth refinement $\mathcal{B}_n$ of $\mathcal{B}$ is

an elementary event relatively to $\tilde{\mathcal{B}}_n$, whereas it corresponds to a sequence of $n+1$ measurements (trials) with respect to $\tilde{\mathcal{B}}$.

Each order-n event is labelled by a string $S = s_1 s_2 \ldots s_n$ of n symbols. Of course, since the B_i's are disjoint, the events at any order are mutually exclusive: for example, $B_j \cap B_k = \emptyset$ implies that the two conditions $\mathbf{x}_n \in B_i \cap \mathbf{F}^{-1}(B_j)$ and $\mathbf{x}_n \in B_i \cap \mathbf{F}^{-1}(B_k)$ cannot hold simultaneously for $j \neq k$.

The *probability* $P(S)$ that the system is found at any time in the state S (where S is the finite label of $B_S \in X$) is given by the measure $m(B_S)$ of B_S and represents the "mass" of the n-cylinder $c_n(S)$ (Eq. (4.6)). If the motion is ergodic (see Section 5.5), $P(S)$ can be evaluated by repeating the measurement N times and by taking the limit $P(S) = \lim_{N \to \infty} n_S/N$, where n_S is the number of occurrences of event S in N trials. This definition assumes that the *relative frequency* n_S/N approaches a limit for $N \to \infty$. The measure m can then be interpreted as a histogram over the elements of $\tilde{\mathcal{B}}$. When dealing with probability measures, the normalization condition

$$\sum_S P(S) = \sum_S m(B_S) = 1 \tag{5.1}$$

is always satisfied at each order $n = |S|$. A property $\mathcal{Q}$ concerning the points of a measure space $(X, \tilde{\mathcal{B}}, m)$ is said to be *almost everywhere* (a.e.) true or *typical* if it fails to hold only in a subset of $(X, \tilde{\mathcal{B}}, m)$ having m-measure zero (i.e., the probability that $\mathcal{Q}$ is verified is strictly positive in the physically relevant domains).

Measure preserving transformations exhibit various degrees of "complexity" that are not revealed by the shape of the invariant measure m but rather emerge from a study of the modifications produced in a nonequilibrium measure m' under time evolution. This process is better understood by introducing the concept of density. Given a measure space $(X, \tilde{\mathcal{B}}, m)$, any nonnegative real function (distribution) $\rho(\mathbf{x})$ with unit L^1 norm is called a *density*. Formally, the density expresses the relationship between two measures m and m'. If they satisfy

$$m'(B) = \int_B \rho(\mathbf{x}) dm, \quad \forall B \in \tilde{\mathcal{B}},$$

m' is *absolutely continuous* with respect to m and ρ is the associated density. Usually, densities are referred to the Lebesgue measure (i.e., $m = m_L$). From the Radon–Nikodym theorem (Lasota & Mackey, 1985), it follows that a measure m' is absolutely continuous with respect to m if $m'(B) = 0$ whenever $m(B) = 0$: e.g., if $m = m_L$, m' has no positive value on zero-volume sets. Two measures m and m' are *mutually singular* if $m(B) = m'(X - B) = 0$ for some $B \in \tilde{\mathcal{B}}$.

Every measure m on $\mathbb{R}^d$ has a unique decomposition as $m = m_{pp} + m_{ac} + m_{sc}$, where m_{pp} is a *pure point* or *atomic* measure (a countable sum of Dirac δ's), m_{ac} is absolutely continuous with respect to Lebesgue's measure m_L and m_{sc}

is continuous and singular with respect to m_L. In the following, absolute continuity will always be referred to the Lebesgue measure, unless otherwise stated. Denoting with ε the diameter of a ball B and setting $\int_B dm \sim \varepsilon^\alpha$, for $\varepsilon \to 0$, one has the following characterization: $\alpha = 0$ a.e. for a *pp*-measure and $\alpha = d$ (the dimension of X) a.e. for an *ac*-measure. A singular–continuous measure, instead, is characterized by intermediate values of the singularity exponent $(0 < \alpha(\mathbf{x}) < d)^1$ which, possibly, depends on the position $\mathbf{x}$ of B in X (see Chapter 6).

Examples:

[5.1] The constant density $\rho(x) = 1$ is invariant under the bare circle map (3.13) for irrational α (quasiperiodic motion).　　　□

[5.2] The same density is preserved by the Bernoulli map (3.9), although the dynamics is completely different in this case (chaotic). A few other examples can be found in Lasota and Mackey (1985), Collet and Eckmann (1980), Adler (1991).　　　□

[5.3] For attractors arising from C^r-diffeomorphisms, any physically meaningful invariant measure must be absolutely continuous on the unstable manifolds. In fact, the density along these curves is expected to be smooth as an effect of stretching. Transversally to the attracting set, instead, the measure may be rather discontinuous because of the accumulation of branches of the unstable manifolds along the contracting directions. Such measures are called SRB (from Sinai, Ruelle, Bowen) and enjoy a number of useful properties, although their existence is hard to prove in general (Eckmann & Ruelle, 1985). Another approach to a physical definition of measure, due to Kolmogorov, prescribes letting the system evolve under the influence of a stochastic perturbation of amplitude ϵ and taking the limit $\rho(\mathbf{x})$ of the resulting density $\rho_\epsilon(\mathbf{x})$ for $\epsilon \to 0$ (*natural measure*).　　　□

[5.4] The normalized sum of n δ-functions centred on the points of a period-n orbit of $\mathbf{F}$ is also an invariant density for $\mathbf{F}$ even in the chaotic regime. Such atomic measures, however, have no physical relevance since they are unstable under perturbations (recall the notion of natural measure). Notice that an MPT $(\mathbf{F}, X, \tilde{\mathscr{B}}, m)$ is *aperiodic* if its periodic points form a set of m-measure 0: i.e., $m(\{\mathbf{x} \in X : \mathbf{F}^n(\mathbf{x}) = \mathbf{x} \text{ for some } n \in \mathbb{N}\}) = 0$.　　　□

[5.5] For a subshift $(\Sigma', \hat{\sigma})$, the measure-preservation property is referred to cylinder sets. The probability $P(S)$ of sequence $S = s_1 s_2 \ldots s_n$ is given by the mass $m(c_n(S))$ of the cylinder $c_n(S)$. The invariance of the measure m under $\hat{\sigma}$

---

1. Some care must be taken for $\alpha = 0$ or d, since corrections to the leading behaviour are crucial for the assessment of the measure type.

is expressed by $m(\hat{\sigma}^i(c)) = m(c)$, for all $i \in \mathbb{Z}$ and all cylinders c at any order n. The Bernoulli map $B(p_0, p_1)$ (Eq. 3.9) yields a full shift with an invariant measure m_{p_0} which assigns the mass $p_0^k p_1^{n-k}$ to each n-cylinder with k zeros and $n - k$ ones. For $n \to \infty$, the proportions of zeros and ones in the sequence tend to p_0 and p_1, respectively. Measures m_p and m_q with $p \neq q$ are, hence, mutually singular: sets of full measure for m_p have zero measure for m_q and vice versa. The shift transformation, however, preserves all of them. □

5.2   Stochastic processes

If no information is *a priori* available about a given physical system, the probabilistic approach, based on the assumption that the outcome of a measurement is a *random variable*, turns out to be very fruitful. The coordinates of the representative point $\mathbf{x}$ of a physical system in phase space or the corresponding energy $E(\mathbf{x})$ can be treated as random variables. A *stochastic process* $\{\mathbf{x}_t\}$ is a family of random variables which depend on a parameter t, usually interpreted as time: a trajectory $\{\ldots, \mathbf{x}_1, \mathbf{x}_2, \ldots\}$ or a symbolic signal $\mathscr{S} = \ldots s_1 s_2 \ldots$ are typical examples thereof.

A stochastic process $\{\mathbf{x}_t\}$ is *stationary* if its statistical properties are invariant under a shift of the time origin. In particular, the probability $P(\mathbf{x}_1, \mathbf{x}_2, \ldots, \mathbf{x}_n; t_1, t_2, \ldots, t_n)$ of observing the values $\mathbf{x}_1$, $\mathbf{x}_2$, ..., $\mathbf{x}_n$ at times t_1, t_2, ..., t_n, respectively, satisfies

$$P(\mathbf{x}_1, \ldots, \mathbf{x}_n; t_1, \ldots, t_n) = P(\mathbf{x}_1, \ldots, \mathbf{x}_n; t_1 + \tau, \ldots, t_n + \tau), \qquad (5.2)$$

for any $\tau \in \mathbb{R}$ (if $\mathbf{x}$ is a continuous variable, the density ρ must be considered instead of P). The stationarity of a signal is synonymous with the existence of an invariant measure for the associated map. In the following, we shall almost exclusively deal with stationary processes. Nonstationary behaviour is observed in bounded dynamical systems when some type of intermittency occurs; it is also common in the spatial configurations of cellular automata generated from particular initial conditions and in formal languages. Physically, stationarity ensures the feasibility of a statistical analysis, since the probabilities are well-defined as limits of occurrence frequencies.

Example:

[5.6]   The outcome $\{s_1, s_2, \ldots\}$ of a coin-tossing experiment (i.e., a Bernoulli shift) is stationary. Its "integral" $z_n = \sum_{i=1}^{n} s_i$, called a *random walk* or discrete *Brownian motion*, is nonstationary as a function of n. If $s_i \in \{-1, 1\}$, the returns of z_n to the initial condition $z_0 = 0$ may occur for even $n = 2m$ only and their probability p_{2m} decreases as $(\pi m)^{-1/2}$ for large m (Feller, 1970). □

The degree of interdependence of successive observations can be quantified by means of *conditional probabilities*. If some event v (a finite symbol-sequence) has just occurred, the conditional probability of observing the new event w, given v, is defined by

$$P(w|v) \equiv \frac{P(vw)}{P(v)}, \tag{5.3}$$

where $P(vw)$ is the *joint* probability for the compound event vw. For a dynamical system $(\mathbf{F}, X)$, $P(vw) = m\big(B_v \cap \mathbf{F}^{-1}(B_w)\big)$ (the notation vw means that v occurs before w). Clearly, $0 \le P(w|v) \le 1$, so that conditional probabilities $P(w|v)$, for all w, enjoy the same properties as ordinary probabilities (to which they reduce if one considers a new sample space, defined by the hypothesis event v). Relation (5.3) is easily extended to n events: for $n = 3$, one has

$$P(uvw) = P(vw|u)P(u) = P(w|uv)P(v|u)P(u) .$$

If $\{s_1, \ldots, s_b\}$ is a set of mutually exclusive events, one of which necessarily occurs (i.e., the s_i span the whole sample space),

$$\sum_{i=1}^{b} P(s_i|S) = 1 , \tag{5.4}$$

for any possible past sequence S. Furthermore, the probability of S satisfies

$$P(S) = \sum_{i=1}^{b} P(s_i)P(S|s_i) . \tag{5.5}$$

When a combination of trials is carried out, either simultaneously or successively, the possible outcomes need not be independent of each other. In general, one expects that the conditional probability $P(w|v)$ does not equal the absolute probability $P(w)$. For events occurring in a sequence, knowledge of the past (v) can be used to improve the prediction on the future outcome w. When, however, $P(w|v) = P(w)$ the uncertainty on w cannot be reduced by the observation of v. In such a case, $P(vw) = P(v)P(w)$ and w is said to be (statistically) independent of v (or vice versa, by symmetry). The general definition can be stated inductively: the events s_1, s_2, ..., s_n are *mutually independent* if any k ($2 \le k < n$) of them are independent and

$$P(s_1 s_2 \ldots s_n) = \prod_{i=1}^{n} P(s_i) . \tag{5.6}$$

The conditional probabilities contain information about the dependence of an experimental observation on the previous ones. The simplest generalization of sequences of independent events is represented by the so-called *Markov chains*, in which the probability of an outcome w depends on that of the previous trial as well, but not on remoter ones. The properties of a Markov process are, therefore, completely specified by one-step conditional probabilities $P(s_{n+1}|s_n)$:

$$P(s_{n+1}|s_1 s_2 \ldots s_n) = P(s_{n+1}|s_n),$$
$$P(s_1 s_2 \ldots s_n) = P(s_1)P(s_2|s_1) \cdot \ldots \cdot P(s_n|s_{n-1}). \qquad (5.7)$$

The probability $P^{(n)}(w|u)$ of observing the string w n time units after u is given by a sum over all possible paths of length n starting at u and ending at w: $P^{(n)}(w|u) = \sum_v P(w|v)P^{(n-1)}(v|u)$. This formula can be further generalized to the *Chapman–Kolmogorov* equation

$$P^{(n+m)}(w|u) = \sum_v P^{(m)}(w|v)P^{(n)}(v|u). \qquad (5.8)$$

Stationary Markov chains are frequently used as models for physical systems, possibly with continuous time and continous space (chemical reactions, nucleation, population dynamics, tunnel diodes, growth phenomena, genetics). In the discrete case, the process can be interpreted as a sequence of *transitions* between *states* of the system, labelled by the symbols of an alphabet $A = \{1, 2, \ldots, b\}$. The corresponding *transition probabilities* are usually disposed in a *transition matrix*

$$\mathbf{M} = \begin{pmatrix} P(1|1) & P(2|1) & \ldots & P(b|1) \\ P(1|2) & P(2|2) & \ldots & P(b|2) \\ \cdot & \cdot & \ddots & \cdot \\ P(1|b) & \cdot & \ldots & P(b|b) \end{pmatrix} \qquad (5.9)$$

which has nonnegative entries and unit row-sums. Matrix $\mathbf{M}$ generalizes the corresponding topological transition matrix introduced in Chapter 4. Together with an initial distribution $P_0 = \{P_0(1), P_0(2), \ldots, P_0(b)\}$, $\mathbf{M}$ completely specifies the Markov chain. A distribution P satisfying Eq. (5.5), for all S, is at *equilibrium* (stationary). A nonequilibrium distribution P_τ evolves in time τ according to

$$P_{\tau+1}(w) = \sum_v P(w|v)P_\tau(v), \qquad (5.10)$$

for all v, $w \in A$. The n-step transition probabilities $P^{(n)}(w|v)$ can be arranged in a matrix $\mathbf{M}^{(n)}$, analogously to the case $n = 1$. Inspection of Eq. (5.8) shows that $\mathbf{M}^{(n)} = \mathbf{M}^n$ is the nth power of the one-step matrix $\mathbf{M}$ ($\mathbf{M}^{(0)}$ being the identity).

State j can be reached from state i if there exists an $n \geq 0$ such that $M_{ij}^n = P^{(n)}(j|i) > 0$. Notice that there may be closed sets of states, from which no transition to outside states is possible. Each closed set corresponds to a distinct Markov process. A Markov chain is called *irreducible* if the only closed set is the set of all states: i.e., if and only if every state can be reached from every other state in a finite number of steps. An even stronger condition is that *all* entries of $\mathbf{M}^n$ are simultaneously nonzero for some n: such a matrix is called *primitive*. Then, no matter which initial condition has been chosen, after time

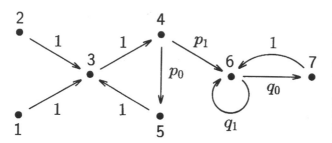

Figure 5.1 A directed graph with attached transition probabilities: $p_0 + p_1 = 1$, $q_0 + q_1 = 1$. Notice the transient states 1 and 2, the transient periodic set (3,4,5) and the closed aperiodic set (6,7).

n the system can be found in any of the available states. A state k is *transient* if and only if $p_{kk} \equiv \sum_{n=0}^{\infty} P^{(n)}(k|k) < \infty$. Furthermore, k has period $n_k \geq 1$ if $P^{(n)}(k|k) = \delta_{n, l n_k}$, with $l \in \mathbb{N}$. If no such n_k exists, the state is aperiodic.

A Markov chain can be represented by means of a directed graph. The difference with the purely topological examples of Chapter 4 lies in the explicit consideration of the transition probabilities appended to each arc, as illustrated in Fig. 5.1. The concept of Markov process can be further generalized by taking into account higher-order conditional probabilities. A kth-order Markov process is defined by

$$P(s_n|s_1 s_2 \ldots s_{n-1}) = P(s_n|s_{n-k} s_{n-k+1} \ldots s_{n-1}) \tag{5.11}$$

where $n \geq k$. Accordingly, a signal is said to originate from a k-*memory source* if every new symbol depends exclusively on the last k (the case $k = 0$ corresponding to complete independence). In order to describe a k-memory source, it is necessary to specify the probabilities of all words of length $k + 1$: the process is then represented as a walk on a graph containing b^k states (words of length k), characterized by an appropriate word-to-word transition matrix.

Example:

[5.7]   Markov chains can be profitably used to describe and classify the behaviour of cellular automata. In fact, this modellization has led to the formulation of the "local structure theory" (Gutowitz *et al.*, 1987), which yields self-consistent estimates of word probabilities. At the lowest order, this approach reduces to the mean field theory (Wolfram, 1986), which considers only symbol probabilities by neglecting memory effects along the spatial direction.

Given a CA rule with interaction range r, let B_S denote the set of sequences S', of length $n' = 2r + n$, that yield the n-block S under the action of the automaton. The probability $P(S)$ of S in the limit set can then be expressed as

$$P(S) = \sum_{S' \in B_S} P(S') . \tag{5.12}$$

With an order-n Markov approximation, all correlations over distances larger than n are ignored, so that the open hierarchy in Eq. (5.12) is turned into a closed set of self-consistent equations. For $n = 1$, $P(S')$ is a product of independent terms which, in the binary case $b = 2$, reads

$$P(S') = P(1)^{n_1}(1 - P(1))^{2r+1-n_1} ,\tag{5.13}$$

where n_1 denotes the number of 1s in S' and $P(1)$ is the probability of symbol 1. Indicating with $N(i)$ the number of preimages S' of symbol 1 which contain i 1s and substituting (5.13) into (5.12), one obtains the relation

$$P(1) = \sum_{i=0}^{2r+1} N(i)P(1)^i(1 - P(1))^{2r+1-i} ,$$

which shows that all rules with the same value of $N(i)$ admit the same mean field theory. This observation led Gutowitz (1990) to view the local structure theory not only as a tool for the estimation of the spatial invariant measure of a CA, but also as a natural way for the parametrization of the rules.

At order n, the correlations forbid a simple factorization such as in Eq. (5.13). By the assumed Markov property, $P(S')$ is approximated as

$$P(S') = P(s'_1 \ldots s'_{n-1})P(s'_n|s'_1 \ldots s'_{n-1}) \cdots P(s'_{2r+n}|s'_{2r+1} \ldots s'_{2r+n-1}) .$$

The conditional probabilities (see Eq. (5.3)) read

$$P(s'_{k+n}|s'_{k+1} \ldots s'_{k+n-1}) = \frac{P(s'_{k+1} \ldots s'_{k+n})}{P(s'_{k+1} \ldots s'_{k+n-1})} .\tag{5.14}$$

While the numerators are unknown, the denominators can be deduced by the probabilities of shorter sequences. Substitution of all terms of the form (5.14) with $k = 0, 1, \ldots, 2r$ into Eq. (5.12) gives a closed set of equations for the nth order approximation of the invariant measure.   □

5.3 Time-evolution operators

The time evolution of a nonequilibrium density can follow rather different routes for different systems. Its study presents several advantages over that of single trajectories: in particular, distributions are not usually strongly sensitive to initial conditions. For a Markov process in the variable $\mathbf{x}$, the density $\rho_n(\mathbf{x})$ at time n can be formally expressed by means of the iteration

$$\rho_{n+1}(\mathbf{x}) = \mathscr{P}\rho_n(\mathbf{x}) ,\tag{5.15}$$

where $\mathscr{P}$ is the *Frobenius–Perron* (FP) operator. Given a measure-preserving transformation $(\mathbf{F}, X, \tilde{\mathscr{B}}, m)$, $\mathscr{P}$ is uniquely defined by

$$\int_B \mathscr{P}f(\mathbf{x})dm = \int_{\mathbf{F}^{-1}(B)} f(\mathbf{x})dm , \ \forall B \in \tilde{\mathscr{B}} ,\tag{5.16}$$

for any function $f \in L^1$. The FP operator $\mathscr{P}$ is linear and belongs to the class of *Markov operators* which act on functions $f \in L^1$ and enjoy the following properties: $\mathscr{P}f \geq 0$ for $f \geq 0$ (*m*-almost everywhere), and $\|\mathscr{P}f\| \leq \|f\|$ (the symbol $\|f\|$ denoting the L^1-norm). The second property (*contractivity*) implies that $\|\mathscr{P}^n f_1 - \mathscr{P}^n f_2\| \leq \|\mathscr{P}^{n-1} f_1 - \mathscr{P}^{n-1} f_2\|$: that is, the distance between two functions cannot increase upon iteration under $\mathscr{P}$. This is known as the *stability property* of Markov operators. A fixed point density $\rho = \mathscr{P}\rho$ of $\mathscr{P}$ is called *stationary*. A few examples of stationary densities for simple maps have been shown in Section 5.1.

The FP operator $\mathscr{P}_n$ associated with the nth iterate $\mathbf{F}^n$ of a map $\mathbf{F}$ is just $\mathscr{P}^n$. The action of $\mathscr{P}$ on a nonequilibrium distribution ρ can be easily visualized for a one-dimensional map $F(x) : [a, b] \to [a, b]$. If $F(x)$ is differentiable and piecewise invertible, with (possibly multivalued) continuous inverse $F^{-1}(x)$, differentiation of Eq. (5.16) yields

$$\mathscr{P}\rho(y) = \sum_{x = F^{-1}(y)} \frac{\rho(x)}{|F'(x)|} , \qquad (5.17)$$

where the sum extends over all preimages x of the point y at which the density is to be computed. In the d-dimensional case, the derivative F' is replaced by the determinant of the Jacobian matrix $J(\mathbf{x})$.

Examples:

[5.8]   Consider the roof map (4.4) at $a = (3 - \sqrt{3})/4$. The four intervals of the Markov partition are mapped according to: $B_\alpha \to B_\beta \cup B_\gamma$, $B_\beta \to B_\delta$, $B_\gamma \to B_\gamma \cup B_\delta$, $B_\delta \to B_\alpha \cup B_\beta$. Hence, Eq. (5.17) yields

$$\rho_\alpha = \rho_\delta/2 , \qquad\qquad \rho_\beta = \left[\rho_\alpha/(1-a) + \rho_\delta \right]/2 ,$$
$$\rho_\gamma = \left[\rho_\alpha/(1-a) + \rho_\gamma \right]/2 , \quad \rho_\delta = \left[\rho_\beta/(1-a) + \rho_\gamma \right]/2 ,$$

where ρ_i is the density in B_i. The density is constant within each interval B_i: in fact, all points y in any B_i have preimages x in the same source intervals $F^{-1}(B_i)$ so that, by the constancy of $F'(x)$ in B_i, they have equal density values $\rho(y)$. Since the four relations above are linearly dependent, one of them can be eliminated and the solution is determined by resorting to the normalization condition $\rho_\alpha |B_\alpha| + \rho_\beta |B_\beta| + \rho_\gamma |B_\gamma| + \rho_\delta |B_\delta| = 1$. The finiteness of the number of equations is a consequence of the existence of a Markov partition (for this reason, subshifts of finite type, such as the present one, are also called Markov shifts). In general, however, the functional equation (5.17) must be solved for every $y \in X$ (for m_L-almost all a, the roof map yields no finite subshift dynamics). The measure ρ can then be approximated by a sequence of functions having an increasing number of constant pieces with decreasing width.   □

[5.9]   An additional problem to the absence of a finite Markov partition is represented by the intrinsic nonlinearity of physical systems. This may induce an even stronger nonuniformity in the measure than the lack of a Markov partition. In fact, Eq. (5.17) shows that a small value of the Jacobian $J(\mathbf{x})$ at some point $\mathbf{x}$ yields a large value of the density $\rho(\mathbf{y})$ at the image point $\mathbf{y} = \mathbf{F}(\mathbf{x})$. A sequence of points with such a characteristic may eventually build up a singularity in ρ. This occurs, e.g., in the Hénon map (3.12), where the density diverges with a power-law behaviour along the unstable manifold, in the neighbourhood of the homoclinic tangencies. For the logistic map (3.11), a single iterate of Eq. (5.17) is sufficient, since the derivative $F'(x)$ vanishes at the critical point $x_c=0$. At $a = 2$, $\rho(x) = 1 / \left[\pi\sqrt{1 - x^2}\right]$ (Ulam & von Neumann, 1947); for $a < 2$, provided that a chaotic attractor exists, square-root singularities are generated at the forward images $F^n(x_c)$ of x_c (their number is infinite if the map does not admit a Markov partition). □

[5.10]   In one-dimensional maps with an order-η extremum (i.e., $F(x) \approx a + b|x - x_c|^\eta$, for $|x - x_c| \ll 1$), the density satisfies $\rho(x) \approx |x - x_n|^{1/\eta-1}$ around $x_n = F^n(x_c)$, whereas $\rho(x) \approx$ const. (absolutely continuous) at almost all x.   □

[5.11]   The phenomenon of *intermittency*, often observed experimentally (Manneville & Pomeau, 1980; Manneville, 1990; Schuster, 1988), is characterized by long regular (laminar) phases alternating with relatively short, irregular bursts. The so-called type-I intermittency can be modelled by the map $x_{n+1} = x_n + x_n^v + \varepsilon$ (mod 1), where $v > 1$ and $\varepsilon \geq 0$. The small-x behaviour of the invariant measure can be estimated from Eq. (5.17), by setting $\rho(x) \approx x^\alpha$ for $x \ll 1$ (i.e., in the intermittent region) and $\rho(x) \approx$ const. around the *right* preimage $F^{-1}(y) \approx 1$ of every point $y \ll 1$. Expanding for $x, y \ll 1$ and equating the various powers of x yields the characteristic exponent $\alpha = 1 - v$, which is indeed observed for $0 < \varepsilon \ll 1$. When $\varepsilon \to 0$, $\rho(x)$ collapses to Dirac's $\delta(x)$.   □

[5.12]   Intermittency is called *weak* if the invariant density is normalizable (i.e., if $\alpha > -1$) (Großmann & Horner, 1985). In the limit $\varepsilon \to 0$, this happens only if the intermittency-induced singularity at the origin is "compensated" by a suitable reinjection from the chaotic, nonlaminar, region. If an order-η critical point x_c is mapped onto $x = 0$ in two iterations, the density $\rho(x)$ at the right preimage x_2 of any point $y \ll 1$ is no more constant but behaves like $|x - x_2|^{1/\eta-1}$, since $x_2 \approx F(x_c)$ (Ex. 5.10). Equation (5.17) now yields $\alpha = 1/\eta - v$. The importance of these characteristic indices will be discussed in Chapter 6. □

[5.13]   For the vector field (3.4), the evolution of a time-dependent density $\rho(\mathbf{x}; t) \equiv \mathscr{P}^{(t)}\rho(\mathbf{x}; 0)$ is described by the (generalized) *Liouville equation*

$$\frac{\partial \rho}{\partial t} = -\nabla \cdot (\rho \mathbf{f}) .$$

The Lebesgue measure is preserved by any divergence-free $\mathbf{f}$ (Eq. (3.6)). When $\mathbf{f}$ represents a Hamiltonian flow, ρ can be any function of the Hamiltonian

H satisfying $\rho(H) \geq 0$ and $\int \rho(H)dH = 1$. Considering, instead, a compact connected invariant component Γ_E of the constant-energy surface Σ_E, where $|\nabla H| \neq 0$, a canonical Liouville measure is given by

$$m_E(B) = \frac{1}{Z} \int_B \frac{d\sigma_E}{|\nabla H|} \, , \, \forall \, B \in \tilde{\mathscr{B}} \, , \tag{5.18}$$

where σ_E denotes the Lebesgue measure on Γ_E and $Z = \int_{\Gamma_E} d\sigma_E/|\nabla H|$ is the normalization constant.   □

5.4   Correlation functions

A fundamental tool in the analysis of stochastic processes (which we take as stationary, scalar, and real, for simplicity) is the (auto)*correlation function*

$$C(t) \equiv R(t) - \bar{x}^2 = \lim_{T \to \infty} \frac{1}{2T} \int_{-T}^{T} x(t')x(t + t')dt' - \bar{x}^2 \, , \tag{5.19}$$

where

$$\overline{f(x)} = \lim_{T \to \infty} \frac{1}{2T} \int_{-T}^{T} f(x(t))dt$$

is the *mean* of $f(x)^2$. The correlation function is often normalized with respect to the *variance* $C(0) = \sigma^2 = \overline{x^2} - \bar{x}^2$ by defining the *correlation coefficient* $r(t) = C(t)/C(0)$. From Eq. (5.19), it is easily seen that $r(t) \leq 1$, since $C(t) = C(-t)$ and

$$R(0) - R(t) = \frac{1}{2}\overline{[x(0) - x(t)]^2} \, . \tag{5.20}$$

If $C(0) = C(T_0)$, for some finite $T_0 > 0$, $C(t)$ is periodic with period T_0; if $C(0) = C(T_1) = C(T_2)$, for two incommensurate positive numbers T_1 and T_2, $C(t)$ is a constant. A process is called *uncorrelated* (or δ-correlated) when $C(t) = 0$ for all $t > 0$: this occurs when the values $x(t')$ and $x(t + t')$ are statistically independent for any time $t' > 0$. A less pronounced, but still sensible, loss of memory occurs in signals with a correlation function that vanishes exponentially fast for $t \to \infty$. This is typical of nondeterministic or chaotic deterministic signals.

A basic classification of stochastic processes is provided by the *power spectrum*, a real nonnegative function of the frequency ω, defined either by

$$S(\omega) = \int_{-\infty}^{\infty} R(t)e^{-i\omega t}dt \, , \tag{5.21}$$

2. If $\{x(t)\}$ is nonstationary, a generalization of Eq. (5.19) is needed (Papoulis, 1984).

or by the squared Fourier transform

$$S(\omega) = \lim_{T \to \infty} \frac{1}{2T} \left| \int_{-T}^{T} x(t) e^{-i\omega t} dt \right|^2 \qquad (5.22)$$

of the signal $\{x(t)\}$. These two definitions coincide under mild restrictions (*Wiener–Khinchin theorem* (Feller, 1970)). The area under the curve $S(\omega)$ equals $\overline{x^2}$ and can thus be interpreted as the average power of $\{x(t)\}$. Hence, $S(\omega)$ is a normalizable measure for processes with bounded variance.

The power spectrum of a periodic signal with period $T_0 = 2\pi/\omega_0$ has Dirac δs at $\omega_k = k\omega_0$, $k \in \mathbb{Z}$: i.e., it is atomic. Quasiperiodic signals also have atomic spectra, although with a more intricate structure. If $(\omega_1, \ldots, \omega_d)$ are the basic incommensurate frequencies, $S(\omega)$ displays peaks at these values and at all linear combinations of them with integer coefficients. The set of all these frequencies is, therefore, dense and countable. In general, the amplitude coefficients of the δ-peaks decrease quite rapidly with the order of the linear combination (the area under $S(\omega)$ is, in fact, bounded), so that only a finite number of them can be resolved when the power spectrum is directly computed from a finite-length signal[3].

Conversely, a progressive decay of correlations is usually associated with absolutely continuous spectra. Common examples are provided by chaotic and stochastic processes which exhibit *broadband spectra* with approximately Lorenzian peaks of the form $S(\omega) \approx A/[\gamma^2 + (\omega - \omega_i)^2]$ around one or more frequencies ω_i.

Finally, between the above two broad classes, one finds singular continuous spectra, which are associated with correlation functions that, while decaying on the average, exhibit striking recurrent revivals. In Chapter 7, we shall relate *sc*-spectra to temporal hierarchical structures.

Power spectra and correlation functions in the long-time regime provide similar, although not fully equivalent, classifications of signals. As shown by Avron and Simon (1981), for instance, also some *ac*-spectra correspond to correlation functions with occasional revivals. In order to account for this, two subclasses have been introduced: namely, *transient* and *recurrent ac*-spectra. In order to understand the distinction, one needs the concept of *essential interior* $\mathscr{I}_m$ of an *ac*-measure m with *support* $s(m)$ (the complement of the largest open set C with $m(C) = 0$):

$$\mathscr{I}_m = \left\{ \omega \in s(m) : m_L \left([\omega - \Delta, \omega + \Delta] \cap s(m) \right) = 2\Delta \text{ for some } \Delta > 0 \right\}.$$

The complement of $\mathscr{I}_m$ in $s(m)$ is called the *essential frontier*: frontier points can induce correlation revivals, even when they are extremely rarefied (Avron & Simon, 1981). Such behaviour is observed, for example, when the support is

---

3. Indeed, the peaks have a finite width, proportional to $1/T$, where T is the length of the available signal.

a *fat Cantor set* (i.e., it has finite Lebesgue measure)[4]. These sets are nowhere dense and coincide with their essential frontier.

Notwithstanding the usefulness of the spectral analysis, sharper indicators are needed in more complex situations, as will be discussed in the next sections. For example, $S(\omega)$ is insensitive to the phase information of the Fourier transform and cannot provide a full characterization of the dynamics. In addition to this, the spectra of chaotic signals are usually indistinguishable from those of noisy (nondeterministic) ones.

Notice that strict "randomness" is often taken as a synonym of a constant spectrum, that is, of a delta-like correlation function. For a signal to be really random, however, it is necessary that all higher-order correlations, such as

$$C^{(n)}(t_1,\ldots,t_n) = \lim_{T\to\infty} \frac{1}{2T} \int_{-T}^{T} x(t'+t_1)\cdot\ldots\cdot x(t'+t_n)dt' - \bar{x}^n,$$

vanish whenever their arguments are not all equal. The power spectrum is related just to the ordinary two-point correlation (5.19). Loosely speaking, however, a signal is "random" if it has an absolutely continuous spectral component.

Examples:

[5.14] The bare circle map (3.13) yields a quasiperiodic spectrum for any irrational α, with a ratio $\omega_1/\omega_2 = \alpha$ between the frequencies of the main peaks.
□

[5.15] Any map with a quadratic maximum yields, at the period-doubling accumulation point, the same language as that of the substitution ψ_{pd} introduced in Ex. 4.7. The power spectrum of the limit sequence has peaks at all odd multiples $2k+1$ of the frequencies $\omega_n = (2\pi)2^{-n}$, for $n \in \mathbb{N}$, and displays a hierarchical structure: the amplitudes drop by a factor 4 in passing from order n to order $n+1$ and are constant for each n. The power spectrum of the trajectory of the logistic map has been analysed in Feigenbaum (1979b) and Nauenberg & Rudnick (1981).
□

[5.16] Several physical systems display a characteristic low-frequency spectral divergence of the type $S(\omega) \approx \omega^{-\theta}$, with $0.8 < \theta < 1.4$, called $1/f$ *noise* (Weissman, 1988), where $f = \omega/2\pi$. This behaviour, if persistent down to $\omega = 0$, presupposes a nonintegrable spectrum in some interval $0 \le \omega \le \omega_0$ for $\theta \ge 1$ and, in turn, the nonstationarity of the signal (recall that the area under the curve $S(\omega)$ is the average "energy" of $\{x(t)\}$). Hence, it is expected that the singularity of $S(\omega)$, observed in real experiments, be smoothed off below some frequency $\omega_c \ll \omega_0$ which lies beyond the achievable resolution. Nevertheless, the observation of $1/f$ noise in a wide, although finite, frequency range still

4. A fat Cantor set can be constructed by removing a fraction $f_k \sim o(1/k)$ of the currently remaining total length (Lebesgue measure) at the kth generation step.

implies slowly decaying correlations (a manifestation of long-range memory): in fact, if $0 < \theta < 1$, $C(t) \sim t^{\theta-1}$. Logarithmic corrections may also be present (Manneville, 1980). The ubiquity of $1/f$ noise (observed in signals of electronic, magnetic, meteorological, biological, economic, and even musical origin) and its puzzling nature have so far defeated any attempt at a general theoretical explanation. Several models have been proposed but none of them is sufficiently "universal" and robust against perturbations. In one of the commonest views, the signal is interpreted as the superposition of infinitely many signals with low-frequency Lorenzian spectrum, each characterized by a different correlation time τ: $1/f$ noise is obtained for a suitable distribution $P(\tau)$ (Dutta & Horn, 1981). An even simpler, deterministic mechanism is represented by type-I intermittency (Ex. 5.11), in which laminar phases alternate with chaotic bursts: in the limit $\varepsilon \to 0$, the process becomes nonstationary and the spectrum displays various types of low-frequency divergencies, depending on the exponent v (Schuster, 1988). Perturbations can be assimilated to a finite displacement ε and introduce a finite cutoff frequency ω_c. □

[5.17] Slowly decaying correlations with a power-law behaviour are present also in DNA molecules. In order to evaluate the power spectrum of a signal consisting of symbols from a b-nary alphabet, a translation table to numerical values must be provided. A method that avoids introducing spurious correlations assigns the value 1 to one of the four symbols $\{A, C, G, T\}$ and 0 to all others. In this way, the DNA sequence S is mapped to four distinct signals, each selecting the positions of a given nucleotide (in the first, the 1s correspond to As, in the second to Cs, and so on). Indicating with $S_k(\omega)$ the spectrum of the kth signal, an overall spectral indicator can be defined as $S(\omega) = \sum_{k=1}^{4} S_k(\omega)$. In Fig. 5.2 we display two averaged power spectra (referring to A and G) for the human cytomegalovirus strain AD 169 which is composed of 229 354 bases. The sharp peak at $\omega = 2\pi/3$ (period-3) may be related to the codon structure of DNA. A low-frequency noisy component of the type $1/f^{\theta_A}$ with $\theta_A \approx 0.76$ is observed in $S_1(\omega)$ in a small range of frequencies. The exponent θ changes slightly with the characteristic base: $\theta_C \approx 0.83$, $\theta_G \approx 0.92$, $\theta_T \approx 0.77$. Extensive calculations (Voss, 1992), performed on several different species, have shown no direct connection between the values of θ and the "complexity" of the associated species. Moreover, different portions of a DNA sequence yield different exponents, a sign of nonstationarity, and the frequency range in which $1/f$-behaviour is observed is not very large (at most two decades). Highly repetitive subsequences, frequently occurring in introns, have been conjectured to contribute to the observed long-range correlations (Li & Kaneko, 1992). The low-frequency component, however, is most likely caused by the (random) alternation of exons and introns in eukaryotes (Borštnik et al., 1993). Exons usually have lengths between 100 and 1000 bases, much smaller than those of introns. □

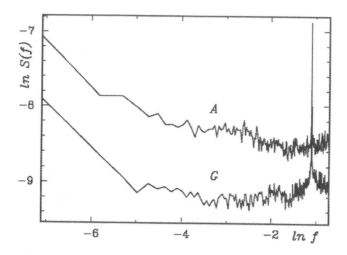

Figure 5.2 Power spectra $S(f)$ versus the frequency $f = \omega/2\pi$ of the human cytomegalovirus strain AD 169, in a doubly logarithmic scale. The labels A and G refer to the positions $\{A, C, G, T\} = \{1, 0, 0, 0\}$ and $\{0, 0, 1, 0\}$, respectively. The curves are vertically displaced to facilitate the comparison. Although a low-frequency component is visible, no clear evidence of $1/f$-behaviour is found.

5.5   Ergodic theory

The study of the long-time average behaviour of dynamical systems and, in particular, of measure-preserving transformations is the central topic of ergodic theory. Different degrees of irregular behaviour can be defined with a set of mathematical tools which yield a classification scheme well beyond the results of a simple estimation of invariant measures and power spectra. For instance, the mere existence of a large number (possibly a continuum) of frequencies in the power spectrum does not imply, by itself, interesting dynamical behaviour: they are indeed shared by many rather well-understood systems, as we have already seen.

The first basic question concerns the existence of the *time average*

$$\bar{f}(\mathbf{x}) = \lim_{n \to \infty} \frac{1}{n} \sum_{i=0}^{n-1} f\left(\mathbf{F}^i(\mathbf{x})\right), \qquad (5.23)$$

where $f : X \to \mathbb{R}$ is a measurable function representing a physical observable and $\mathbf{F} : X \to X$ is a measure-preserving transformation with respect to $(X, \tilde{\mathscr{B}}, m)$: *Birkhoff's ergodic theorem* states that the limit (5.23) exists a.e. and that $\bar{f}(\mathbf{x}) = \bar{f}(\mathbf{F}(\mathbf{x}))$. This problem is related to the formulation of the conditions under which the time mean (5.23) coincides (a.e. in X) with the space or *ensemble* average

$$\langle f \rangle \equiv \int_X f(\mathbf{x}) dm , \qquad (5.24)$$

which can be interpreted as the weighed sum of the observable f over all possible states of the system. The equality holds if and only if the orbit $\{\mathbf{F}^n(\mathbf{x}), n \in \mathbb{Z}\}$ visits each set of positive measure, for m-almost every initial point $\mathbf{x}$: i.e., $m(B) > 0$ and $m(B') > 0$ implies that $m(\mathbf{F}^n(B) \cap B') > 0$ for some $n > 0$ and all $B \neq B' \in \tilde{\mathcal{B}}$. Systems enjoying this fundamental property are called *ergodic*. By setting $B = B'$, it is readily verified that ergodicity presupposes plain *recurrence*: the trajectory returns to the original neighbourhood B indefinitely often (recall the definition of nonwandering set in Chapter 3). Hence, for an ergodic MPT, the mean $\bar{f}(\mathbf{x})$ is independent of the initial condition $\mathbf{x}$. For example, the average Lyapunov exponents $\bar{\lambda}_i(\mathbf{x}) = \langle \lambda_i \rangle$ are the same for m-a.e. (almost every) $\mathbf{x}$. A system is (measure-theoretically) chaotic if $\langle \lambda_1 \rangle > 0$.

An immediate consequence of ergodicity is that the fraction of iterates $\{\mathbf{F}^n(\mathbf{x})\}$ belonging to a set $B \in \tilde{\mathcal{B}}$ approaches $m(B)$, in the limit $n \to \infty$, for almost all $\mathbf{x}$. More generally, *Poincaré's recurrence theorem* states that, for each $B \in \tilde{\mathcal{B}}$, almost every point x of B is recurrent with respect to B (i.e., $\mathbf{F}^n(x) \in B$, for some $n \geq 0$): hence, every set of nonzero measure is visited infinitely often by the iterates of a.e. $\mathbf{x}$. Equivalently, $\mathbf{F}$ is ergodic if every invariant set $B \in \tilde{\mathcal{B}}$ is such that either $m(B) = 0$ or $m(X \setminus B) = 0$: i.e., the system $(\mathbf{F}, X, \tilde{\mathcal{B}}, m)$ is *metrically transitive* or *indecomposable*. Indeed, a transformation might be nonergodic on the whole space X but ergodic on subsets of X (*ergodic components*). Since, however, every system can be decomposed into ergodic subsystems, the assumption of ergodicity is usually not very restrictive (in chaotic Hamiltonian systems, e.g., the motion can be ergodic on subsets of certain constant-energy surfaces and periodic on others). If $\mathbf{F}$ is ergodic, there is at most one stationary density $\rho(\mathbf{x})$ for the Frobenius–Perron operator and vice versa, if $\rho(\mathbf{x}) > 0$ a.e. (Lasota & Mackey, 1985). In particular, a unique invariant measure exists for every irreducible Markov shift whose states have finite mean recurrence times. The simplest example of an ergodic transformation is represented by rotation through an irrational angle (bare circle map (3.13)), that is, by quasiperiodic motion.

A definite stochastic character, instead, is exhibited by ergodic systems with the additional *mixing* property. In order to clarify the relationship between the two concepts, it is useful to recall that ergodicity of an MPT $(\mathbf{F}, X, \tilde{\mathcal{B}}, m)$ is equivalent to the following condition:

$$\lim_{n \to \infty} \frac{1}{n} \sum_{i=0}^{n-1} m(\mathbf{F}^{-i}(B) \cap B') = m(B)m(B') \quad \forall B, B' \in \tilde{\mathcal{B}} . \qquad (5.25)$$

Then, $\mathbf{F}$ is *weakly mixing* (w-mixing) if

$$\lim_{n\to\infty} \frac{1}{n} \sum_{i=0}^{n-1} \left| m\left(\mathbf{F}^{-i}(B) \cap B'\right) - m(B)m(B') \right| = 0 \quad \forall\, B, B' \in \tilde{\mathscr{B}} \qquad (5.26)$$

and *strongly mixing* (s-mixing) if

$$\lim_{i\to\infty} m\left(\mathbf{F}^{-i}(B) \cap B'\right) = m(B)m(B') \quad \forall\, B, B' \in \tilde{\mathscr{B}}. \qquad (5.27)$$

By dividing both sides by $m(B') > 0$ in Eq. (5.27), it is easily recognized that s-mixing implies a definite spreading of every set B throughout phase space: the fraction of mass in a remote preimage of B that intersects B' is nearly the same as the relative weight $m(B)$ of B in the whole space X. In other words, the conditional probability of observing B at time 0, given B' at time $-i$, asymptotically becomes independent of the past measurement B'.

Equations (5.25)–(5.27) consitute a hierarchy of increasingly more erratic behaviour, as the term "mixing" suggests: indeed, any s-mixing transformation is w-mixing and any w-mixing system is, in turn, ergodic. Because of this inclusive relation, the terms "ergodic" or "merely ergodic" will be used to exclude any mixing and, similarly, "w-mixing" will signify "not s-mixing" when no misunderstanding can arise. The Bernoulli and baker maps are easily seen to be s-mixing. In general, chaotic transformations are believed to be s-mixing, because of the positivity of the Lyapunov exponents, although a direct verification of condition (5.27) is not easy. Irreducible aperiodic Markov shifts are s-mixing.

The asymptotic factorization of the n-step probabilities $P^{(n)}(j|i) = m\left(\mathbf{F}^{-n}(B_j) \cap B_i\right)$ in strongly mixing systems testifies, on the one hand, to the long-term unpredictability of the detailed dynamics in phase space but secures, on the other, a rather uneventful, quickly converging statistical behaviour. Notice, however, that exponential spreading of phase space regions does not prevent the invariant measure from being highly structured: its support, in fact, may be a fractal set.

Two other classes with even more pronounced stochastic character can be defined: exact systems and K-automorphisms (an automorphism being an invertible MPT). An MPT $(\mathbf{F}, X, \tilde{\mathscr{B}}, m)$, such that $\mathbf{F}(B) \in \tilde{\mathscr{B}}$, for each $B \in \tilde{\mathscr{B}}$ is called *exact* if

$$\lim_{n\to\infty} m\left(\mathbf{F}^n(B)\right) = 1, \text{ for every } B \in \tilde{\mathscr{B}} : m(B) > 0. \qquad (5.28)$$

The image of any positive-measure set B will eventually spread over the whole space X. These transformations are noninvertible and, therefore, have little physical relevance. They can be regarded as an extension of Markov systems with primitive transition matrix. The Bernoulli shift is exact. The definition of K-automorphism (which is an invertible "extension" of an exact system) is more involved and is skipped for simplicity (see Lasota & Mackey, 1985): the typical example in this class is the baker map. Finally, we recall that higher-order notions of strong mixing exist (Petersen, 1989).

In contrast to strong mixing, mere ergodicity does not allow exponential divergence of nearby orbits. The intermediate case of weak mixing can be seen to exhibit some stretching. Indeed, any w-mixing MPT satisfies the s-mixing condition (5.27) provided that the index i is chosen outside a certain zero-density set $J \in \mathbb{Z}^+$ (such as the set $\{n^2\}$ of the squares or the set $\{F_n\}$ of the Fibonacci numbers). In order to satisfy (5.26) but not (5.27), it is sufficient to have $|a_i| = |m(\mathbf{F}^{-i}(B) \cap B') - m(B)m(B')| = p$ if $i \in J$ and $|a_i| = q^{-i}$ if $i \notin J$. This can be interpreted by saying that the sequence $\mathbf{F}^{-i}(B)$ becomes independent of any other set B', except at a few instants of time. Although the definition of w-mixing arises as a quite natural complement of those of ergodicity and s-mixing (they express different convergence criteria for infinite series), a long time was necessary before explicit examples of strictly w-mixing (i.e., not s-mixing) MPTs could be constructed (Petersen, 1989). Notwithstanding this, there are notions of "genericity" according to which weak mixing is the prevalent property, as will be shown later.

Clearly, the ergodic-theoretical classification cannot be ignored in the study of "complexity". Ergodicity incontestably represents the first property to be investigated. On the one hand, the lack of ergodicity, when it does not arise from trivial periodic behaviour, requires the examination of all separately ergodic components: this is the problem discussed in Chapter 3 with reference to phase transitions in statistical mechanics. On the other hand, ergodicity is the first step beyond insignificance: in fact, a quasiperiodic signal, although ergodic, is still simple to describe. It is wise, however, to refrain from asking the question "Which is the most complex of the three classes?", natural though it may come at this point. It is, indeed, both premature, since much more insight into these topics must be gained before definite conclusions can be drawn, and ill-posed, since the degree of complexity (evaluated according to some criterium) need not change monotonically while going from one class to the next.

5.5.1 Spectral theory and isomorphism

A direct application of definitions (5.25)–(5.27) often leads to rather involved calculations. Indeed, the ergodic-theoretical analysis of MPTs is usually performed by means of a linear positive operator $\mathscr{U}_\mathbf{F} : L^2(m) \to L^2(m)$, named after *Koopman* and defined by

$$\mathscr{U}_\mathbf{F} f(\mathbf{x}) = f(\mathbf{F}(\mathbf{x})) , \tag{5.29}$$

which is the adjoint of the Frobenius–Perron operator $\mathscr{P}$: i.e., $(\mathscr{P}f, g) = (f, \mathscr{U}g)$, where $(f, g) = \int_X f(\mathbf{x})g(\mathbf{x})dm$ is the scalar product of f with g (Lasota & Mackey,

1985). If the MPT $\mathbf{F}$ is invertible, $\mathscr{U}_{\mathbf{F}}$ is unitary: in fact, $\mathscr{U}_{\mathbf{F}}^{-1} f(\mathbf{x}) = f(\mathbf{F}^{-1}(\mathbf{x}))$ (if $\mathbf{F}^{-1}$ is defined). A nonzero function $f \in L^2(m)$ satisfying $\mathscr{U}_{\mathbf{F}} f = \lambda f$ is called an *eigenfunction* of $\mathscr{U}$ with *eigenvalue* λ. Since $\mathscr{U}_{\mathbf{F}}$ satisfies $\|f\|^2 = \|\mathscr{U}_{\mathbf{F}} f\|^2$, $|\lambda| = 1$ $\forall \lambda$. Any MPT has $\lambda = 1$ as an eigenvalue with any nonzero constant function as an eigenfunction (Walters, 1985).

An MPT has *discrete spectrum* if there is a set $f_i : \mathscr{U}_{\mathbf{F}} f_i = \lambda_i f_i$ which constitutes an orthonormal basis for $L^2(m)$. Ergodic MPTs with discrete spectrum are conjugate to irrational rotations; moreover, ergodicity implies that every $\mathscr{U}_{\mathbf{F}}$-invariant measurable function f on X is a.e. constant. Conversely, an MPT is said to have *continuous spectrum* if 1 is the only eigenvalue and the only eigenfunctions are the constants. This is the case of weakly mixing transformations (Walters, 1985). Finally, one can consider the extreme situation in which the basis functions can be regrouped into classes within which each function is mapped by $\mathscr{U}$ to another one: maps with this behaviour are said to have *countable Lebesgue spectrum* and lie between strongly mixing and Bernoulli transformations (Petersen, 1989).

In order to understand the relation between the mixing property and the decay of correlations, consider the *cross-correlation*

$$C_{f,g}(n) = \int_X f(\mathbf{F}^n(\mathbf{x})) g(\mathbf{x}) dm - \langle f \rangle \langle g \rangle \,, \qquad (5.30)$$

between two observables f and g, where the ensemble average has been taken, as allowed in the ergodic case. By the definition of the Koopman operator,

$$C_{f,g}(n) = (\mathscr{U}_{\mathbf{F}}^n f, g) - \langle f \rangle \langle g \rangle \qquad (5.31)$$

and, since a system is s-mixing if and only if the scalar product in Eq. (5.31) tends asymptotically to $\langle f \rangle \langle g \rangle$ (Petersen, 1989), the correlation is seen to vanish at infinity. Recalling that s-mixing also implies the asymptotic (long-time) factorization of conditional probabilities (5.27), we have quite a complete view of this phenomenon.

Two transformations with the same spectral properties are called *spectrally isomorphic*: notice that, even when they belong to the same class (i.e., they are both either ergodic or comparably mixing) they need not be similar. Closer equivalence relations can be established through *conjugacy* or MPT-*isomorphism* (Walters, 1985). In order to decide whether two systems are connected via one of the above relations, say $\mathscr{R}$, it is necessary to find an $\mathscr{R}$-invariant attribute that is shared by both of them. A particularly meaningful invariant is the *metric entropy* (to be discussed in the next section) which is often used to distinguish nonisomorphic systems. Ergodicity and mixing (both spectral invariants) are much weaker than conjugacy which, in turn, is implied by MPT-isomorphism. Provided that a precise meaning can be assigned to the term "complexity", it is then natural to ask whether this is also invariant under some relation $\mathscr{R}$ and which indicators can be used for the purpose of classifying complex systems.

5.5.2 Shift dynamical systems

The branch of ergodic theory that studies continuous transformations such as the shift map $\hat{\sigma}$ (4.7) is called *topological dynamics*. The methods illustrated above can be carried over to shift dynamical systems in a straightforward way. As remarked in Chapter 4, the main properties of a smooth dynamical system (periodicity, minimality, ergodicity, mixing) are shared by the corresponding symbolic one. In addition to these, it is useful to recall *unique* and *strict* ergodicity: the former occurs when the invariant measure is unique and the latter when minimality and unique ergodicity hold. Minimality implies ergodicity (on the invariant set Λ, not on the whole space X, if $\Lambda \subset X$).

Several examples of ergodic and mixing systems can be found in Walters (1985), Lasota & Mackey (1985), Petersen (1989). Smooth perturbations of the bare circle map, such as the sine map (3.14) with small K, preserve ergodicity. Hyperbolic chaotic maps are s-mixing.

Examples:

[5.18] The dynamics at the period-doubling transition point, described by either of the substitutions ψ_{pd} and ψ_{MT} (Ex. 4.7), is strictly ergodic (Kakutani, 1986a and 1986b). □

[5.19] Consider the bare circle map (3.13) with irrational α and let $0 < \alpha < \beta < 1$, with $\beta \in \mathbb{R}$. Setting $s_n = \lfloor x_n/\beta \rfloor$, where $\lfloor x \rfloor$ is the integer part of x, one obtains the strictly ergodic dynamical system $(\overline{\mathcal{O}(\mathscr{S})}, \hat{\sigma})$, where $\mathscr{S} = \ldots s_{-1} s_0 s_1 \ldots$ is the orbit. Systems constructed in this way are called *Sturmian*. □

[5.20] By performing the substitution $1 \to 11$ *once* in the sequence $\mathscr{S}$ of the previous example, one obtains a new sequence $\mathscr{S}'$ in which every 1 has been "doubled". The system $(\overline{\mathcal{O}(\mathscr{S}')}, \hat{\sigma})$ is still strictly ergodic. It has been further proved (Kakutani, 1973) that $(\overline{\mathcal{O}(\mathscr{S}')}, \hat{\sigma})$ is weakly mixing but not strongly mixing if α is a transcendental (Liouville) number defined by $\alpha = \sum_{k=1}^{\infty} 10^{-n_k}$, where $\{n_k : k \in \mathbb{N}\}$ is an increasing sequence of positive integers satisfying $\lim_{k \to \infty}(n_{k+1} - 2n_k) = \infty$, and if the fractional part of $10^{n_k}\beta$ lies between 0.5 and 0.6 for all $k \in \mathbb{N}$. In spite of their apparent complication, these conditions are not very restrictive (almost all numbers are transcendental). A less general result, based upon delicate diophantine properties of α and β, was proved by Katok & Stepin (1967). The same substitution $1 \to 11$ applied once to the limit sequence of the Morse–Thue morphism ψ_{MT} also renders it weakly mixing (Kakutani, 1973). □

[5.21] Dynamical systems arising from substitutions are not strongly mixing but some substitutions with variable-length image words are weakly mixing: one of them is $\psi_{DK} : \{0, 1\} \to \{001, 11100\}$ (Dekking & Keane, 1978). Substitutions provide a straightforward algorithm for the generation of minimal and, possibly, w-mixing systems with zero topological entropy. A general construction method for minimal (more precisely, strictly ergodic) shifts over b symbols with arbitrary topological entropy $K_0 < \ln b$ has been introduced by Grillenberger (1973).   □

[5.22] The product $\mathbf{F} = \bigotimes_i \mathbf{F}_i$ of k MPTs $(\mathbf{F}_i, X_i, \tilde{\mathscr{B}}_i, m_i)$ $(i = 1, \ldots, k)$ is defined on $\prod_i X_i$ by $\mathbf{F}(\mathbf{x}_1, \ldots, \mathbf{x}_k) = (\mathbf{F}_1(\mathbf{x}_1), \ldots, \mathbf{F}_k(\mathbf{x}_k))$. Then, if $\mathbf{F}_1$ and $\mathbf{F}_2$ are w-mixing, $\mathbf{F}_1 \times \mathbf{F}_2$ is w-mixing. If $\mathbf{F}_1$ is w-mixing, $\mathbf{F}_1 \times \mathbf{F}_2$ is ergodic for each ergodic $\mathbf{F}_2$. If each $\mathbf{F}_i$ is ergodic, the product map is also ergodic if and only if the Koopman operators $\mathscr{U}_i$ have $\lambda = 1$ as a unique common eigenvalue (Cornfeld et al., 1982). Moreover, products of s-mixing automorphisms are s-mixing.   □

[5.23] Chaotic cellular automata (Section 3.4) are believed to be mixing in both space and time. Moreover, spatial mixing without time mixing is possible as in the example of Fig. 3.5(a). The opposite cannot occur since the time variability of the symbol at a given site i necessary for mixing can only arise from a flow of information from adjacent regions in the pattern. The simplest s-mixing elementary CA is rule 90 which can be written as

$$s_i(t + 1) = s_{i-1}(t) + s_{i+1}(t) \bmod 2 \ . \tag{5.32}$$

Any configuration can be generated at any time. It is, in fact, immediately seen that each block $S = s_1 s_2 \ldots s_n$ has four different preimages $S' = s_0' s_1' \ldots s_{n+1}'$, which begin with all distinct realizations of $s_0' s_1'$. Accordingly, the automaton is surjective and admits a uniform shift-invariant measure (Shirvani & Rogers, 1991). In order to assess the mixing property, consider an infinite random configuration $\mathscr{S}$ with image $\mathscr{S}^{(1)} = f(\mathscr{S})$, and their respective length-n subsequences $S = s_1 s_2 \ldots s_n$ and $S^{(1)} = s_1^{(1)} s_2^{(1)} \ldots s_n^{(1)}$ which start at site $i = 1$. The linearity (modulo 2) of the rule implies that $s_1^{(1)}$ and $s_n^{(1)}$ are independent of the whole string S, since they are determined by the random symbols s_0 and s_{n+1} adjacent to S. In the next image $\mathscr{S}^{(2)} = f^2(\mathscr{S})$, also the symbols at sites 2 and $n - 1$ become independent of S. Hence, the probability of observing S at time $t = 0$ and any given string $S^{(t)}$ at time t factorizes for $t > n/2$. In general, one-dimensional CAs for which the rule is a linear expression as in Eq. (5.32) are s-mixing at all orders (Shirvani & Rogers, 1988). Strong mixing has also been proved for a broader class of one-dimensional CAs (Shirvani & Rogers, 1991) which includes some "nonlinear interactions", such as rule 30 (defined by

$s_0' = s_{-1} + s_0 + s_1 + s_0 s_1$ mod 2). Furthermore, all surjective one-dimensional CAs have been conjectured to be s-mixing.   □

5.5.3   What is the "generic" behaviour?

A property that is shared by all systems in a "massive" set in a suitable measure space will be called "generic". In particular, a *residual* set (see Appendix 3) can be regarded as massive and its complement, a first-category set, as "thin". In ergodic theory, the following spaces are frequently considered: the group G of all automorphisms (invertible transformations) of the unit d-dimensional cube, the group $H(X, m)$ of all homeomorphisms of a compact metric space X which preserve a continuous measure m (as, e.g., shift dynamical systems), the group $\Delta^r(M, m)$ of all C^r diffeomorphisms of a compact manifold M which preserve a smooth measure m. The notion of genericity depends on the choice of the topology. Two definitions prove to be particularly useful (Halmos, 1956; Petersen, 1989). In the *weak topology* on G, a sequence $\{\mathbf{F}_n\}$ of transformations converges to $\mathbf{F}$ if and only if $\lim_{n \to \infty} m(\mathbf{F}_n(B) \triangle \mathbf{F}(B)) = 0$ for all $B \in \mathscr{B}$, where $B \triangle B' = B \cup B' - B \cap B'$. The *strong topology* is induced by either of the metrics

$$d_1(\mathbf{F}, \mathbf{G}) = m(\mathbf{x} \in X : \mathbf{F}(\mathbf{x}) \neq \mathbf{G}(\mathbf{x}))   \text{or}$$

$$d_2(\mathbf{F}, \mathbf{G}) = \sup\{m(\mathbf{F}(B) \triangle \mathbf{G}(B)) : B \in \tilde{\mathscr{B}}\} .$$

According to the weak topology, the set of strongly mixing MPTs is of the first category, whereas weakly mixing MPTs form a residual set (Petersen, 1989). Hence, the difficulty of constructing w-mixing transformations that are not strongly mixing does not imply their uncommonness. On the contrary, weak mixing is the prevailing behaviour. In the strong topology, instead, the set of all ergodic MPTs is nowhere dense, whereas the set of periodic MPTs is everywhere dense in G (Halmos, 1956). These apparently surprising results are in part a consequence of the properties of the topology only and not of the MPTs. In fact, two transformations are not very likely to be near to each other in the strong topology (which is "virtually" discrete). On the other hand, the density of a set of transformations does reflect a structural property of MPTs (Halmos, 1956).

Notice that, when constructing examples explicitly, one cannot choose freely, with "infinite resolution", from the space of all MPTs, but is constrained to shortly defined, simple functions (piecewise linear, polynomial, or involving a few trigonometric, exponential, or logarithmic expressions) which depend on a few real parameters. This is reminiscent of the structure of real numbers: although they are Lebesgue-almost all normal, writing just one down is a formidable task.

On the other hand, rational numbers are dense in $[0, 1)$ but have zero Lebesgue measure. Hence, questions about the genericity of a certain behaviour should be formulated and studied with great care. Similar considerations arise in the discussion of complexity: how generic is it, according to a given definition?

5.5.4   Approximation theories

The genericity of weak mixing is related to the convergence criterion used to evaluate how well a sequence $\{\mathbf{F}_n\}$ of maps approximates the object map $\mathbf{F}$ acting on $(X, \tilde{\mathscr{B}}, m)$. Remarkably, any automorphism can be approximated by periodic ones. The construction of such approximations is a classical topic in ergodic theory (Cornfeld *et al.*, 1982; Halmos, 1944).

In particular, one considers a sequence $\{\mathscr{C}_n\}$ of finite partitions of X, where $\mathscr{C}_n = \{C_i^n : i = 1, 2, \ldots, N(n)\}$ contains $N(n)$ elements, and a sequence $\{\mathbf{F}_n\}$ of transformations, such that each of them *preserves* the corresponding partition $\mathscr{C}_n$: i.e., $\mathbf{F}_n$ sends every element of $\mathscr{C}_n$ into an element of $\mathscr{C}_n$ itself, for all n. It is further required that the σ-algebra $\tilde{\mathscr{C}}_n$, defined on $\mathscr{C}_n$, refines $\tilde{\mathscr{B}}$ arbitrarily for $n \to \infty$, so that every set in $\tilde{\mathscr{B}}$ can be approximated by some set in $\tilde{\mathscr{C}}_n$ (up to zero-measure sets). When these conditions hold, the map $\mathbf{F}$ is said to admit an *approximation of the first type by periodic transformations* (APT$_1$) with *speed* $f(N)$ if

$$\sum_{i=1}^{N} m\Big(\mathbf{F}\big(C_i^{(n)}\big) \triangle \mathbf{F}_n\big(C_i^{(n)}\big)\Big) < f(N), \tag{5.33}$$

where $N = N(n)$ is the cardinality of $\mathscr{C}_n$ (Katok & Stepin, 1967; Cornfeld *et al.*, 1982). If, in addition, $\mathbf{F}_n$ cyclically permutes the elements of $\mathscr{C}_n$, the approximation to $\mathbf{F}$ will be called *cyclic*.

The sum in Eq. (5.33) represents a distance in the space of MPTs. A typical APT for the bare circle map (3.13) consists of a sequence of bare circle maps, the parameters of which are the diophantine approximants $\alpha_n = p_n/q_n \to \alpha$ (Ex. 3.6). In this case, $N(n) = q_n$ grows exponentially fast with n. Obviously, a map that is quickly approximated by periodic ones cannot exhibit ergodic or mixing properties. The following statements, proved in Katok and Stepin (1966), hold:

1. Any automorphism admits an APT$_1$ with speed $f(N) = a_N/\ln N$, where a_N is an arbitrary monotonic sequence of real numbers tending to infinity. Clearly, interesting cases occur when a_N grows less rapidly than $\ln N$.

2. If $\mathbf{F}$ admits a *cyclic* APT$_1$ with $f(N) = \beta/N$, the number of its ergodic components is at most $\beta/2$. If $\beta < 4$, $\mathbf{F}$ is ergodic.

3. If $\mathbf{F}$ admits a more stringent type of periodic approximation, called APT$_2$ (Katok & Stepin, 1967), with speed β/N and $\beta < 2$, it produces no mixing.

Notice that while property (3) is self-explanatory, result (2) is somewhat coun-
terintuitive. The case $f(N) = h/\ln N$, with h a constant, will be discussed in the
next section. It should be remarked that the approximation method is hierarchical, being based on a sequence of maps defined on a number $N(n)$ of intervals
which increases exponentially with the resolution level n. Furthermore, it need
not make use of periodic transformations but may be carried out with maps of
the Markov type or belonging to other classes of MPTs (Alpern, 1978): notice
that periodic maps are special Markov maps (i.e., without any branching).

Equivalently, the analysis of complexity must be based on a clear-cut definition
of the class of systems under investigation and on a precise notion of metric,
in order to judge about the agreement between the object and the models that
are supposed to describe it. At this point of the discussion, it can be safely
stated that complexity must be "distant" from periodicity and quasiperiodicity
since these properties admit a finite, exact description. On the other hand, it
is tempting to say that complexity should be likewise regarded as distinct from
finite-memory Markov processes (although some complexity measures disagree
with this remark). Bernoulli shifts and, more generally, chaotic maps with
a finite number of prohibitions yield rather "bulky", uninteresting languages
which admit an exponentially increasing number of sequences of all orders.
As in natural languages, it is the strong nonuniformity of the measure and,
possibly, the absence of a large fraction of symbol concatenations that make
the "message" worthwhile of study.

These observations suggest that one should characterize complexity through
an exclusion procedure that eliminates systems with increasingly more "structure", as soon as they can be explained by compact and accurate mathematical
models: periodic and completely random objects are excluded first, quasiperiodic signals and subshifts of finite type (Markov processes) next, and so on.
The resulting "negative" definition of a complex pattern (not periodic, not
quasiperiodic, not an SFT, not sofic, ..., not random) would then be a statement
of *simplicity* for all well-understood systems.

5.6   Information, entropy, and dimension

Information theory, originally formulated to deal with compression and transmission of data (Shannon & Weaver, 1949), successively developed outside the
domain of communication theory with significant contributions to statistical
mechanics, computer science (algorithmic complexity), probability theory and
statistics. Indeed, the central concept of this theory, the *entropy*, was carried
over from thermodynamics where it constitutes the base of the Second Law and
was later extended for application in ergodic theory. In each setting, entropy is
a measure of randomness or unpredictability.

Consider a source emitting a string S of symbols s_i drawn from an alphabet $A = \{1, 2, \ldots, b\}$ with a given probability $P(s)$, $\forall\, s \in A$. The *uncertainty* associated with each event obviously depends on its probability. If $P(s) = 1$, for some s, the outcome of the experiment is univocal and the source conveys no information. If, on the contrary, $P(s) = 1/b\ \forall\ s$, the uncertainty is maximal (provided that the source has no memory). Each observation conveys the information that is associated with the display of the current symbol itself, thereby removing the preexisting uncertainty. The average information over all outcomes $s_i \in A$ of an experiment $\mathscr{A}$ is conveniently measured by the *entropy*

$$H(\mathscr{A}) = -\sum_{i=1}^{b} P(s_i) \ln P(s_i)\,, \tag{5.34}$$

where $P = \{P(1), \ldots, P(b)\}$ is the probability distribution for $\mathscr{A}$. Although the entropy can be defined also for a continuous distribution $\rho(\mathbf{x})$, as $H = -\int_X \ln \rho(\mathbf{x}) dm$, we shall deal only with discrete random variables.

The entropy H enjoys three fundamental properties. Firstly, $H = 0$ if and only if one of the symbols has probability 1 (under the convention that $x \ln x = 0$ for $x = 0$, which implies that adding zero-probability terms does not change the entropy). Secondly, since $h(x) = -x \ln x$ is continuous, nonnegative and concave in $(0, 1]$, Jensen's inequality implies

$$H = \sum_{i=1}^{b} h(p_i) \le b\, h\left(\frac{1}{b} \sum_{i=1}^{b} p_i\right) = b h(1/b) = \ln b\,, \tag{5.35}$$

for any probability vector $P = \{p_1, \ldots, p_b\}$. The maximum value $\ln b$ is achieved in the most uncertain case (i.e., when $p_i = 1/b$, $\forall\ i$), as desired. Finally, the information behaves additively for independent compound events, represented by n-blocks $S = s_1 \ldots s_n$ obeying Eq. (5.6): in such a case, in fact, $\ln P(S) = \ln\left(\prod_i P(s_i)\right) = \sum_i \ln P(s_i)$. The function $H(p_1, \ldots, p_b)$ (with $\sum_i p_i = 1$) of Eq. (5.34) is the only continuous one satisfying these three requirements (Khinchin, 1957). In general, the quantity $I(S) = -\ln P(S)$ can be interpreted as the information content of the event $S = s_1 \ldots s_n$. If the base-2 logarithm is taken, as usually done in information theory, $I(S)$ represents the number of *bits* (the binary unit of information) necessary to specify S: this is particularly evident for the symmetric Bernoulli shift $B(1/2, 1/2)$, for which $P(S) = 2^{-|S|}$ for any sequence S. An initial condition given with a precision of n binary digits produces a perfectly predictable sequence for the first n iterates, whereas symbol s_{n+1} is completely unknown.

For a general source, the probability of any symbol or word depends on the previous history. Memory effects can be evaluated by considering a sequence of n repetitions $\{\mathscr{A}_i\}$ of the same experiment $\mathscr{A}$. The *n-block entropy* is defined as

$$H_n = H(\mathscr{A}_1, \mathscr{A}_2, \ldots, \mathscr{A}_n) = -\sum_{S:|S|=n} P(S) \ln P(S), \qquad (5.36)$$

where the sum is taken over all sequences S of length n. Choosing $n = 2$ and indicating with $\mathscr{A}$ and $\mathscr{B}$ the experiments carried out at times 1 and 2, respectively, we have

$$H(\mathscr{A}\mathscr{B}) = -\sum_{i,j} P(s_i)P(s_j|s_i)\ln[P(s_i)P(s_j|s_i)] = -\sum_i P(s_i)\ln P(s_i)$$

$$-\sum_i P(s_i)\sum_j P(s_j|s_i)\ln P(s_j|s_i), \qquad (5.37)$$

since $\sum_j P(s_j|s_i) = 1, \forall i$ (Eq. (5.4)). The quantity

$$H(\mathscr{B}|s_i) = -\sum_j P(s_j|s_i)\ln P(s_j|s_i) \qquad (5.38)$$

represents the entropy of experiment $\mathscr{B}$, with the condition that experiment $\mathscr{A}$ has yielded the event s_i. Relation (5.37) can be rewritten as

$$H(\mathscr{A}\mathscr{B}) = H(\mathscr{A}) + H(\mathscr{B}|\mathscr{A}), \qquad (5.39)$$

where $H(\mathscr{B}|\mathscr{A}) = \sum_i P(s_i)H(\mathscr{B}|s_i)$ is the *conditional entropy* of experiment $\mathscr{B}$ relative to $\mathscr{A}$. Notice that the additivity property of H is recovered for independent events. Moreover, by the concavity of $h(x)$, we have

$$\sum_i p_i h(q_i) \le h\left(\sum_i p_i q_i\right), \qquad (5.40)$$

for any probability vector $P = \{p_i\}$ and $q_i \in \mathbb{R}$. By setting $p_i = P(s_i)$, $q_i = P(s_j|s_i)$ and recalling Eq. (5.5), we obtain

$$H(\mathscr{B}|\mathscr{A}) \le H(\mathscr{B}). \qquad (5.41)$$

Hence, the entropy behaves *subadditively* for general compound events:

$$H(\mathscr{A}\mathscr{B}) \le H(\mathscr{A}) + H(\mathscr{B}), \qquad (5.42)$$

with the obvious interpretation that the previous knowledge of the outcome of experiment $\mathscr{A}$ cannot increase the uncertainty on the outcome of experiment $\mathscr{B}$. Memory (or experience, in common parlance) reduces perplexity. If the result of $\mathscr{B}$ is in part determined by that of $\mathscr{A}$, less information is conveyed to the observer than in the completely uncorrelated case. The *chain rule* (5.37) can be generalized to *n*-blocks as

$$H_n = H(\mathscr{A}_1, \ldots, \mathscr{A}_n) = \sum_{i=1}^n H(\mathscr{A}_i|\mathscr{A}_0, \ldots, \mathscr{A}_{i-1}), \qquad (5.43)$$

with the convention that $H(\mathscr{A}_1|\mathscr{A}_0) = H(\mathscr{A}_1)$. The average information per symbol $\langle I \rangle_n = H_n/n$ quantifies the dependence of the words of length n on those of length $n-1$. Relation (5.42) implies that $\langle I \rangle_n$ is monotonic nonincreasing. The entropy of the source or *metric entropy* K_1 is then defined as

$$K_1 = \lim_{n \to \infty} H_n/n \,, \tag{5.44}$$

where the limit exists for a stationary source (the reason for the index 1 will be given later). Since there are at most b^n sequences of length n, $0 \leq K_1 \leq \ln b$. Random signals have a large entropy: that is, they convey a high amount of information per symbol. Conversely, natural languages, with their grammatical and logical rules, restrict severely the number of possible concatenations: they exhibit a certain degree of predictability and, hence, achieve low information values. This property is called *redundancy* (notice, for example, that the letter "u" is superfluous in the pair "qu" in the English language, since $P(\mathrm{u}|\mathrm{q}) = 1$) and it greatly facilitates the comprehension of the language. In fact, it permits recognition of errors or reconstruction of a sentence from a few of its parts. On the contrary, a misprint during the compilation of a computer program usually has catastrophical consequences.

Any measure-preserving system $(\mathbf{F}, X, \tilde{\mathscr{B}}, m)$ acts as a source of a finite-state stationary process $\{s_i\}$ when endowed with an arbitrary partition $\mathscr{C}$ of finitely many measurable sets. The experiment is carried out by recording the labels s_i of the elements $C_{s_i} \in \mathscr{C}$ visited by the trajectory at times $i = 1, 2, \ldots, n$. Hence, we use the same symbol $\mathscr{C}$ to designate both the partition and the experiment. As a result, the entropy $H(\mathscr{C})$ of the partition is given by Eq. (5.34), where $P(s_i) = m(C_{s_i})$ is the mass in the element $C_{s_i} \in \mathscr{C}$. By taking the n-th refinement $\mathscr{C}_n = \mathscr{C} \vee \mathbf{F}^{-1}\mathscr{C} \vee \ldots \vee \mathbf{F}^{-n}\mathscr{C}$ of $\mathscr{C}$, one obtains the $(n+1)$-block entropy $H_{n+1} = H(\mathscr{C}_n)$ since each probability $P(S)$ coincides with the measure of an element of $\mathscr{C}_n$. The limit

$$K_1(\mathscr{C}, \mathbf{F}) = \lim_{n \to \infty} \frac{H(\mathscr{C}_n)}{n} \,, \tag{5.45}$$

called the *entropy of the transformation* $\mathbf{F}$ *with respect to the partition* $\mathscr{C}$, exists for each countable measurable partition (it is possibly $+\infty$) (Petersen, 1989). It quantifies the average uncertainty per time step about the partition element that is currently visited by the orbit. Since inappropriate choices of $\mathscr{C}$ can eliminate this uncertainty even though the motion is chaotic, it is necessary to take the supremum over all possible partitions. This leads to the definition of the *metric* or *Kolmogorov–Sinai* entropy K_1 of the transformation $\mathbf{F}$ as

$$K_1(\mathbf{F}) = \sup_{\mathscr{C}} K_1(\mathscr{C}, \mathbf{F}) \,, \tag{5.46}$$

which represents the maximum information that can be obtained from an experiment performed on $(\mathbf{F}, X, \tilde{\mathscr{B}}, m)$ by taking into account all possible finite partitions $\mathscr{C}$. If $\mathscr{C}$ is chosen so carefully that $K_1(\mathbf{F}) = K_1(\mathscr{C}, \mathbf{F})$, the entropy can be computed directly from Eq. (5.44) without need of taking the supremum (5.46). This result, known as the *Kolmogorov–Sinai theorem* (Cornfeld *et al.*, 1982), is achieved by the generating partitions (already defined in Chapter 4): upon infinite refinement, they reproduce the full σ-algebra $\mathscr{B}$. Furthermore, Krieger's theorem (Krieger, 1972) states that every ergodic MPT with $K_1 < +\infty$ has a finite generator with a number b of elements satisfying $e^{K_1} < b \le e^{K_1} + 1$. The theorem also establishes an equivalence between ergodic MPTs and shift homeomorphisms.

The topological analogue of the entropy K_1 for a shift dynamical system is the *topological entropy* K_0, briefly mentioned in Chapter 4 (Eqs. (4.9) and (4.10)), and defined as (Adler *et al.*, 1965)

$$K_0 = \lim_{n \to \infty} \frac{\ln N(n)}{n}, \tag{5.47}$$

where $N(n)$ is the number of n-blocks in the language $\mathscr{L}$. Both K_0 and K_1 are special cases of a generalized entropy function K_q which will be discussed in Chapter 6.

As already remarked, the concept of information cannot be captured completely by the entropy alone. The necessity to take into account the influence of the past history on the present state of the system led us to consider the conditional entropy (5.39). More generally, the dependence of a distribution $P = \{p_1, \dots, p_b\}$ on a test distribution $Q = \{q_1, \dots, p_b\}$ is expressed by the *relative entropy*

$$H(P \| Q) = \sum_{i-1}^{b} p_i \ln \frac{p_i}{q_i}, \tag{5.48}$$

where the conventions $0 \ln(0/q) = 0$ and $p \ln(p/0) = \infty$ are adopted. The quantity $H(P \| Q)$ measures the inadequacy of Q as a possible substitute for the true distribution P. Jensen's inequality, applied to the function $\ln x$, yields

$$H(P \| Q) \ge \ln \sum q_i = 0, \tag{5.49}$$

where the equality holds if and only if $q_i = p_i$, that is, when the two distributions coincide. Therefore, $H(P \| Q)$ is sometimes interpreted as a *distance* (named after *Kullback* and *Leibler*) between probability distributions. By setting $q_i = 1/b, \forall i$, in Eq. (5.48), we recover the upper bound $H \le \ln b$ for the entropy. Taking the two-symbol probability $p(s_i s_j)$ and making the factorization assumption $q(s_i s_j) = p(s_i) p(s_j)$ ($\forall i, j$) in $H(P \| Q)$, one obtains the so-called *mutual information*: it measures the deviation of the process from an uncorrelated one.

Examples:

[5.24] The entropies K_0 and K_1 of periodic and quasiperiodic signals are zero.

The number $N(n)$ of n-blocks in the language is constant in the former case and proportional to n in the latter.   □

[5.25] The metric entropy is a measure of unpredictability. The difference $h_n = H_{n+1} - H_n$ between block entropies at consecutive hierarchical levels can be interpreted as the information needed to specify the symbol $n + 1$, given the previous n. It can be easily verified that $\lim_{n \to \infty} h_n = K_1$. For an order-$k$ Markov process, $h_n = K_1$ for all $n > k$: the limit is reached when the memory of the process is exhausted. The entropy of a Bernoulli process $B(p_1, \ldots, p_N)$ is $K_1 = -\sum_i p_i \ln p_i$.   □

[5.26] The metric entropy of the product $\mathbf{F}_1 \times \mathbf{F}_2$ of two independent MPTs satisfies $K_1(\mathbf{F}_1 \times \mathbf{F}_2) = K_1(\mathbf{F}_1) + K_1(\mathbf{F}_2)$, by the (sub)additivity property (5.42). The same holds for the topological entropy K_0: in fact, the set of allowed n-blocks for $\mathbf{F}_1 \times \mathbf{F}_2$ is just the product of the analogous sets for the two individual maps.   □

[5.27] Random (i.e., nondeterministic) systems with continuous-valued variables have infinite entropy: no finite generator exists. This is the case of (physical) noise. Deterministic systems affected by a small amount of noise yield finite entropy values as long as the estimate is performed with limited resolution. The outcomes of a coin tossing experiment (Bernoulli shift) form a *discrete* random process which, hence, has finite entropy ($\ln 2$).   □

[5.28] The positivity of the metric entropy is not equivalent to the strong-mixing property. There are s-mixing systems with entropy 0: for a stationary Gaussian process, see Girsanov (1958), for horocycle flows, Parasjuk (1953) and Gurevich (1961), and, for a two-dimensional subshift of finite type, Mozes (1992). Zero-entropy MPTs are generic in the weak topology (Walters, 1985). □

[5.29] Although strong mixing implies spreading of phase-space domains, it does not imply the positivity of the largest Lyapunov exponent λ_1 or the exponential decay of correlations (as in the case of "periodic-chaos", where the orbit visits a number p of distinct domains periodically but is chaotic when restricted to any one of these). There are examples of s-mixing transformations with $\lambda_1 = 0$ or $K_1 = 0$ and of nonmixing ones with $\lambda_1 > 0$ (Oono & Osikawa, 1980; Courbage & Hamdan, 1995). Not much is known about the genericity of these systems. Similarly, positive metric entropy K_1 (or positive λ_1), even in the strongly mixing case, does not imply the exponential decay of correlations, as shown for a certain one-dimensional gas of particles by Niwa (1978).   □

[5.30] Shannon (1951) estimated the entropy of English prose by letting people guess the next letter in a text, having seen the previous n. The result was $K_1 \approx 1.3 \ln 2$, much less than the maximum value $\ln 26$, as expected for a random source. This redundancy can be eliminated by a suitable *coding* which permits compression of the information contained in a text to achieve optimal data transmission (see Chapter 8).   □

[5.31]   Any MPT **F** with finite entropy K_1 admits an APT_1 with speed $f(N) = [2K_1 + \delta]/\ln N$, where $\delta > 0$ is arbitrary. If **F** admits an APT_1 with speed $f(N) = \gamma/\ln N$, then $K_1 \leq \gamma$; if, in addition, **F** is ergodic, $K_1 \leq \gamma/2$. Finally, an MPT has zero metric entropy if and only if it admits an APT_1 with speed $o(1/\ln N)$. These and a few other related results were established by Katok & Stepin (1967). Schwartzbauer (1972) has later proved that $K_1 = \gamma/2$ for any automorphism.   □

[5.32]   The time evolution of cellular automata provides a useful illustration of the concept of entropy. With the exception of a few particular rules, the dynamics modifies the relative abundance of subsequences, so that the spatial probability distribution is asymptotically nonuniform (some sequences may be forbidden altogether). Given a uniformly random initial configuration, the spatial metric entropy usually decreases with time. The consequent natural interpretation that order is created from disorder (self-organization) is, however, misleading. In fact, there are special configurations (e.g., homogeneous except for just one symbol) which may evolve into structures characterized by increasing spatial entropy. This is displayed in Fig. 5.3. The metric entropy can also be used to characterize the irregularity of the pattern in the time direction, by choosing a fixed site i and considering the signal $s_i(t)$ (where t is the discrete time), provided that stationarity holds.   □

The information-theoretic approach permits also investigation of the dimension of a fractal measure m. Let the support $s(m)$ of m be covered with a collection of disjoint domains which, for the moment, are assumed to have fixed diameter ε. The covering is then iteratively refined by further splitting each of the elements into subsets with equal diameter. Indicating with $N(\varepsilon)$ the number of nonempty elements at the resolution ε, the *fractal dimension*[5] D_0 of the support $s(m)$ is defined as (Mandelbrot, 1982)

$$D_0 = \lim_{\varepsilon \to 0} -\frac{\ln N(\varepsilon)}{\ln \varepsilon} . \qquad (5.50)$$

A set Λ is called *fractal* if $D_0 > d_t$, where d_t is the ordinary topological dimension of Λ (d_t is zero for a point, one for a line, ..., n for a set having $(n-1)$-dimensional sections: for a precise definition, see Hurewicz & Wallman (1974)). For smooth (absolutely continuous) measures, D_0 coincides with d_t.

---

5. This quantity is often referred to under different names, such as *capacity* or *self-similarity dimension*. The denomination of *box dimension*, however, is becoming the predominant one.

x

t

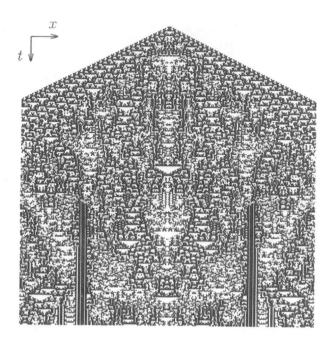

Figure 5.3
Spatio-temporal pattern
generated by the totalistic
cellular automaton 1024,
with $b = 3$ and $r = 2$, of
Fig. 3.7, starting with the
initial configuration
having $s_i = \delta_{i0}$.

Evidently, D_0 is the "static" counterpart of the topological entropy K_0 (5.47),
where the scaling parameter n has been replaced by $-\ln \varepsilon$.

The scaling properties of the measure m can be characterized by the *information dimension* (Renyi, 1970)

$$D_1 = \lim_{\varepsilon \to 0} \frac{\sum_{i=1}^{N(\varepsilon)} P_i(\varepsilon) \ln P_i(\varepsilon)}{\ln \varepsilon} , \tag{5.51}$$

which is the analogue of the metric entropy K_1 (5.44). Hence, K_1 can be
interpreted as an information dimension in the space of symbolic sequences S
when they are mapped to points x of the unit interval as $x(S) = \sum_i s_i b^{-i}$ (the
resolution ε being b^{-n}).

In general, $D_0 \geq D_1$: the whole set of generalized dimensions D_q will be
defined in Chapter 6. In the trivial case $P(\varepsilon) = 1/N(\varepsilon)$, D_1 reduces to D_0.
The simplest example of a self-similar fractal is perhaps the ternary Cantor
set: it consists of all real numbers $x = \sum_{i=1}^{\infty} s_i 3^{-i}$ for which s_i is either 0 or 2.
At the resolution $\varepsilon_n = 3^{-n}$, the set can be covered with 2^n segments, so that
$D_0 = D_1 = \ln 2/ \ln 3$.

The covering procedure can be generalized to include collections of domains
having different diameters ε_i. This yields a more rigorous definition of dimension.
Consider the *Hausdorff measure* $m_\gamma(\varepsilon)$ of a set Λ of points in $\mathbb{R}^d$

$$m_\gamma(\varepsilon) = \inf \sum_i \varepsilon_i^\gamma , \tag{5.52}$$

where the infimum is taken over all coverings of Λ that satisfy $\varepsilon_i < \varepsilon, \forall i$ (this avoids choosing unnecessarily large domains). The *Hausdorff dimension* d_h of Λ is then defined as the value of γ such that $m_\gamma(\varepsilon) \to \infty$ for $\gamma < d_h$, and $m_\gamma(\varepsilon) \to 0$ for $\gamma > d_h$. Since $m_\gamma(\varepsilon) \leq N(\varepsilon)\varepsilon^\gamma$, $d_h \leq D_0$ although the two numerical values are usually the same for strange attractors arising from diffeomorphisms or experimental systems (Mandelbrot, 1982; Mayer-Kress, 1989).

Examples:

[5.33] At the period-doubling transition, the first 2^n images of the critical point (consider a one-dimensional map, for simplicity) define 2^{n-1} intervals which represent a covering of the attractor (the subsequent 2^n points falling inside these segments). In the lowest-order approximation, the attractor is a two-scale Cantor set with contraction rates $|\alpha|^{-1}$ and α^{-2} for the segment lengths, where $\alpha = -2.5029\ldots$ is Feigenbaum's scaling constant (Ex. 3.3). Application of Eq. (5.52) leads to the *self-similarity condition*

$$|\alpha|^{-d_h} + |\alpha|^{-2d_h} = 1$$

which is satisfied by $d_h \approx 0.525$. A more refined estimate, which takes into account more length scales (i.e., the lack of an exact self-similarity), yields $d_h \approx 0.538$ (Grassberger, 1981). $\qquad\qquad\qquad\qquad\qquad\qquad\square$

[5.34] The attractor of the generalized baker map (3.10) is the product of a line (along the expanding y-direction) and a Cantor set (along the contracting x-direction). Its overall dimension D (defined according to any of the variants (5.50–5.52)) satisfies $D = 1 + D^{(2)}$, where the *partial dimension* $D^{(2)}$ (see Appendix 4) is the corresponding dimension of the Cantor set. At the nth level of approximation, the Cantor set can be covered with 2^n segments having lengths $r_1^k r_2^{n-k}$, for all $k \in \{0, 1, \ldots, n\}$, and carrying a mass $p_1^k p_2^{n-k}$ (k is the number of 0s in the symbolic encoding of the generic segment). Its Hausdorff dimension $D_h^{(2)} = d_h - 1$, which coincides with the partial dimension $D_0^{(2)} = D_0 - 1$ of the support, satisfies

$$r_1^{d_h-1} + r_2^{d_h-1} = 1$$

and its information dimension $D_1^{(2)}$ is explicitly given by the ratio

$$D_1^{(2)} = \frac{p_1 \ln p_1 + p_2 \ln p_2}{p_1 \ln r_1 + p_2 \ln r_2} = -\frac{K_1}{\langle \ln r \rangle}$$

between the entropy K_1 and the average logarithmic contraction rate $\ln \varepsilon = \langle \ln r \rangle$ (this result will become clear with the introduction of the generalized dimensions D_q in Chapter 6). $\qquad\qquad\qquad\qquad\qquad\qquad\qquad\qquad\qquad\square$

[5.35] Dimensions and entropies of experimental systems can be estimated from a scalar time series $\{x(\Delta t), x(2\Delta t), \ldots, x(N\Delta t)\}$ obtained by recording some observable x at equally spaced time itervals. For this to be possible, resolution, number, and noise "content" of the data must satisfy certain criteria (Grassberger *et al.*, 1991). The attractors are first reconstructed by *embedding* the time series in an E-dimensional space, the points of which have the form $\mathbf{x}_i = (x_{i+\tau}, \ldots, x_{i+E\tau})$ (Packard *et al.*, 1980; Takens, 1981; Sauer *et al.*, 1991), where $x_k = x(k\Delta t)$: the integer τ is called the "delay time". For an overview of recent results, see Mayer-Kress (1989), Abraham *et al.* (1989), and Drazin & King (1992).

□

The interesting topic from the point of view of complexity is not so much the fractality of the invariant set or a possibly large value of dimensions and entropies but the inhomogeneity of the singularities of the measure: namely, a difference among the various dimensions and entropies or the lack of an exact self-similarity. These phenomena, which constitute the overwhelming majority of the cases, will be the subject of the next chapter.

Further reading

Arnold and Avez (1968), Ash (1965), Billingsley (1965), Edgar (1990), Falconer (1990), Sinai (1994).

Chapter 6

Thermodynamic formalism

The concepts of memory and interactions are fundamental for the understanding of complexity and, indeed, they naturally arise in statistical mechanics, especially in connection with phase transitions. In this chapter, we illustrate the extension to general symbolic patterns of the thermodynamic formalism developed for spin lattice systems, which provides a common background to different disciplines.

A nonuniform occurrence of certain blocks of symbols in the configuration can be ascribed to an interaction potential that favours the patterns with lowest "energy". This heuristic observation has various rigorous formulations, corresponding to different equivalent thermodynamic schemes. We select a few which have a direct connection with physical observables such as probabilities, fractal dimensions, and Lyapunov exponents. The nature of the resulting interactions induces a classification that represents a characterization of complexity. Slowly decaying interactions, for example, are suggestive of some degree of organization.

6.1 Interactions

Consider a shift dynamical system $(\Sigma', \hat{\sigma})$, with $\Sigma' \subseteq \Sigma_b$, arising from a doubly-infinite symbolic sequence $\mathscr{S}$ over a b-symbol alphabet A and preserving the measure m. The signal $\mathscr{S}$ can be interpeted as a spin configuration over the lattice $\mathbb{Z}$. In order to pursue the parallel with a statistical mechanical system, the lattice is coarse-grained into n-blocks $S = s_1 s_2 \ldots s_n$ (the *microstates*) and the value $O(S)$ of a physical observable $\mathcal{O}$ is assigned to each of them. Because of the existence of a stationary measure, the most natural choice for $O(S)$ is the weight $P(S)$ associated with the n-cylinder $c_n(S)$. Accordingly, one defines (Sinai, 1972) the Hamiltonian as

$$\mathcal{H}(s_1, s_2, \ldots, s_n) = -\ln P(s_1 s_2 \ldots s_n). \tag{6.1}$$

This position is justified by the observation that, in mixing shifts, the probability exhibits a typical leading exponential decay for increasing n: $P(S) \sim \exp[-n\kappa(S)]$, with $\kappa(S) > 0$ for m-almost all S. As already remarked, this is not always the case: the breakdown of this behaviour is responsible for a number of interesting phenomena.

Notice that the situation is reversed with respect to "physical" statistical mechanics, since in this formulation the probabilities are given (either analytically or through a numerical evaluation) for every finite sequence S and the Hamiltonian (6.1) is an unknown function of the spin variables which is to be fitted to the data.

Examples:

[6.1] In the Bernoulli shift, each sequence composed of k 0s and $n - k$ 1s (in any order) appears with probability $P(s_1 s_2 \ldots s_n) = p_0^k p_1^{n-k}$ (see Ex. 5.5). Hence,

$$\mathcal{H} = -k \ln p_0 - (n - k) \ln p_1 = -n \ln p_0 - \ln(p_1/p_0) \sum_{i=1}^{n} s_i, \tag{6.2}$$

since $k = \sum_i (1 - s_i)$ and $n - k = \sum_i s_i$. This is a lattice-gas Hamiltonian in which $s_i = 1$ indicates that site i is occupied and $s_i = 0$ that it is empty. The usual spin-Hamiltonian picture is recovered by performing the substitution

$$\sigma_i = 2s_i - 1. \tag{6.3}$$

As expected, a sequence of *independent*, uncorrelated events corresponds to *noninteracting* particles. The last term in Eq. (6.2) represents the contribution of an external field which vanishes for $p_0 = p_1$. □

[6.2] If the ith symbol depends on the previous one in a Markovian way, the probability $P(s_0 s_1 \ldots s_n) = P(s_0) \prod_i P(s_i|s_{i-1})$ yields

$$\mathcal{H} = -\ln P(s_0) - \sum_{i=1}^{n} \ln P(s_i|s_{i-1}). \tag{6.4}$$

The chain is defined by the b^2 possible values of $P(s_i|s_{i-1})$. Each term $p_{\alpha|\beta} \equiv P(\alpha|\beta)$, where α and β are either 0 or 1 in the binary case, appears a number $n_{\alpha\beta}$ of times in the sum. The values n_{00}, n_{01}, n_{10} and n_{11} can be written as sums involving the occupation numbers s_i: for example, $n_{01} = \sum_i (1 - s_i) s_{i-1}$. Substituting the four expressions into Eq. (6.4), one obtains

$$\mathcal{H} - -n \ln p_{0|0} - \ln \frac{p_{0|1} p_{1|0}}{p_{0|0}^2} \sum_{i=1}^{n} s_i - \ln \frac{p_{0|0} p_{1|1}}{p_{0|1} p_{1|0}} \sum_{i=1}^{n} s_i s_{i-1} , \qquad (6.5)$$

where "surface terms" such as $- \ln P(s_0)$ have been neglected. This Hamiltonian clearly describes a lattice gas with nearest-neighbour interactions (Huang, 1987). The equivalent Ising Hamiltonian (3.25) is recovered with the substitution (6.3). A positive (negative) sign of the coefficient of the quadratic term corresponds to ferromagnetic (antiferromagnetic) interactions. $\qquad\qquad\qquad\qquad\quad \square$

As the above examples show, the memory of the stochastic process directly translates into an interaction. In the study of symbolic dynamical systems, the Hamiltonian is deduced from probabilities generated by given (possibly unknown) dynamical rules. In general, it can be written as a sum of terms involving multi-spin interactions of the type

$$\mathcal{H} = -\sum_i J_1 \sigma_i - \sum_{i>j} J_2(i-j) \sigma_i \sigma_j - \sum_{i>j>k} J_3(i-j, j-k) \sigma_i \sigma_j \sigma_k - \dots , \quad (6.6)$$

where the coefficients J_ℓ are *translationally invariant*. Certain restrictions must be imposed on them in order to guarantee the existence of the *thermodynamic limit*: namely, the convergence of particle-, energy- and entropy-densities to a finite limit when the number of particles n tends to infinity. For pair interactions (i.e., $J_\ell = 0$, for $\ell > 2$), this condition reads

$$\sum_j |J_2(i-j)| < \infty \qquad\qquad\qquad\qquad\qquad (6.7)$$

for any fixed i. This is tantamount to saying that surface effects can be neglected, so that only bulk properties survive. Having defined the Hamiltonian as in Eq. (6.1), for some measure m, inequality (6.7) implies that the conditional probability $P(s_1 s_2 \dots s_n | s_0)$ becomes essentially independent of s_0 for sufficiently large n.

   This approach is by no means restricted to probabilities but it is directly extendible to other observables that can be associated with symbol sequences: for example, the size $\varepsilon(s_1 s_2 \dots s_n)$ of each element of the covering of a fractal set at the nth level of resolution or the velocity v of an eddy in a turbulent fluid. The concept of interaction is accordingly generalized. For a "multiplicative" process, such as the construction of a fractal set or a turbulent cascade, it is natural to define $\mathcal{H}$ as the logarithm of the observable O[1]. Clearly, the choice of $\mathcal{H}$ must lead to a consistent thermodynamic formulation. It cannot be excluded that the

---

1. The term "multiplicative" usually means that $0 < c_1 \leq |\ln O(S)|/|S| \leq c_2 < +\infty$ uniformly in S in the limit $|S| \to \infty$, where c_1 and c_2 are constants.

observable, even if correctly identified, behaves in such an irregular fashion that the n-spin Hamiltonian fails to be an extensive function, "proportional" to the number n of particles for large n. The existence of the thermodynamic limit is usually discussed with reference to the total potential energy $U(X)$ of a finite set X of $n = |X|$ particles located at (distinct) lattice sites $\mathbf{x}_1, \mathbf{x}_2, \ldots, \mathbf{x}_n \in X$ (two particles may not occupy the same site). Configurations for which $U = \infty$ have zero thermodynamic weight and are, therefore, forbidden. It is further assumed that U is invariant under permutation of its arguments and under translations:

$$U(X + r) = U(X), \quad \forall r \in L.$$

Sufficient conditions for the existence of the limit require that the interactions do not cause the collapse of infinitely many particles into a bounded region of L and that they become negligible at large distances (Ruelle, 1974):

1. *Stability:* $U(X) \geq -an(X)$ for all X and some constant $0 < a < \infty$ (a restriction on the attractive parts of the potential);

2. *Tempering:*

$$W_{n,m} = U(x_1, \ldots, x_n, y_1, \ldots, y_m) - U(x_1, \ldots, x_n)$$
$$-U(y_1, \ldots, y_m) \leq Anmr^{-\gamma}, \tag{6.8}$$

where $\{x_i\}$ and $\{y_j\}$ are the coordinates of two distinct groups of particles, $A \geq 0$, $|x_i - y_j| \geq r \geq R_0 > 0$ for all $i \in [1, 2, \ldots, n]$ and $j \in [1, 2, \ldots, m]$, and $\gamma > 1$ ($> d$ on a d-dimensional lattice). This condition limits the mutual repulsive interaction between separated groups of particles.

Since U is essentially the Hamiltonian, rewritten as a function of the positions of the particles, the quantity $W_{n,m}$ represents the logarithmic difference between the conditional probability of the configuration $(x_1, \ldots, x_n)$, given $(y_1, \ldots, y_m)$, and its unconditioned probability, when definition (6.1) is adopted.

The potential energy U can be seen as the superposition of ℓ-body potentials $\Phi_\ell = \Phi(Y) = \Phi(y_1, y_2, \ldots, y_\ell)$, where $Y = \{y_1, y_2, \ldots, y_\ell\}$ is a generic set of ℓ particles. Because of translational invariance, Φ_1 is a finite constant, conventionally set to 0. The real-valued functions Φ_ℓ satisfy the same invariance properties as U. Accordingly $U(X) = U(x_1, \ldots, x_n)$ is expressed by

$$U(X) = \sum_{\ell \geq 2} \sum_{1 \leq i_1 < \ldots < i_\ell \leq n} \Phi(x_{i_1}, \ldots, x_{i_\ell}) = \sum_{Y \subset X} \Phi(Y), \tag{6.9}$$

in analogy with the sum over 1-, 2-, $\ldots$, n-body interaction energies of Eq. (6.6). A class of interaction spaces can then be conveniently specified by the condition

$$\|\Phi\|_\rho = \sum_{Y \ni y_0} \frac{|\Phi(Y)|}{|Y|^{1-\rho}} < \infty, \tag{6.10}$$

where $|Y| = \ell$ is the number of particles in Y and ρ is a convergence exponent. Expression (6.10) represents a bound on the total interaction of *one* particle (situated at y_0) with all others. For $\rho = 0$, one obtains a Banach space $\mathcal{B}_0$, the functions of which need not decrease very rapidly to zero for large values of their arguments. This condition implies stability (Fisher, 1972). The choice $\rho = 1$ defines a space $\mathcal{B}_1 \subset \mathcal{B}_0$ and imposes a more stringent bound on the long-distance decay of the interaction.

Excluding the cases in which the thermodynamic limit does not exist, the most interesting behaviour is expected to be associated with the spontaneous emergence of long- (infinite-) range *correlations* in the equilibrium configurations. It is well-known that they can be produced by short-range interactions, provided that the lattice is two- or higher-dimensional. In the one-dimensional case, instead, this phenomenon is possible only if the interactions themselves are long-ranged (see Sections 6.3 and 6.4). Notice that this can occur even if the dynamical laws responsible for the generation of the pattern obey local rules as in the case of cellular automata. Other interesting properties are exhibited by systems belonging to the interaction class $\mathcal{B}_0 \setminus \mathcal{B}_1$ (Fisher, 1972).

6.2   Statistical ensembles

The formal analogy between one-dimensional statistical mechanics and dynamical systems was originally developed by Sinai (1972), Ruelle (1978) and Bowen (1975) for one-dimensional expanding maps and Julia sets. Several different formulations are possible (Bohr & Tél, 1988; Badii, 1989; Beck & Schlögl, 1993). We discuss a few equivalent sets of definitions of "thermodynamic" quantities which have a close connection with coarse-graining procedures commonly employed in the theory of dynamical systems. In particular, we concentrate on entropies and dimensions.

6.2.1   Generalized entropies

Recalling Eq. (6.1), we construct the *canonical* "partition function"

$$Z_{\text{can}}(n, q) = \sum_{\{s_1, s_2, \ldots, s_n\}} P^q(s_1 s_2 \ldots s_n) \sim e^{-n(q-1)K_q}, \tag{6.11}$$

where the symbol $\sim$ indicates the leading scaling behaviour for large n and the sum is taken over all distinct sequences S of length n. The function K_q (Renyi, 1970), called *generalized entropy*, accounts for the fluctuations of the *local entropy*

$$\kappa(s_1 s_2 \ldots s_n) = -\frac{1}{n} \ln P(s_1 s_2 \ldots s_n) \tag{6.12}$$

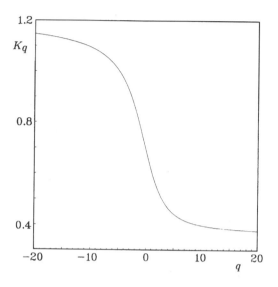

Figure 6.1 Generalized entropy K_q for the Bernoulli shift $B(1/3, 2/3)$ as a function of q.

"around" its expectation value $\sum \kappa(S)P(S)$ which converges to the Kolmogorov–Sinai entropy K_1 (Eq. (5.44)) for $n \to \infty$, as can be verified by expanding Eq. (6.11) to first order in $q - 1$. The curve K_q vs. q for the Bernoulli shift $B(1/3, 2/3)$ is shown in Fig. (6.1). The expression

$$I_q = \frac{1}{1-q} \ln \sum P^q \sim n K_q ,\tag{6.13}$$

an extension of the usual Shannon information H (see Eq. (5.36)), is known as *Renyi information* of order q. By differentiating K_q with respect to q and defining $\pi(S) = P^q(S) / \sum P^q(S)$, one obtains

$$K_q' = -\frac{1}{n(q-1)^2} \sum_{\{S\}} \pi(S) \ln \frac{\pi(S)}{P(S)}$$

which is never positive for the property (5.49) of the relative entropy: K_q is a monotonic nonincreasing function of q. In the limit $q \to 0$, with the convention that $P^0 = 0$ if $P = 0$, the topological entropy K_0 is recovered: in this way, the sum in Eq. (6.11) yields the number $N(n)$ of allowed n-blocks. In the limit $q \to +\infty$ $(-\infty)$, one obtains the smallest (largest) local entropy. The curve K_q is a constant in the so-called uniform case, in which all probabilities asymptotically scale with the same rate κ.

In the thermodynamic approach, the generalized entropy K_q can be formally identified with the free energy $\mathscr{F}(\mathscr{V}, T)$, where the volume $\mathscr{V}$ is simply n and the temperature T is proportional to the inverse of the exponent $q - 1$, so that

$$\begin{aligned} \mathscr{F}(\mathscr{V}, T) &\Longleftrightarrow n K_q \\ \mathscr{E} &\Longleftrightarrow n\kappa \\ \mathscr{V} &\Longleftrightarrow n \\ k_B T &\Longleftrightarrow (q-1)^{-1} , \end{aligned}\tag{6.14}$$

where $\mathcal{E}$ is the energy and k_B the Boltzmann constant. Besides the extensive quantities $\mathcal{F}$ and $\mathcal{E}$, it is customary to introduce the analogous free energy $F(T) \iff K_q$ and energy $E \iff \kappa$ per unit volume. While the first three assignments in Eq. (6.14) are quite harmless, the last one poses a problem of interpretation when negative temperatures ($q < 1$) are involved. This inconvenience is immediately removed by a simultaneous change of sign for both energy and temperature across the infinite-temperature point $q = 1$. It must be stressed that, in this application of the thermodynamic formalism, the temperature is an entirely free parameter, the variation of which is equivalent to focusing on subsystems (sets of n-cylinders) characterized by close energy values κ.

The "asymmetry" in the exponents (q versus $q-1$) involved in the definition (6.11) of the "free energy" K_q flaws the parallel with ordinary thermodynamics. This problem can be resolved with a careful definition of the ensemble. Notice that the sum $\sum P^q$ runs over all the b^n possible distinct n-blocks and that it equals the expectation $\langle P^{q-1} \rangle$. This average can be evaluated by dividing the configuration $\mathcal{S}$ into N sequences S of length n and by performing the sum $\sum' P^{q-1}(S)/N$, where each sequence S appears a number $N(S)$ of times proportional to its own probability $P(S) \approx N(S)/N$. This is the canonical ensemble. Sums performed as in Eq. (6.11), instead, are based on a uniform partition of phase- or sequence-space (box-counting). The items are hence sampled with respect to the Lebesgue measure, so that the extra weight P must be explicitly included to compute the required average. The choice of the canonical ensemble restores the symmetry (the reason for having $q-1$ as an exponent instead of β is purely historical).

The microcanonical description is obtained by considering the ensemble consisting of all sequences with an energy E in the range ($\kappa, \kappa + d\kappa$). Accordingly, the microcanonical partition function $Z_{\mathrm{mic}}(\kappa)$ reads

$$Z_{\mathrm{mic}}(\kappa) \equiv e^{n\mathsf{S}(\mathsf{E})/k_B} = e^{n[g(\kappa)-\kappa]} , \tag{6.15}$$

where $\mathsf{S}(\mathsf{E})$ is the microcanonical entropy and $g(\kappa) - \kappa$ is its dynamical counterpart. The function $g(\kappa)$ is called *spectrum of entropies*. The reader should notice that κ is a local *dynamical* entropy which plays the role of an energy in the thermodynamic construction: its meaning is absolutely distinct from that of the *thermodynamic* entropy $\mathsf{S}(\mathsf{E})$. The relationship between the canonical and microcanonical formulations can be obtained by noticing that

$$\sum' P^{q-1} \sim \int e^{-n\kappa(q-1)} e^{n[g(\kappa)-\kappa]} dk ,$$

for $n \to \infty$. Recalling the definition (6.11) of K_q, one obtains

$$\int e^{-n[q\kappa-g(\kappa)]} d\kappa \sim e^{-n(q-1)K_q} \tag{6.16}$$

which, in the limit $n \to \infty$, can be solved by the saddle-point approximation. This yields the two equivalent Young–Fenchel conditions

$$(q - 1)K_q = \inf_{\kappa} [q\kappa - g(\kappa)] \ , \quad g(\kappa) = \inf_{q} \left[q\kappa - (q - 1)K_q\right] \ . \qquad (6.17)$$

In particular, if $g(\kappa)$ and K_q are differentiable, the above conditions reduce to the Legendre transforms

$$g(\kappa) = q\kappa(q) - (q - 1)K_q \ , \quad \kappa(q) = \frac{d\left[(q - 1)K_q\right]}{dq} \ , \quad q = \frac{dg(\kappa)}{d\kappa} \ , \qquad (6.18)$$

where $\kappa(q)$ denotes the value of κ at which the infimum is reached in Eq. (6.17). The first of equations (6.18) can be rewritten as

$$F(T) = E(T) - TS(E) \ ,$$

by recalling definitions (6.14) and (6.15). All usual thermodynamic relations hold. For example

$$\left(\frac{\partial \mathscr{S}}{\partial \mathscr{E}}\right)_{\mathscr{V}} = \frac{1}{T} \ , \quad \left(\frac{\partial \mathscr{S}}{\partial \mathscr{V}}\right)_{\mathscr{E}} = \frac{P}{T} \ , \quad \left(\frac{\partial \mathscr{F}}{\partial \mathscr{V}}\right)_{T} = -P \ , \qquad (6.19)$$

where the pressure P formally corresponds to $= -K_q$. In this approach, in fact, there are no fluctuating quantities besides the energy E. Moreover, there is no interaction with the environment which could be modelled by a term like $P d\mathscr{V}$ (or $H d\mathscr{M}$, in a magnetic language where H is the external field and $\mathscr{M} = \sum \sigma_i$ the magnetization) in an infinitesimal transformation. In the next subsection, a bivariate case will be discussed, in which consideration of volume and pressure is essential.

The meaning of the function $g(\kappa)$ can be understood with reference to Eq. (6.15). Operationally, it can be evaluated through a histogram: in fact, $e^{ng(\kappa)} d\kappa$ is the number of *distinct* sequences of length n characterized by a local entropy in the range $(\kappa, \kappa + d\kappa)$. For the Bernoulli shift $B(p_0, p_1)$, where $e^{-n\kappa} = p_0^k p_1^{n-k}$,

$$e^{ng(\kappa)} = \binom{n}{k}$$

is the number of n-blocks containing k zeros. Assuming, without loss of generality, $p_0 < p_1$, one obtains $g(\kappa)$ (see Fig. 6.2) as

$$g(\kappa) = -\frac{\kappa_+ - \kappa}{\kappa_+ - \kappa_-} \ln \frac{\kappa_+ - \kappa}{\kappa_+ - \kappa_-} - \frac{\kappa - \kappa_-}{\kappa_+ - \kappa_-} \ln \frac{\kappa - \kappa_-}{\kappa_+ - \kappa_-} \ ,$$

where $\kappa_+ = -\ln p_0 = K_{-\infty}$ and $\kappa_- = -\ln p_1 = K_{+\infty}$ are, respectively, the largest and the smallest value of the local entropy. They mark the borders of the support of the entropy spectrum $g(\kappa)$. Since they occur for just one sequence each (0^n and 1^n, respectively), $g(\kappa)$ is zero at these points. The value of $g(K_{\pm\infty})$, however, may be positive in other systems: for example, in the ternary Bernoulli

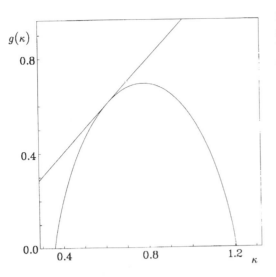

Figure 6.2 Entropy spectrum $g(\kappa)$ vs. κ for the Bernoulli shift $B(1/3, 2/3)$.

shift $B(p_0, p_1, p_2)$ with $p_0 = p_1$, all the 2^n sequences containing no 2s exhibit the same rate p_0 and contribute to $g(\kappa_-)$ (assuming $p_0 > p_2$).

The relation between $g(\kappa)$ and K_q is simply illustrated by recalling the geometrical interpetation of the Legendre transforms (6.18). The tangent to $g(\kappa)$ with slope q intersects the bisectrix $g(\kappa) = \kappa$ at the point (K_q, K_q). Accordingly, the maximum of $g(\kappa)$ is the topological entropy K_0. At $q = 1$, $g(\kappa)$ equals the metric entropy K_1 and is tangent to the bisectrix. The function $F(q) = (q - 1)K_q$ is always concave, even when $g(\kappa)$ is not: in fact, its second derivative is

$$\frac{d^2 F}{dq^2} \propto -\sum_{i > j}(P_i P_j)^q \left(\ln \frac{P_i}{P_j}\right)^2 \leq 0 \,,$$

where the indices i and j are a shorthand notation for the sequences $S = s_1 s_2 \ldots s_n$. Hence, when the entropy spectrum $g(\kappa)$ is computed from the Legendre transform (6.18), the resulting $\tilde{g}(\kappa)$ is automatically concave since

$$\frac{d^2 \tilde{g}}{d\kappa^2} = \left(\frac{d^2 F}{dq^2}\right)^{-1} \,. \tag{6.20}$$

When directly estimated from a histogram, however, $g(\kappa)$ need not be concave: in such a case, the Legendre transform $\tilde{g}$ is the concave envelope of g (i.e., the smallest possible concave function greater than g). In the uniform case, when all probabilities scale with the same rate κ^*, the support of g reduces to the single point $\kappa = \kappa^*$. Notice that the two-scale Bernoulli process $B(p_0, p_1)$ produces an infinity of scaling exponents $\kappa(S)$.

As already remarked, the curve $g(\kappa) - \kappa$ versus κ is analogous to the thermodynamic entropy $S(E)$: its derivative is, in fact, the "inverse temperature" $q - 1$. Since $g(\kappa) \leq \kappa$, almost all sequences (with respect to the natural measure m) exhibit the same local entropy $\kappa = K_1$ (where $g(\kappa) = \kappa$) in the limit $n \to \infty$. This is easily understood by noticing that $Z_{mic}(\kappa)$ (see Eq. (6.15)) is proportional to the probability of finding an n-sequence with entropy κ in the spin configuration. Expanding $g(\kappa)$ around $\kappa = K_1$ to second order, $Z_{mic}(\kappa)$ reduces to a Gaussian distribution with variance

$$\sigma^2(n) = \frac{1}{n|g''|} ,$$

provided that the derivative $g'' = d^2g/d\kappa^2$ is nonzero. Hence, finite-volume fluctuations can be neglected in the thermodynamic limit. The same result is obtained by making this approximation in the canonical formulation Eq. (6.16), where g'' is evaluated at $\kappa(q)$ for each q. Recalling the definition of the specific heat

$$c_{\mathscr{V}} = \left(\frac{\partial \mathscr{E}}{\partial \mathsf{T}}\right)_{\mathscr{V}} = -(q-1)^2 \frac{d^2F}{dq^2} \tag{6.21}$$

and equation (6.20), one recovers the well-known connection between specific heat and energy fluctuations. The coincidence of the local entropy $\kappa = -\ln P(S)/n$ with the metric entropy K_1 for m-almost all S, in the limit $n \to \infty$, is known as the *Shannon–McMillan–Breiman theorem* (Cornfeld *et al.*, 1982).

6.2.2   Generalized dimensions

The thermodynamic formalism has been so far developed to account for the fluctuations of a single quantity: the probability. In the case of fractal measures, the relevant observables are both the mass P and the size ε of the elements B of a suitable covering. Recalling the procedure adopted for the definition of the Hausdorff dimension (Section 5.6), one notices that the domains B need not be identical to each other nor contain the same mass: their sizes have an upper bound ε, which sets the resolution level, and are adjusted in order to achieve the infimum of the measure (5.52). The scale ε can be made exponentially dependent on an integer n in a hierarchical approach to the limit. Accordingly, each domain B at the nth level of resolution is labelled with an n-symbol sequence S, as illustrated in Fig. 6.3.

The scaling behaviour that relates the probability P to the size ε (see Eq. (A4.1)) is expressed by

$$P(\varepsilon) \sim \varepsilon^{\alpha(S)} , \tag{6.22}$$

for $\varepsilon \ll 1$. The exponent $\alpha(S)$ can be interpreted as the *local dimension* of the element B_S and, in general, fluctuates with the position (labelled by S), in analogy with the entropy κ. At variance with the previous subsection, where the resolution e^{-n} was fixed, the scaling variable $\varepsilon = \varepsilon(S)$ is here a fluctuating quantity, just as the probability $P(S)$. This property is taken into account by

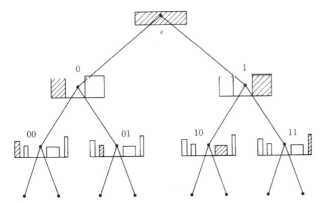

Figure 6.3 Illustration of a hierarchical coarse-graining of a fractal measure on a binary Cantor set. The mass carried by each interval is represented by a box with appropriate height. Each symbol sequence refers to the shaded box nearby and the empty string ϵ to the whole unit interval.

introducing two parameters, q and τ, and constructing the following *isothermal-isobaric* partition function

$$Z_{\mathrm{iso}}(n, q, \tau) = \sum_{S\,:\,|S|=n} P^q(S)\varepsilon^{-\tau}(S)\,, \tag{6.23}$$

where the sum is taken over all sequences S of length n, in analogy with the usual thermodynamic expression

$$Z_{\mathrm{iso}}(\mathsf{P}, \mathsf{T}) = e^{-\mathscr{G}(\mathsf{P},\mathsf{T})/k_B\mathsf{T}}\,,$$

where $\mathscr{G}(\mathsf{P}, \mathsf{T})$ is the Gibbs free energy and

$$
\begin{aligned}
\mathscr{E} &\iff -\alpha(S)\ln\varepsilon \\
\mathscr{V} &\iff -\ln\varepsilon \\
k_B\mathsf{T} &\iff (q-1)^{-1} \\
\mathsf{P} &\iff -\tau/(q-1)\,.
\end{aligned}
\tag{6.24}
$$

Hence, the energy per unit volume $\mathsf{E} = \mathscr{E}/\mathscr{V}$ equals the local dimension $\alpha(S)$. A relation between τ and q, obtained by requiring that the partition sum (6.23) remains bounded away from 0 and ∞ in the limit $n \to \infty$, defines the *generalized dimension* D_q as

$$D_q = \tau(q)/(q-1)\,. \tag{6.25}$$

In fact, recalling Eq. (6.22), the r.h.s. of Eq. (6.23) can be rewritten as the expectation value

$$\left\langle P^{q-1}\varepsilon^{-\tau}\right\rangle = \left\langle\varepsilon^{(q-1)(\alpha-D_q)}\right\rangle\,,$$

where D_q appears as the "average dimension" necessary to balance the expression. From relation (6.24), one obtains $D_q \iff \mathsf{F}(\mathsf{T}) = -\mathsf{P}$.

More precisely, following Hausdorff's criterion (5.52), the most economic covering is defined by a new partition function Z'_{iso} which equals the infimum of Z_{iso} over all choices of $\varepsilon(S) < \varepsilon$, for negative τ, and the supremum of Z_{iso}, for positive τ. Accordingly, $\tau(q)$ is the value of τ such that

$$\lim_{n\to\infty} Z'_{\text{iso}}(n, q, \tau) = \begin{cases} \infty\, , & \text{if } \tau > \tau(q), \\ 0\, , & \text{if } \tau < \tau(q). \end{cases} \tag{6.26}$$

The selection of the infimum (supremum) is motivated by the need of avoiding "marginal" intersections between the elements of the covering and the fractal set, which would lead to the underestimation of the local density and, in turn, to the overestimation of the dimension α. This necessity may prevent a q-independent optimal covering from existing, thus rendering the thermodynamic interpretation more problematic. This difficulty is obviously absent for a self-similar Cantor set, in which both spatial scales and probabilities are given by a memory-less multiplicative process: $\varepsilon(s_1 \ldots s_n) = \prod_i r_{s_i}$, $P(s_1 \ldots s_n) = \prod_i p_{s_i}$, where $r_{s_i} \in \{r_0, r_1, \ldots, r_{b-1}\}$ and $p_{s_i} \in \{p_0, p_1, \ldots, p_{b-1}\}$ are the scaling rates for lengths and masses, respectively. In general, however, the construction of a hierarchy of partitions which respect the intrinsic, unkown rules of the process at each level may be a forbidding task. This is another aspect of complexity which, moreover, appears in the course of the coding procedure itself, i.e., before intrinsic properties of the system manifest themselves through atypical scaling with effects on the shape of the thermodynamic functions K_q and D_q.

The general formula (6.23) can be simplified in two extreme situations: namely, when the covering consists of either constant-mass (Badii & Politi, 1984 and 1985) or constant-size (Renyi, 1970; Grassberger & Procaccia, 1983; Halsey *et al.*, 1986) elements. As a consequence, the dependence of the sum on the scaling parameter can be made explicit. In the former case, the partition function is rewritten as

$$\left\langle \varepsilon^{-\tau}(P) \right\rangle \sim P^{1-q(\tau)}\, , \tag{6.27}$$

for $P \to 0$, where τ is now the independent parameter and $q = q(\tau)$ is the inverse function of $\tau(q)$. The constant-size partition function reads

$$\left\langle P^{q-1}(\varepsilon) \right\rangle \sim \varepsilon^{\tau(q)}\, , \tag{6.28}$$

for $\varepsilon \to 0$. The two previous expressions are completely analogous to the metric-entropy partition function (6.11), the main difference being that the scaling parameter is either e^{-n}, or P, or ε, in the three cases. In fact, as already remarked at the end of Chapter 5, the metric entropy K_1 can be interpreted as a dimension in sequence space

$$X_n = \left\{ x : x = \sum_{i=1}^{n} s_i b^{-i}, s_i \in \{0, 1, \ldots, b-1\} \right\} \to [0, 1]\, , \tag{6.29}$$

for $n \to \infty$. This immediately suggests the possibility of generalizing the covering procedure for the entropy by choosing a variable-size subdivision of the unit interval, which can be obtained by means of cylinder sets $c_n(S)$ of different length n. Hence, the problem of finding an optimal covering arises also in this case. The appropriate statistical mechanical framework then requires the grand canonical ensemble.

In general, the "symmetry" between P and ε, for fractal dimensions, and between P and n, for metric entropies, leaves the freedom of choosing any of them as the energy and its dual as the volume in the thermodynamic formulation. In the constant-mass approach (6.27), for example, one might define $\mathscr{E} = -\ln \varepsilon$, $k_B T = -1/\tau$ and $\mathscr{V} = -\ln P$ (Badii, 1989). The most natural assignment for constant-size coverings (6.28), instead, is given by Eq. (6.24), which yields $\mathscr{F}(\mathscr{V}, T) = -D_q \ln \varepsilon$. In complete analogy with the treatment of the metric entropy, the microcanonical partition function $Z_{\mathrm{mic}}(\alpha)$ takes the form

$$Z_{\mathrm{mic}}(\alpha) \equiv \varepsilon^{\alpha - f(\alpha)}, \tag{6.30}$$

where $f(\alpha) - \alpha$ plays the role of the entropy $S(E)$ and the function $f(\alpha)$ is called the *dimension spectrum*. The quantity $Z_{\mathrm{mic}}(\alpha)$ is proportional to the probability of finding a domain in the covering with a local dimension in the range $(\alpha, \alpha + d\alpha)$. Following the same procedure as for the metric entropy (see Eq. (6.16)), one obtains the Legendre transforms

$$f(\alpha) = q\alpha(q) - \tau(q), \quad \alpha(q) = \frac{d\tau(q)}{dq}, \quad q = \frac{df(\alpha)}{d\alpha}, \tag{6.31}$$

where $\alpha(q)$ denotes the value of α at which the extremal condition is attained. The maximum of the $f(\alpha)$ curve equals the support dimension D_0. The tangency point $f(\alpha) = \alpha$ occurs at the information dimension $\alpha = D_1$. Almost all ε-domains exhibit the same local dimension D_1 in the limit $\varepsilon \to 0$.

In order to preserve the bivariate character of the statistics for a general covering, the microcanonical partition function ought to be written as

$$Z_{\mathrm{mic}}(\kappa, \lambda) \equiv e^{nh(\kappa, \lambda)}, \tag{6.32}$$

where $P(S) = e^{-n\kappa(S)}$ and $\varepsilon(S) = e^{-n\lambda(S)}$. The intensive variables κ and λ satisfy $\kappa(S) = \alpha(S)\lambda(S)$, because of Eq. (6.22). These assignments are appropriate only if it is possible to construct a sequence $\{\mathscr{C}_n\}$ of coverings that allows associating $\kappa(S)$ and $\lambda(S)$ with suitable system observables. In particular, when $\mathscr{C}_n$ represents the nth refinement of the generating partition of a two-dimensional measure preserving map, κ and λ turn out to be the entropy and the positive Lyapunov exponent computed over the finite-time trajectory corresponding to S (Grassberger *et al.*, 1988). The thermodynamic entropy $S(E) = f(\alpha) - \alpha$ is hence determined as

$$f(\alpha) - \alpha = \sup_{\lambda : \kappa = \alpha\lambda} \frac{h(\kappa, \lambda)}{\lambda}.$$

Example:

[6.3] Let us investigate the dimension spectrum $f(\alpha)$ for the logistic map (3.11) at the Ulam point $a = 2$ (Ex. 5.8). The invariant measure $\rho(x) = 1/\left[\pi\sqrt{1-x^2}\right]$ presents two regions with distinct scaling indices: the smooth "internal" one, characterized by a local dimension $\alpha = 1$, and the two borderpoints (-1 and 1) exhibiting a square-root singularity ($\alpha = 1/2$). By covering the interval $[-1, 1]$ with $N(\varepsilon) = 1/\varepsilon$ cells of length ε, the mass $P(\varepsilon; x)$ in the cell centred at x is

$$P(\varepsilon; x) = \frac{1}{\pi} \int_{x-\varepsilon/2}^{x+\varepsilon/2} \frac{dx}{\sqrt{1-x^2}} \propto \sqrt{1 - x + \varepsilon/2} - \sqrt{1 - x - \varepsilon/2}\,,$$

for $\varepsilon/2 \ll 1 - x \ll 1$. Setting $1 - x = M\varepsilon$ and recalling Eq. (6.22), one obtains

$$\varepsilon^\alpha \sim \sqrt{\varepsilon}\left(\sqrt{M + 1/2} - \sqrt{M - 1/2}\right) \sim \sqrt{\frac{\varepsilon}{M}}\,,$$

where the symbol $\sim$ indicates the double limit $\varepsilon \ll 1$, $M \gg 1$. Clearly, M is the number of cells between x and 1. The local dimension α, for any finite ε, grows monotonically from $1/2$ to 1 while moving x from 1 to 0. Hence, the number of cells contributing an index in $(\alpha, \alpha + \delta\alpha)$ (equal to $\varepsilon^{-f(\alpha)}\delta\alpha$) is given by the variation δM that must be made in order to observe an increment $\delta\alpha$ in α. This yields

$$\varepsilon^{-f(\alpha)} = \frac{dM}{d\alpha} \sim |\ln \varepsilon| \varepsilon^{-(2\alpha - 1)}\,.$$

The function $f(\alpha)$ is hence $2\alpha - 1$ in the interval $J = [1/2, 1]$ and is by definition $-\infty$ outside this range: in fact, no cell carries an exponent $\alpha \notin J$. With this extension, $f(\alpha)$ can be considered as a concave function, in spite of its linearity in J. This result, therefore, is not just an outcome of the Legendre transform but, as the above calculation has shown, is obtained directly from the histogram. Although there cannot be "punctual" exponents other than $1/2$ and 1, a continuous range J of α-values is observed: no matter how small ε is, there is always a number $\varepsilon^{-f(\alpha)}$ of cells which have not yet settled in the asymptotic scaling regime with $\alpha = 1$.   □

6.3   Phase transitions

It is well known that several physical systems undergo a deep structural modification when an external parameter is increased beyond a "critical" value. Among them, we mention smectic vs. nematic liquid crystals, para- vs. ferromagnets, superconductors, and superfluids. The best known of these phenomena is probably the evaporation of water. At atmospheric pressure, H_2O exists as

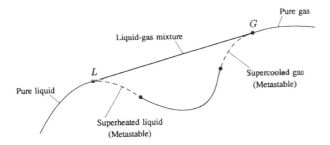

Pure gas

G

Liquid-gas mixture

L

Pure liquid

Supercooled gas
(Metastable)

Superheated liquid
(Metastable)

Figure 6.4 Schematic plot of the entropy-energy diagram for a fluid at the liquid–gas transition. The segment connecting points L and G is the concave envelope of the entropy curve. Metastable states are found along the dashed line.

a liquid for temperatures less than the boiling point $T = 100°C$ and as a gas beyond it. The change of state is called a *phase transition* and is characterized by an abrupt decrease of the fluid density $\rho = n/\mathscr{V}$ ($\rho_l \approx 1600\rho_g$, where the indices refer to liquid and gas, respectively). The boiling temperature $T(P)$ increases with the pressure P, while the density difference between the two phases decreases until it vanishes at the *critical point* $(T_c, P_c) = (374°C, 218\,atm)$. At still higher temperatures, a single fluid phase is left. The curve $P(T)$ is called the *transition line*. In general, the distinction between the phases is expressed by means of one or more thermodynamic quantities: the *order parameters* $\mathbf{Q} = \{Q_1, \dots, Q_p\}$. For water, the scalar $Q = \rho_l - \rho_g$ is appropriate. In ferromagnetic materials, Q is the total magnetization. Usually, $\mathbf{Q}$ is identically zero in one phase. The change of the order parameter across the transition can be either discontinuous, as in the liquid-gas case, or continuous, as in magnets. Transitions of the former type are said to be of the *first order*, whereas all others are called *continuous* (Fisher, 1967a). The order parameter, an extensive thermodynamic variable, is usually the derivative of a thermodynamic potential with respect to a *conjugate field*, such as the pressure or the external magnetic field (Huang, 1987). Hence, phase transitions are associated with the occurrence of nonanalytic behaviour in some thermodynamic function. Discontinuity in the first derivative characterizes a first order transition, whereas continuous transitions may be distinguished on the basis of the discontinuity or divergence of some higher-order derivative. In typical mathematical models, this behaviour appears only in the limit of an infinite system since the partition function is, by definition, analytic for any finite n. Therefore, the thermodynamic limit $n \to \infty$, although an idealization of real physical systems, appears as a fundamental tool to catch the essence of the phenomenon, irrespective of finite-size corrections.

The distinguishing feature of first order transitions is the coexistence of different phases. In this situation, equilibrium conditions imply that the intensive variables (T and P) are the same for all phases (Gibbs, 1948). Extensive variables, instead, assume different values. Suppose, for example, that the system possesses two *pure states*, characterized by the values (S_1, E_1) and (S_2, E_2) of entropy and energy. Phase coexistence manifests itself with the appearance of a linear dependence of S on E in the interval (E_1, E_2) (see Fig. 6.4). Accordingly, the linear combinations $S = rS_1 + (1 - r)S_2$ and $E = rE_1 + (1 - r)E_2$, for $r \in (0, 1)$, define the thermodynamic variables for the *mixture* of the two phases coexisting at the temperature $T = \partial \mathscr{E} / \partial \mathscr{S}$ and at the pressure $P = -\partial \mathscr{E} / \partial \mathscr{V}$. The prototype for this behaviour is the above mentioned liquid–gas transition which is usually illustrated by displaying its equation of state[2] $f(P, V, T) = 0$ in the (P, V) plane, where $V = 1/\rho$ is the volume per unit mass. By the properties of the Legendre transform, the same transition appears in the conjugate function $F(T)$ as a cusp at the transition temperature T_t, where the left and right derivatives $F'_{\mp}(T_t)$ do not coincide.

Example:

[6.4]   For the logistic map (3.11) at the Ulam point (Ex. 6.3), the free energy

$$D_q = \begin{cases} 1, & \text{for } q \leq 2, \\ \dfrac{q}{2(q-1)}, & \text{for } q > 2, \end{cases} \tag{6.33}$$

presents a discontinuous derivative at $q = q_c = 2$, typical of a first order phase transition. The two phases correspond to the open interval $-1 < x < 1$, where the local dimension is $\alpha = 1$, and to the pair of points $x = \pm 1$, contributing with $\alpha = 1/2$.

$\square$

This is not the only possible kind of a first order transition. A systematic classification has been proposed by Fisher & Milton (1986) who identified nine different scenarios. Among them, we recall the complementary phenomenon to the liquid–gas condensation, in which the isothermal curve $P(V)$ presents a discontinuous jump. This is equivalent to the occurrence of a cusp in the $S - E$ representation. This quite bizarre behaviour is actually observed in a lattice model (Fisher, 1967b and 1972). An example arising in the context of nonlinear dynamics is discussed in the next section. Transitions of this type can occur only as a result of long-range interparticle forces which, however, need not violate the tempering condition (6.8). The interactions in the model considered by Fisher (1972) are in the Banach space $\mathscr{B}_0 \setminus \mathscr{B}_1$.

6.3.1   Critical exponents, universality, and renormalization

As already mentioned, the order parameter Q varies along the transition line until it vanishes at the critical point $\mathbf{P}_c$. Since Q is nothing but the "width" of the coexistence region, the distinction between the phases is then lost. At $\mathbf{P}_c$, the transition becomes continuous. The approach to $\mathbf{P}_c$ is signalled by the

---

2. Notice that the information conveyed by the equation of state is not sufficient to deduce all thermodynamic properties of the system as it can be done, instead, from the entropy–energy–volume characteristics $S = S(E, V)$ (Gibbs, 1873; Wightman, 1979).

appearance, in the system configuration, of "coherent" subregions that may be viewed as the remnants of previously ordered phases. Their sizes and evolution times vary in broad ranges which become infinite at P_c. This phenomenon is quantitatively described by the divergence of coefficients like the isothermal compressibility $k_T = -(\partial \mathcal{V}/\partial P)/\mathcal{V}$ or the specific heat $c_{\mathcal{V}}$ (6.21) when the reduced temperature $t = (T - T_c)/T_c$ tends to zero. Indicating with χ any of these quantities, one has

$$\chi(t) \sim \begin{cases} t^{-\gamma_+}, & \text{for } t \to 0^+, \\ |t|^{-\gamma_-}, & \text{for } t \to 0^-, \end{cases} \qquad (6.34)$$

where γ_+ and γ_- are called *critical exponents*. The vanishing of the order parameter at T_c is characterized by yet another exponent β: $Q(t) \sim (-t)^\beta$, for $t \to 0^-$, and is identically 0 above T_c. Additional critical exponents can be defined for the correlation function

$$C(l - k) = Z_{\text{can}}^{-1} \sum_{\{\sigma\}} \sigma_k \sigma_l e^{-\beta \mathcal{H}}, \qquad (6.35)$$

where the sum is over all spin configurations $\{\sigma\} = \{\sigma_1, \ldots, \sigma_n\}$. Far from the critical region, $C(l - k)$ decays exponentially as $\exp(-|l - k|/\xi)$. The onset of the multi-scale structure is expressed by the divergence of $\xi = \xi(t)$ as

$$\xi(t) \sim |t|^{-\nu}, \quad \text{for } t \to 0. \qquad (6.36)$$

At the critical point, the correlation function decays algebraically as

$$C(l - k) \sim |l - k|^{-\theta}.$$

The divergence of the correlation length ξ indicates that the details of the interparticle potentials are irrelevant to the critical behaviour of the system. This observation suggests the *universality* of critical phenomena: for example, the liquid–gas and the para–ferromagnetic transitions exhibit the same critical exponents, within the experimental errors (Patashinskii & Pokrovskii, 1979). More specifically, physical systems can be organized into *universality classes*, each characterized by the same critical properties which mostly depend on the dimensionality of the problem and on the symmetry group of the order parameters. An exhaustive theoretical explanation has been achieved by the *renormalization group* approach (Wilson, 1971; Wilson & Kogut, 1974) which combines scale-invariance and coarse-graining concepts. The former is formalized by the *scaling hypothesis* (Widom, 1965) which asserts that thermodynamic functions like the Gibbs potential are, in the neighbourhood of the critical point

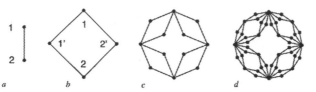

Figure 6.5 The first four steps in the construction of a hierarchical lattice. Each bond is replaced by a suitably rescaled rhombic cluster.

$\mathbf{P}_c$, homogeneous functions of the form

$$g(x_1, x_2, \ldots) \sim |t|^{a_g} g\left(x_1 |t|^{-a_1}, x_2 |t|^{-a_2}, \ldots\right), \qquad (6.37)$$

where $\mathbf{x} = (x_1, x_2, \ldots)$ is a set of *scaling fields*, such as the interaction coefficients $\{J_i\}$ (Eq. (6.6)). The key point is that the structure of typical configurations of a critical system is left essentially unaltered by a sequence of coarse-graining steps, apart from an overall scale factor (Kadanoff, 1966 and 1976a). The postulated *scale-invariance* suggests the construction of an effective Hamiltonian $\mathcal{H}_e^{(r)}$ in order to describe the interactions between subdomains having the size r used in the coarse-graining. Together with a rescaling step (a change in both length and energy scales), this procedure constitutes the renormalization group transformation $\mathcal{R}$.

Upon infinite iteration of $\mathcal{R}$, $\mathcal{H}_e^{(r)}$ usually converges to a fixed-point Hamiltonian $\mathcal{H}^*$. This means that the spatial structure appears to be statistically invariant under a (repeated) change of scale. Fully ordered and totally random patterns are the most elementary examples of this behaviour; a more interesting case is the fixed point found in correspondence of the critical point $\mathbf{P}_c$.

Hierarchical lattices, because of their exact self-similarity, provide a simple testing ground for the theory. Indeed, an exact recursive relation in a finite-dimensional space is obtained. We illustrate some of the relevant features of renormalization group (RG) dynamics with reference to a lattice constructed as in Fig. 6.5. In the first step, the bond $1-2$ is replaced by a rhombic cluster of $B = 4$ new bonds (Figs. 6.5(a) and (b)). Upon iteration, each new bond is analogously replaced by a properly rescaled cluster (Figs. 6.5(c) and (d)).

A hierarchical spin model results from placing a spin on each lattice node with an appropriate coupling with all others. Here, we choose Ising-like (i.e., nearest-neighbour) interactions by attributing the energy $-J\sigma_i\sigma_j$ to the bond $i-j$.

The RG-transformation $\mathcal{R}$ is just the inverse of the hierarchical construction of the lattice: each cluster of the last generation is removed and replaced by a bond with an effective coupling such that the canonical partition function is left unchanged. In the lattice of Fig. 6.5, an RG-step involves the decimation of two spins in each of the innermost clusters (as, for instance, $1'$ and $2'$ in Fig. 6.5(b)). Since the coupling involves only the two nn-sites in the cluster, the contributions of all decimated clusters to the partition function can be treated separately and independently. Letting $K = J/\kappa_B T$, the Boltzmann factor $\mathcal{H}/(\kappa_B T)$ is made to depend on the temperature T and on the coupling strength J through the variable K only. The partition function of a single cluster is hence

$$Z_{cl} = \exp\left[2K(\sigma_1 + \sigma_2)\right] + 2 + \exp\left[-2K(\sigma_1 + \sigma_2)\right],$$

which can be formally rewritten as

$$Z_{cl} = \exp\left[K'\sigma_1\sigma_2 + G(K')\right], \tag{6.38}$$

where

$$K' \equiv \mathscr{R}(K) = \ln\cosh 2K \tag{6.39}$$

and

$$G(K') = K' + \ln 4. \tag{6.40}$$

Inspection of Eq. (6.38) reveals that the energy contribution of the renormalized bond $1-2$ is still of the Ising type with a coupling strength K' given by the RG-map (6.39) plus an additional bond energy $G(K')$. Since the forms of both the interaction and the lattice are left unchanged by the RG-transformation, the procedure can be iterated indefinitely, so that the RG-dynamics is determined by just one variable, K, to be interpreted as either the coupling constant or, equivalently, as an inverse temperature. It is precisely the one-dimensionality of the RG-space that makes this model particularly appealing: in general, instead, $\mathscr{R}$ acts in a functional space with an infinity of interaction terms to be simultaneously renormalized.

Hierarchical lattices do not just yield a class of artificial models for testing scaling methods. Some of them, in fact, arise in the construction of approximate RG-transformations on Bravais lattices, as the one depicted in Fig. 6.5. The transformation discussed above for this lattice is an exact implementation of the Migdal–Kadanoff renormalization scheme (Migdal, 1976; Kadanoff, 1976b) which was originally developed for a two-dimensional square lattice. The relationship is better understood on an even simpler model, in which only one new site is introduced at each step, so that each bond (such as $1-2$ in Fig. 6.5(b)) is replaced by only two bonds (such as $1-1'$ and $1'-2$). It is not difficult to discover that this procedure leads to a perfectly renormalizable one-dimensional lattice. More generally, by formally associating an expansion factor c with each step of the construction, the effective dimensionality d of the lattice can be determined from the relation $c^d = B$, where B is the number of bonds in each cluster. The natural choice $c = 2$ ensures that $d = 1$ when $B = 2$ (as in the simplified model) and $d = 2$ for the lattice of Fig. 6.5.

The term $G(K')$ in Eq. (6.40), which appears at each renormalization step, represents the bond contribution to the free energy F. Upon iteration, the total number of bonds decreases exponentially as B^{-n}, where n is the number of RG-steps. In the thermodynamic limit, $F(K)$ is given by the infinite sum

$$F(K) = \sum_{n=1}^{\infty} G[K(n)]B^{-n},\tag{6.41}$$

where $K(n) = \mathcal{R}[K(n-1)]$ and $K(0) = K$. Formula (6.41) holds for general RG-transformations, for suitable G and B (Niemeijer & van Leeuwen, 1976).

An important result of the theory of critical phenomena is that even if G is an analytic function, the infinite sum in Eq. (6.41) can give rise to a singularity. Let us first discuss the case in which $K(n)$ converges to a fixed point K^*. If K^* is stable, a change in the initial coupling $K(0)$ becomes less and less pronounced in the higher order addenda of Eq. (6.41), so that we cannot expect any breaking of the analyticity. Indeed, a stable fixed point, in general, identifies a single phase (as, e.g., solid, liquid, etc.) of a statistical mechanical system. Conversely, the amplification of an initial deviation occurring in the vicinity of an unstable fixed point, although counterbalanced by the exponential drop of the volume B^{-n}, may lead to the divergence of some derivative of $F(K)$ and thus to singular behaviour. This is better illustrated by rewriting Eq. (6.41) as

$$F(K) = G(K) + F(\mathcal{R}(K))/c.$$

If $F(K)$ contains a singular component $F_s \sim (K - K^*)^\psi$, the following relation must hold (remember that $G(K)$ is assumed to be analytical):

$$F_s(K) = F_s(\mathcal{R}(K))/c.$$

Thus,

$$\psi = \frac{\ln c}{\ln|\mu|},$$

where $\mu = d\mathcal{R}/dK$ is the (only) eigenvalue of the RG-map at K^*. Therefore, a singularity is expected to occur as soon as $|\mu| > 1$[3].

Let us illustrate the above implications again with reference to the model in Fig. 6.5. The RG-map (6.39) possesses an unstable fixed point $K_c = 1/T_c \approx 0.60938$ and a stable fixed point $K^* = 0$. It is immediately seen that all temperatures above T_c are eventually mapped to K^*: they correspond to the "disordered", high-temperature phase. Conversely, all temperatures below T_c are mapped to $K = \infty$, the ordered, zero-temperature phase. Thus, the critical temperature T_c corresponds to the para–ferromagnetic phase transition. This analysis can be seen as an approximate solution of the two-dimensional Ising model.

---

3. It should be noticed, however, that the free energy remains an analytic function for some special choices of $G(K)$ (Derrida et al., 1983).

In general systems, the Hamiltonian $\mathscr{H}$ can be linearly expanded around $\mathscr{H}^*$, in terms of the eigenvectors of $\mathscr{R}$. The deviations of $\mathscr{H}$ from $\mathscr{H}^*$ associated with contracting eigenvalues correspond to *irrelevant interactions*, which have no influence at the critical point $\mathbf{P}_c$. The existence of unstable directions explains the type of scale invariance that holds at $\mathbf{P}_c$. If the system is not exactly at $\mathbf{P}_c$, in fact, the correlation length is finite and the iteration of $\mathscr{R}$ drives the system away from the fixed point exponentially fast. The number of unstable directions coincides with the number of independent parameters that must be tuned in order to reach the critical point: for the ferromagnetic transition, these are the temperature T and the external field H. The exponents v and δ, associated with the correlation length ξ (Eq. (6.36)) and with the magnetization $M \sim H^{-1/\delta}$, are determined directly from the unstable eigenvalues of $\mathscr{R}$. All others are obtained from these two through a set of "scaling laws" which follow from general dimensional arguments (Huang, 1987). In the absence of the external field H, a system at an initial distance t from $\mathbf{P}_c$ is mapped to a distance $t' = t\mu_1^n$ in n steps ($|\mu_1| > 1$). Simultaneously, the correlation length changes from ξ to $\xi' = \xi L^n$, where $L \in (0,1)$ is the rescaling factor of the RG-map $\mathscr{R}$. By substituting into Eq. (6.36), one finds $v = -\ln L / \ln |\mu_1|$.

In general, the scaling relation (6.37) can be rewritten in terms of the relevant fields only:

$$g(t, H) = L^{\ d} g(\mu_1 t, \mu_2 H) ,$$

where μ_2 is the second expanding eigenvalue and d is the dimension of the system. Hence, the homogeneity of g directly follows from the transformation $\mathscr{R}$. Furthermore, the renormalization group approach elucidates the concept of universality. Critical properties depend essentially on four factors:

1. the dimensionality d of the system;
2. the number of components of the order parameter Q;
3. the symmetry properties of the Hamiltonian;
4. the range of the interactions.

An instructive example is provided by the Ising model which has different exponents in two- and three-dimensional lattices, while it does not exhibit any phase transition in one dimension (except for the singular behaviour at zero temperature). In fact, as proved by Dyson (1969), long-range order occurs at nonzero temperatures in one-dimensional lattice systems if the pair interaction $J(i - j)$ decays as $|i - j|^{-\alpha}$ with $\alpha < 2$. In the presence of such long-ranged interactions, irrespective of the dimensionality d, the application of the renormalization group map $\mathscr{R}$ cannot eliminate the infinite-memory nature of the forces. The fixed point $\mathscr{H}^*$ that is eventually reached depends on the form chosen for the interactions (i.e., on the initial condition $\mathscr{H}_0$).

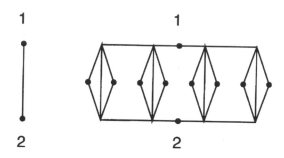

Figure 6.6 The first step in the construction of a hierarchical lattice. The bond 1-2 is replaced by the structure on the right. The four top (bottom) vertices must be considered as a single node: namely, the one indicated by a full dot on the same line. The four rhombs may be imagined to lie on the surface of a sphere, with the upper and lower vertices emanating from the north (1) and south (2) poles, respectively.

For a generic system, $\mathscr{R}$ need not admit only distinct fixed points, but also periodic or even chaotic trajectories. In fact, it has been shown that more elaborated hierarchical lattices can give rise to a chaotic dynamics (McKay *et al.*, 1982; Švrakić *et al.*, 1982; Derrida *et al.*, 1983). One such example is the lattice depicted in Fig. 6.6, where $q = 4$ equal sub-clusters are introduced in parallel at each RG-step. The corresponding RG-map can be straightforwardly derived from the previous example (see Eq. (6.39)) as

$$K' = \mathscr{R}(K) = q(K + \ln \cosh 2K) . \tag{6.42}$$

The factor q accounts for the multiplicity of the sub-clusters involved in the transformation (in the following we allow it to be noninteger), while the linear term K follows from the existence of an extra-link connecting sites 1 and 2, which was not present in the previous construction. The state $K = 0$ is still the infinite-temperature fixed point. Moreover, for $q \in [1, 4.658\ldots]$, there exists an interval to the left of $K = 0$ (corresponding to anti-ferromagnetic interactions) which is mapped into itself. Therefore, one expects to find the typical scenario of the logistic map, including chaotic dynamics for suitable q-values[4].

For what concerns the analyticity properties of the free energy, it is not difficult to see that the same arguments developed for a fixed point apply to the case of periodic or even chaotic trajectories, provided that $\ln|\mu|$ is replaced by the Lyapunov exponent of the orbit. A chaotic RG-flow, however, has deeper physical implications. Firstly, topological chaos alone (i.e., the positivity of the topological entropy) is sufficient to yield an infinity of phase transitions even if the trajectory asymptotically reaches an unstable fixed point $K_c = 1/T_c$. In fact, the same singular behaviour can be observed in the vicinity of the temperatures corresponding to the infinitely many preimages of K_c. Secondly, if a strange

---

4. Notice that, for the integer value $q = 4$ of Fig. 6.6, the asymptotic evolution consists of a period-4 solution in the middle of the period-doubling process.

attractor exists, the set of singularities ψ of the free energy F becomes dense (recall that changing K is tantamount to changing the map's initial condition $K(0)$ and that the set of the unstable periodic orbits, each carrying its own Lyapunov exponent, is dense). Hence, F is nowhere analytic (at least in some temperature range).

Since these results appear to be rather anomalous, it is natural to ask whether they hold for more realistic models of physical systems. In two dimensions, a theorem by Zamolodchikov (1986) forbids the existence of a chaotic RG-flow for translationally invariant systems. An extension of the theorem to three dimensions, however, has so far resisted the efforts of many researchers (Damgaard, 1992). Moreover, Zamolodchikov's theorem does not apply to disordered systems, where translational invariance holds only in a statistical sense. In fact, as we shall see in the next section, the glassy phase may provide an example of a chaotic renormalization.

From the point of view of complexity, a single stable phase, whether periodic as in a crystal or disordered as in a gas, is quite uninteresting since it can be simply described: the system is the collection of identical subunits which are either rigidly bound to one another or completely independent. Systems at a first order transition represent a higher degree of complexity, with the possible coexistence of phases. Finally, critical systems, being devoid of an absolute scale, are suggestive of a still higher type of complexity. In particular, notice the remarkable onset of macroscopic correlations from short-range forces in dimensions greater than one. The importance of critical phenomena is, however, weakened by their lack of genericity: in fact, two or more parameters must usually be tuned to achieve criticality.

6.3.2   Disordered systems

In the previous applications of thermodynamic formalism, we have considered "ordered" Hamiltonians, characterized by a set of coupling constants which are either fixed or prescribed by a deterministic procedure. In such a case, the irregularity of the spin configurations arises from the stochasticity of the interaction with a heat bath. There are physical phenomena, however, in which a further level of randomness operates, in that the model (Hamiltonian) is itself intrinsically stochastic. Examples of this behaviour are random walks in random media (such as directed polymers) and spin glasses (see Chapter 3). Random fractals with scaling rates which fluctuate along the hierarchical tree can also be assimilated to statistical mechanical systems with quenched disorder. They are encountered, for example, both in the modelling of fully developed turbulence (Section 2.2) and in the evolution of random dynamical systems in which the chaotic rule varies stochastically in time. A beautiful example of the latter phenomenon is given by the surface motion of a periodically pumped

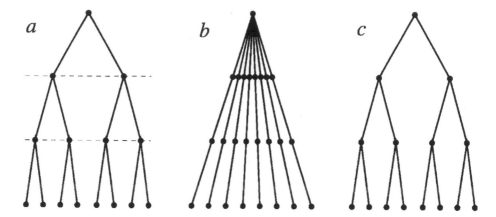

a   *b*   *c*

Figure 6.7 Tree representations of three different stochastic processes: (a) a coherent process, where the same rate is attributed to all homologous branches throughout each hierarchical level; (b) a fully decorrelated process, where the sequence probabilities are all independent of one another (apart from the normalization condition); (c) a random Cayley tree.

fluid (Sommerer & Ott, 1993a): the evolution on the surface is driven by the bulk dynamics which is intrinsically irregular even in the presence of a perfectly periodic forcing. For this reason, the system can be modelled with a dissipative random map (Yu *et al.*, 1990).

In these examples, the Hamiltonian is known only in a probabilistic sense, through the distribution of the interaction coefficients. Therefore, the extensive thermodynamic variables depend on the specific realization of the disorder (see Section 3.5) and the application of the thermodynamic formalism is subordinated to a suitable averaging over the ensemble of all possible realizations, and can give meaningful answers only for those variables the fluctuations of which vanish in the thermodynamic limit (called *self-averaging*). Furthermore, the answer may depend on the type of averaging, in analogy with the dependence of the generalized entropies K_q on q.

Let us illustrate the phenomenology with reference to three similar models characterized by a different type of disorder. The first consists of a chain of free spins in a random magnetic field (RMF). This is definitely the simplest conceivable example, with the energy being the sum $\mathscr{H} = \sum_i h_i \sigma_i$ of independent variables with no mutual interactions. A specific instance of this model is given by a binary Bernoulli process with random weights, in which the probability of a sequence Ss of $n+1$ symbols factorizes as $P(Ss) = P(S)p_n(s)$ and $p_n(\cdot)$ is a random variable with its own probability distribution. In the tree representation of Fig. 6.7(a), a dichotomic disorder is assumed: namely, $p_n(s)$ equals either $\pi_a(s)$ or $\pi_b(s)$ (where $\pi_a(0) + \pi_a(1) = \pi_b(0) + \pi_b(1) = 1$) with probabilities W_a and W_b ($W_a + W_b = 1$), respectively. It is easily seen that this process is equivalent to a free spin system with $h_i = -\ln p_i(1)/\ln p_i(0)$. Although random, this process is coherent in so far as the same rate $p_n(s)$ is applied to *all* configurations exhibiting symbol s at the nth place. An example of this structure is the phase-space measure for a chaotic dynamical system, where the mapping is randomly chosen at each time step. The above mentioned coherence is here guaranteed by the simultaneous application of the same rule throughout the whole phase-space (i.e. on all configurations). If the alternative is restricted to two Bernoulli maps, we have precisely the case just described.

In disordered systems, the partition function $Z(n, q)$ (6.11) is a fluctuating quantity itself which is useful to average over all realizations of the disorder. Accordingly, one usually introduces the *effective* partition function

$$Z(n, q; r) = \left[\sum_j W_j Z^r(n, q) \right]^{1/r} , \tag{6.43}$$

where r is an averaging exponent and W_j is the probability of the jth realization. On the one hand, r can be interpreted as an inverse "disorder-temperature", by analogy with q (or, rather, $q-1$). On the other hand, the rth integer power of $Z(n, q)$ is nothing but the partition sum for a system consisting of r *replicas* of the same disorder realization. In the dichotomic RMF model of Fig. 6.7(a), $Z(n, q; r)$ factorizes as

$$Z(n, q; r) = \left[W_a Z_a^r(q) + W_b Z_b^r(q) \right]^{n/r} , \tag{6.44}$$

where $Z_a(q) = \pi_a^q(0) + \pi_a^q(1)$ (and analogously for $Z_b(q)$). The generalized free energy density, hence, simplifies to $F(q, r) = -\ln Z(1, q; r)$. The typical thermodynamic properties are given by the *quenched average*, defined by the limit $r \to 0$ (i.e., by the number of replicas tending to 0!). This is the starting point of the so-called *replica method*, which is based on the analytic continuation of $Z(n, q; r)$. By the linearity of the Legendre transform, the entropy $S(E)$ associated with Eq. (6.44) is simply obtained by taking the logarithmic average of the partition functions in the two alternatives a and b. The result, plotted in Fig. 6.8(a), shows that the function is analytic and rather featureless (the choice of the weights is irrelevant). By studying its dependence on r, information about the sample-to-sample fluctuations could be extracted.

As a second example, let us consider the random energy model (REM), introduced and solved by Derrida (1981), which exhibits some of the properties of the spin-glass transition. The probability $P(s_1 \ldots s_n) = \prod_{i=1}^n p(s_i)$ is the product of n randomly chosen numbers $p(s_i)$: that is, a trial is performed at each *leaf* of the hierarchical tree and no longer at each *level* only. In particular, the sum $P(S0) + P(S1)$ need not equal $P(S)$: now one may have, e.g., $P(S0) = P(S)\pi_a(0)$ and $P(S1) = P(S)\pi_b(1)$ (still using a dichotomic disorder). Hence, all "parental

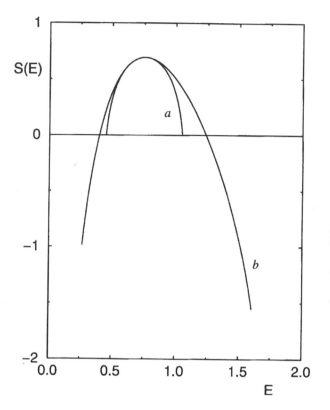

Figure 6.8
Energy-entropy diagrams
(E, S) of two
doubly-random processes.
The conditional
probability $P(Ss)/P(S)$ is
randomly chosen between
the two alternatives $\pi_a(s)$
and $\pi_b(s)$, for $s \in \{0, 1\}$.
In (a), the choice is made
at each *level* of the tree as
in Fig. 6.7(a); in (b), at
each *leaf* of the tree
independently (the total
probability is then
normalized at each level).
The parameter values are:
$\{\pi_a(0), \pi_a(1)\} = \{0.8, 0.2\}$,
with weight $W_a = 0.2$, and
$\{\pi_b(0), \pi_b(1)\} = \{0.6, 0.4\}$,
with weight $W_b = 0.8$.

relations" on the tree are removed: the energy $\ln P(S)$ of each of the 2^n n-blocks
is the result of n independent random trials, as schematized in Fig. 6.7(b),
where all paths are disconnected from one another. This model possesses the
maximum extent of statistical independence: the free energy landscape results
from a completely incoherent assembly of minima and maxima. Such a wealth
of components is a consequence of the huge amount of information needed to
specify a single instance of the interactions: in fact, 2^n independent variables
are needed at level n in the REM, in contrast to the n variables of the RMF
model. A physical realization of this process is achieved by a system involving
n-body interactions. If the coupling constants are randomly distributed, the
free-energy configurations are totally random as well. One might say that each
configuration feels a different (and independent) realization of the disorder.
This, in turn, implies frustration (see Section 3.5), since the energy cannot be
minimized simultaneously for all spin pairs.

Because of the statistical independence of the configuration energies, the
entropy S is obtained from the product of the number of n-blocks, 2^n, with the
average number of times the energy E is observed over all realizations of the
disorder. Thus, indicating with $P(\mathscr{D})$ the probability of a disorder instance $\mathscr{D}$
and with $N(E, n, \mathscr{D})$ the number of n-blocks with energy E in instance $\mathscr{D}$, one

gets

$$e^{nS(E)} \sim 2^n \sum_{\mathscr{D}} N(E, n, \mathscr{D}) P(\mathscr{D}) \,.$$

If disorder is again dichotomic, the fraction W_a of a-choices defines a realization uniquely and the sum can be evaluated as an integral in dW_a. A standard application of the steepest descent method on W_a permits determination of the entropy as the maximum value of the logarithm of the r.h.s in the above expression. The resulting curve $S(E)$ for the same parameter values as in Fig. 6.8(a) is reported in Fig. 6.8(b). Two observations are in order: the upper part of the spectrum is almost the same as in the previous case, indicating that correlations do not play a relevant role at high temperatures; significant differences are noticed in the tails, where the REM exhibits even negative entropies. Since each energy value above the ground state must be obviously observed at least once in a single realization of the process, the corresponding entropy must be positive. Therefore, the energy spectrum in Fig. 6.8(b) must be truncated at the energy values $E_\pm$ at which $S(E) = 0$: the corresponding slopes can be interpreted as critical temperatures separating the paramagnetic from the glassy phase (Derrida, 1981).

   While the truncation is perfectly justified when studying thermodynamic properties of a physical system, it is more controversial for symbolic dynamical systems. In the latter, in fact, the temperature is not an observable but just a formal parameter of the theory. Hence, nothing prevents defining and computing averages over many samples and thereby measuring negative thermodynamic entropies. This standpoint is indeed implicitly assumed in the study of fully developed turbulence, where the time evolution of the dissipation field is modelled through a fractal measure in a one-dimensional space: time-series recorded at different positions in space are interpeted as outcomes of different realizations of the disorder and the number of configurations with the same "energy" (i.e., the same probability in that context) is arithmetically averaged over such realizations (Chhabra & Sreenivasan, 1991). Alternatively, one can split a single long time-series into many independent pieces (Cates & Witten, 1987). Negative entropies are also meaningful in that they permit the recovery of hidden information about improbable events. Consider a set C embedded in a plane and having a dimension spectrum $f(\alpha)$ which extends below $\alpha = 1$. There is no chance to reveal that spectral component by studying the intersection of C with a randomly placed straight line, if a single cut is analyzed. By averaging over many cuts, however, that information is eventually recovered from the negative part of the $f(\alpha)$ curve (Mandelbrot, 1989).

   As a third example of a doubly-random process, consider the so-called Cayley tree with disorder (CTD) which has been solved exactly by Derrida and Spohn (1988). As in the RMF model, the energy of each configuration is obtained by following a path departing from the vertex and terminating at one of the leaves (Fig. 6.7(c)). The conditional probabilities $\pi_a(s)$ or $\pi_b(s)$ are now chosen at each *node* independently rather than at each level. Accordingly, the number of independent variables necessary to specify a given realization grows exponentially with n as in the REM, so that the two models are expected to exhibit similar statistical behaviour (in the CTD, however, $P(S0)+P(S1) = P(S)$ holds again, at variance with the REM). In fact, they are characterized by exactly the same thermodynamic functions (Derrida & Spohn, 1988). The CTD is a useful testing ground for mean-field approximations and, although rather crude, it is a realistic model for polymer growth (whereby each configuration S is mapped to the shape of the polymer in space). Moreover, it is applicable to the generation of random fractals of the type introduced in the theory of turbulence. In the dichotomic case, the tree can be decomposed into two independent subtrees having the same structure. Accordingly, the partition function satisfies the recursion

$$Z(n+1,q) = p^q[Z'(n,q) + Z''(n,q)]$$

where p is the random weight attributed to the left subtree, with a probability distribution $W(p)$, and Z' and Z'' are the partition functions of the two subtrees. Since they are mutually independent, their distribution $\rho_n(Z)$ turns out to obey the simple recursive relation

$$\rho_{n+1}(Z) = \iiint \rho_n(Z')\rho_n(Z'')W(p)\delta[Z - p^q(Z'+Z'')]dZ'dZ''dp \quad (6.45)$$

with initial condition

$$\rho_0(Z) = \delta(Z-1)\,.$$

An analogous equation can be written for the effective partition function (6.43) by multiplying Eq. (6.45) by Z^r and integrating over Z. Clearly, the resulting recursion involves lower-order moments for integer $r > 1$, while it is self-consistent for $r = 1$. Therefore, an analytic solution for the free energy $F(q,r)$ can be found in all such cases, yielding

$$F(q,r) = \begin{cases} \ln(2\mu_1(q))\,, & \text{for } q < q_c, \\ (\ln 2 + \ln \mu_r(q))/r\,, & \text{for } q > q_c, \end{cases} \quad (6.46)$$

where $\mu_r(q)$ is defined as

$$\mu_r(q) = \int W(p)p^{rq}dp$$

and the critical value q_c is satisfies the relation $\mu_r(q_c) = (2\mu_1)^r/2$. Thus, depending on the temperature (i.e., on q), one of two phases is selected. Interestingly, the high-temperature regime is independent of r and corresponds to the *annealed average*, i.e., to the first moment of the partition function ($r = 1$). Unfortunately, the most relevant free-energy $F(q,0)$ cannot be obtained by analytically continuing Eq. (6.46). This is the typical difficulty encountered in the study of spin glass models. Its solution requires the introduction of a new formalism for the study of the glassy phase. For a general discussion of this subject, we direct the reader to specialized books (Mézard et al., 1986; Fischer & Hertz, 1991; Crisanti et al., 1993). Here, we just recall that a solution of the CTD can be found by writing an equation for a special generating function (Derrida & Spohn, 1988): namely, the average of $\exp(e^{-qx}Z(n,q))$. It turns out that the high-temperature phase persists up to a critical temperature $1/q_c$, below which the free energy is strictly proportional to q. This is nothing but a rephrasing of the results obtained for the REM.

From the study of these three doubly-random models, we conclude that the quenched average yields, in some cases, the same results as the much more tractable annealed average (see, e.g., the REM and the CTD beyond the critical temperature). Whenever this does not verify, much care is needed, since the quenched average need not be the one that best characterizes random symbolic dynamical systems. Moreover, in general physical systems, the situation is not as simple as in the above three models where we just truncate the entropy spectrum $S(E)$ at 0. As the Sherrington–Kirkpatrick model (Sherrington & Kirkpatrick, 1975) shows, the glassy phase is much more intricate: not all replicas give rise to the same configurations, and even the ground state is characterized by a strictly positive entropy (Mézard et al., 1986; Fischer & Hertz, 1991).

In more realistic models, like Edwards–Anderson's (EA) in three dimensions (see Chapter 3), it has been conjectured that the glassy phase is characterized by a chaotic renormalization flow. Let us first reformulate the arguments used to prove the existence of a phase transition in the Ising–Lenz model to conclude that a glassy phase indeed exists. Consider a cubic portion of the lattice with linear size L, in which a certain realization of the disorder occurs. Let the spins belonging to two opposite faces be randomly chosen and kept fixed, while periodic boundary conditions are assumed for the other pairs of opposite faces, and take a copy of this configuration in which all spins on one of the fixed faces are reversed. The free energy difference ΔF between the two subsystems can be interpreted as an estimate of the effective coupling strength J' on the scale L. In fact, when performing an RG-step with a spatial scale factor L and an approximate Ising-type interaction at the new scale, the terms σ_1 and σ_2 appearing in the effective Hamiltonian, analogously with Eq. (6.38), can be seen as representative of the spin configurations of the differing faces. Obviously, reversing the configuration on one of the two faces is equivalent to reversing the corresponding σ_i. In general, it is assumed that $J' \sim JL^y$. If $y < 0$, the effective coupling renormalizes to 0, so that the RG-flow converges to the high-temperature phase. Conversely, the divergence of J', occurring when $y > 0$, indicates the existence of a low-temperature phase. In the Ising–Lenz model at low-temperatures, ΔF is essentially determined by the energy of the boundary walls separating different homogeneous regions, so that $y = d - 1$: this once more confirms the existence of a phase transition only for $d > 1$. In the EA model, numerical simulations and an approximate RG-transformation suggest that $y < 0$ for $d = 2$ and $y > 0$ for $d = 3$, so that a glassy phase should exist only in the latter case.

The peculiarity of the glassy phase is the conjectured extreme sensitivity of its spatial pattern to temperature changes (in analogy to what is seen in the hierarchical model of Fig. 6.6). Let us perturb the ground state of a spin glass by reversing the spins in a droplet of linear size L. According to the previous analysis, the energy of the system increases by a typical amount $J' = E = JL^y$. Bray and Moore (1987) have suggested that this energy contribution is localized along some domain wall with fractal dimension d_s, separating the inner from the outer region of the droplet. While a domain wall in a ferromagnetic system is just an ordinary "curve" separating regions with spins pointing in opposite directions, here frustration makes the concept of domain wall much more undefined. In fact, its (fractal) dimension can be computed only in an indirect way, by adding a small random perturbation of size δJ to the bonds and estimating the resulting energy variation δE. Since the number of the sites in the domain wall is of the order of $N \sim L^{d_s}$ and the total contribution of the independent perturbations scales like $N^{1/2}$, one expects that $\delta J' = \delta E = (\delta J) L^{d_s/2}$ (numerical simulations indicate that $d_s = 1.26$ in the two-dimensional EA model). By combining the two results, one finds $\delta J'/J' \sim (\delta J/J) L^{d_s/2-y}$. Accordingly, if $\zeta = d_s/2 - y$ is larger than 0, any small change of the coupling strength (i.e., the temperature) is amplified, and ζ can be interpreted as the Lyapunov exponent of a suitable RG-map (apart from a multiplicative factor accounting for the change of scale involved in each step). This is precisely the case of the EA model in any dimension (Fischer & Hertz, 1991).

6.4   Applications

The thermodynamic formalism not only provides a detailed characterization of the asymptotic behaviour of symbolic dynamical systems but also helps in establishing useful relationships among various indicators of complexity. In this section, we demonstrate the utility of this approach on a few examples.

6.4.1   Power spectral measures and decay of correlation functions

The frequency domain provides one of the possible contexts for the study of fractal features of signals and, in turn, for the application of the thermodynamic formalism. Let us consider the Fourier spectrum

$$S_N(\omega) = \Big| \sum_{n=1}^{N} s_n e^{-i\omega n} \Big|^2 , \tag{6.47}$$

for a discrete-time signal $S = \{s_1, s_2, \ldots\}$ of length N. If, at some frequency $\omega = \omega_0$, $S_N(\omega_0) \sim N^2$ for $N \to \infty$, the spectrum has a pure-point component at ω_0; if $S_N(\omega_0) \sim N$, ω_0 lies in the *ac*-component of the spectrum; if, finally,

$$S_N(\omega_0) \sim N^{2\gamma} \tag{6.48}$$

with the spectral exponent $\gamma(\omega_0)$ bounded in the interval $(1/2, 1)$, ω_0 is in the singular continuous set. Such intermediate values of γ do not exhaust the phenomenology of singular continuous points (absence of a limit behaviour for S_N, dependence of γ on N, or scaling laws of the form $S_N \sim N^\eta / \ln N$ are all examples thereof). The "normal" classification can be summarized by defining the singularity exponent $\alpha(\omega_0)$ as

$$\int_{\omega_0}^{\omega_0 + \epsilon} S_N(\omega) d\omega \sim N \epsilon^{\alpha(\omega_0)} , \tag{6.49}$$

where $\epsilon = \epsilon(N) = 2\pi/N$ is the separation between two consecutive frequencies, at resolution N, and the factor N on the r.h.s. accounts for the normalization of the spectral measure. Substituting $S_N \sim N^{2\gamma}$ into Eq. (6.49), one obtains

$$\alpha = 2(1 - \gamma) . \tag{6.50}$$

Hence, $\alpha = 0$ and $\alpha = 1$ are suggestive of a pure point and an absolutely continuous component, respectively.

Fractal features of singular–continuous power spectra can be shown to be related to the long-time decay properties of the mean squared correlation

$$\langle C^2 \rangle(t) \equiv \frac{1}{t} \int_0^t C(t')^2 dt' , \tag{6.51}$$

which converges to 0 for $t \to \infty$ even when $C(t)$ presents revivals which mask its overall decay. This is particularly evident for the map (Grebogi *et al.*, 1984)

$$x_{t+1} = \sigma \tanh x_t \cos(2\pi \theta_t)$$
$$\theta_{t+1} = \theta_t + \omega \bmod 1 . \tag{6.52}$$

As its skew-product structure (see Ex. 3.2) indicates, the equation for the *x*-variable is quasiperiodically driven when the angular frequency ω is irrational. In such a case, the solution is quasiperiodic for $\sigma < 2$. At $\sigma = 2$, the solution $x_t = 0$, $\theta \in [0, 2\pi]$, becomes unstable and the motion asymptotically unfolds on a so-called *strange nonchaotic attractor*[5] (Grebogi *et al.*, 1984). The behaviour is

5. An attractor which, even in the absence of positive Lyapunov exponents, displays an anomalous geometrical structure.

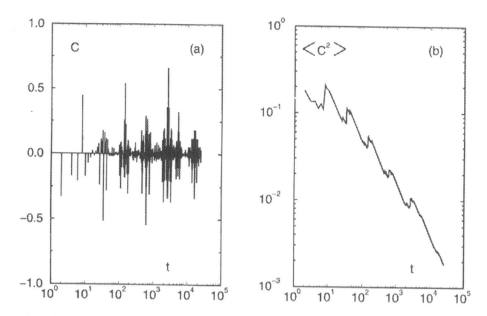

Figure 6.9 The correlation function $C(t)$ for the variable x_t in model (6.52) for $\sigma = 3$ (a) and the corresponding mean square correlation (b) versus $\ln t$.

thereby qualitatively different, as it is possible to infer from the autocorrelation function of the variable x_t (Pikovsky & Feudel, 1994), reported in Fig. 6.9(a). While $C(t)$ resurges nearly periodically (on a logarithmic scale), $\langle C^2 \rangle$ presents the power-law decay

$$\langle C^2 \rangle(t) \sim t^{-\beta} , \tag{6.53}$$

as shown in Fig. 6.9(b), where $\beta = 0.65 \pm 0.05$.

Assuming that Eq. (6.53) has general validity, Ketzmerick *et al.* (1992) have shown that

$$\beta = D_2 , \tag{6.54}$$

where D_2 is the order-2 Renyi dimension (Eqs. (6.25) and (6.28)), also called the *correlation dimension* (Grassberger & Procaccia, 1983), of the spectral measure $S(\omega)$ associated with $C(t)$. In fact, in the limit $\Delta \to 0$,

$$\Delta^{D_2} \sim \int_0^\pi \left[\int_{\omega-\Delta}^{\omega+\Delta} dS(\omega') \right] dS(\omega) , \tag{6.55}$$

where the inner integral yields the "mass" $P(\Delta; \omega)$ contained in the interval $[\omega - \Delta, \omega + \Delta]$ and the outer one is an average over all ωs. Defining $A = [-\Delta, \Delta]$, Eq. (6.55) can be rewritten as

$$\Delta^{D_2} \sim \int \int \chi_A(\omega - \omega') dS(\omega) dS(\omega') , \tag{6.56}$$

where $\chi_A(\cdot)$ is the characteristic function of A. The convolution theorem permits computation of the second integral by multiplying the inverse Fourier transforms of $S(\omega')$ and $\chi_A(\omega')$ together and taking the direct Fourier transform. For simplicity, the signal is shifted by the amount $\langle x \rangle$ into $y_t = x_t - \langle x \rangle$, so that $C(t) = R(t)$ (Eq. (5.19)). Hence,

$$\Delta^{D_2} \sim \int dS(\omega) \int_{-\infty}^{+\infty} dt\, e^{-i\omega t} C(t) \frac{\sin(\Delta t)}{t} .$$

Performing the remaining Fourier transform and recalling that $C(t) = C(-t)$ finally yields

$$\Delta^{D_2} \sim \int_0^\infty C^2(t) \frac{\sin(\Delta t)}{t} dt . \qquad (6.57)$$

In the limit $\Delta \to 0$, the r.h.s. essentially coincides with the definition (6.51) of $\langle C^2 \rangle$, up to some irrelevant multiplicative factor. Relation (6.54) follows by substituting the scaling assumption (6.53) in Eq. (6.57) and scaling the dummy variable t by a factor Δ^6.

The case $D_2 = 0$ corresponds to a pure-point spectrum: as expected, correlations persist in the infinite-time limit. Conversely, for $D_2 = 1$, the spectral measure becomes absolutely continuous: the $1/t$ decay law for $\langle C^2 \rangle(t)$ is easily understood by realizing that correlations are significantly different from zero only in a finite initial time interval.

6.4.2 Thermodynamics of shift dynamical systems

[6.5] **Intermittency.** Type-I intermittency (Ex. 5.11), modelled by the map $x' = f(x; \varepsilon) = x + x^\nu + \varepsilon \pmod 1$ in the limit $\varepsilon \to 0$, is one of the simplest mechanisms that yield a phase transition. The alternation of long "laminar" motion and "turbulent" bursts suggests the analogy with a two-phase system at a first-order transition. The map admits a binary generating partition with the Markov property, defined by $B_0 = [0, x_c)$ and $B_1 = [x_c, 1]$, where $x_c = f^{-1}(1)$ is the left preimage of 1. By the continuity of the invariant measure, the local dimension $\alpha(S)$ is 1 a.e. (and equals $2 - \nu$ at $x = 0$, with $1 < \nu < 2$). Hence, the local entropy $\kappa(S)$ coincides with the local Lyapunov exponent $\lambda(S) = -\ln \varepsilon(S)/n$, where $\varepsilon(S)$ is the length of the interval B_S labelled by the n-sequence S. Noting that the order-n image $f^n(B_S)$ of B_S is the whole unit interval $[0, 1]$, it is possible to estimate $\varepsilon(S)$ as $\varepsilon(S) \sim \prod_{i=1}^n 1/f'(x_i)$, where $x_i = f^{-1}(x_{i-1}) \in B_{s_1 \dots s_i}$ and $x_0 = x_c$. Sequences composed of a large number of consecutive zeros are characterized by a small value of λ, since the system spends a long time in the laminar region where $f'(x)$ is close to 1. Conversely, the turbulent component of the dynamics is characterized by a large local

---

6. For a rigorous proof, the reader is recommended to consult Holschneider (1994).

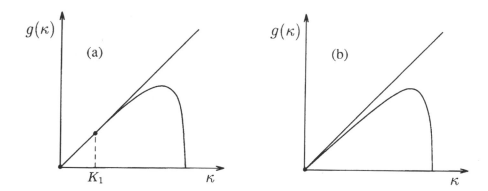

Figure 6.10 Graph of the local entropy spectra $g(\kappa)$ vs. κ for type-I intermittency: (a) $1.5 < v < 2$, (b) $v \geq 2$. In (a), the curve $g(\kappa)$ is tangent to the bisectrix $g(\kappa) = \kappa$ in the whole range $0 \leq \kappa \leq K_1$; in (b), only at $\kappa = 0$.

Lyapunov exponent. The cases $v < 2$ and $v \geq 2$ correspond to weak and strong intermittency, which yield qualitatively different behaviour. In the former, characterized by the existence of a normalized invariant measure, one must further distinguish the interval $1.5 < v < 2$ in which the number N_n of laminar trajectories of length n exhibits non-Gaussian fluctuations (at variance with the case $1 \leq v < 1.5$) and $1/f$-like power spectra appear (Schuster, 1988).

The entropy spectrum $g(\kappa)$ for this parameter region is displayed in Fig. 6.10(a). The curve $g(\kappa)$ is tangent to the bisectrix $g(\kappa) = \kappa$ between $\kappa = 0$ ("periodic" or laminar phase, corresponding to the fixed point $x = 0$) and $\kappa = K_1$ (lower border of the turbulent state, at which the local entropy coincides with the metric entropy K_1). The "states" in this range are mixed ones, coexisting at the inverse temperature $q_c - 1 = 0$. In the strongly intermittent case, $v \geq 2$, the singularity of the invariant measure at $x = 0$ is so strong that the metric entropy vanishes, as illustrated in Fig. 6.10(b). For a detailed investigation of this phase transition, see Wang (1989a and b) who introduces a countable Markov partition for the map and illustrates its relationship with Fisher's droplet model (Fisher, 1967b; Felderhof & Fisher, 1970).   □

[6.6] **Hénon map.** A "generic" example of a first order phase transition in the dimension spectrum $f(\alpha)$ is provided by the Hénon map (3.12). Although the strange attractor is not hyperbolic, it is the product of a continuum with a Cantor set almost everywhere with respect to the natural measure m. We describe the overall shape of the $f(\alpha)$ function following the discussion of Politi et al. (1988). Since the unstable manifold W^u is tangent to the stable one at the homoclinic points, contraction may occur along W^u for an arbitrarily large number of iterates. This enhances the probability density in the vicinity of such orbits, in analogy with the formation of square-root singularities in the logistic

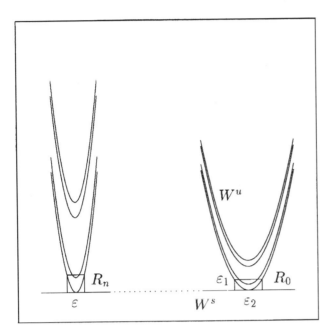

Figure 6.11 Schematic representation of the stable and unstable manifolds (W^s and W^u, respectively) of the Hénon map in the vicinity of a homoclinic tangency (inside rectangle R_0) and of its nth iterate (inside R_n).

map. The typical direct-product structure of the measure in hyperbolic systems is thereby locally destroyed.

With reference to Fig. (6.11), consider a rectangle R_0 lying on the stable manifold W^s, with sides of length ε_2 along W^s and ε_1 in the transversal direction. These lengths are fixed by requiring that the nth image of R_0 is a square R_n of size ε placed similarly to R_0 with respect to W^s, in such a way as to cover exactly the tangency region enclosed by the folds of W^u (it is assumed, without loss of generality, that the corresponding curvature is 1). This choice captures the global features of the attractor in the nonhyperbolic domains, in so far as it avoids resolving separately the continuous and the Cantor structure with different precision. The mass in the rectangle R_n, equal to that in R_0, scales as

$$P(\varepsilon) \sim \varepsilon^\alpha \sim \varepsilon_1^{\alpha_2} \varepsilon_2 \,, \tag{6.58}$$

where α is the local dimension associated with R_n and α_2 the local partial dimension (see Appendix 4) in R_0 transversal to the two manifolds. The partial dimension α_1 along the unstable manifold is assumed to be 1 in R_0, as appropriate for hyperbolic attractors, since R_0 contains a "primary" tangency point (D'Alessandro *et al.*, 1990): the singularity is expected to build up only after a large number n of iterates in the nonhyperbolic regions. Clearly, $\varepsilon_i = \varepsilon \exp(-n\lambda_i)$, where $\lambda_1 > 0$ is the expansion rate in the direction perpendicular to W^s and $\lambda_2 < 0$ the contraction rate along W^s (see Appendix 4). Assuming a

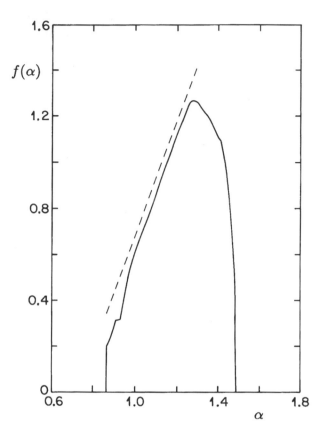

Figure 6.12 Dimension spectrum $f(\alpha)$ vs. α for the Hénon attractor at $(a,b) = (1.4, 0.3)$ evaluated from a set of 10^5 points (Broggi, 1988).

parabolic shape for W^u in R_0, one has $\varepsilon_1 \approx \varepsilon_2^2$. Substituting into Eq. (6.58), we finally obtain

$$\alpha = (1 + 2\alpha_2)\frac{1 + \alpha_2}{2 + \alpha_2} \, , \qquad (6.59)$$

where the Kaplan–Yorke relation $\alpha_2 = -\lambda_1/\lambda_2$, Eq. (A4.8), has been used. Fluctuations of the Lyapunov exponents λ_i (and, consequently, of α_2) have been neglected for simplicity. Expression (6.59) gives the typical value of the dimension α in the nonhyperbolic phase. When the Jacobian b tends to 0, the logistic map is recovered and $\alpha_2 \to 0$: hence, $\alpha = 1/2$, as expected. Conversely, for $b \to 1$ (the conservative case), $\alpha_2 \to 1$ and $\alpha \to 2$. At the usual parameter values ($a = 1.4$, $b = 0.3$), identifying λ_i with the ith average Lyapunov exponent yields $\alpha \approx 0.76$. The value of $f(\alpha)$ at this point has been evaluated to be about 0.29. Inclusion of the fluctuations of the local indices with the position $\mathbf{x}$ in the *hyperbolic* phase induces analogous fluctuations in the *nonhyperbolic* part of the dimension spectrum $f(\alpha)$ as well.

The whole dimension spectrum (Fig. 6.12) can be therefore interpreted as the superposition of two distinct concave functions, corresponding to the hyperbolic and nonhyperbolic points, connected by a tangent segment as for a usual first

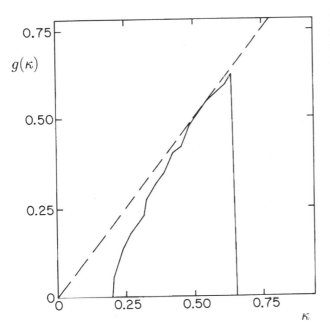

$g(\kappa)$

Figure 6.13 Entropy spectrum $g(\kappa)$ vs. κ for the elementary CA rule 22.

order phase transition. The identification of the two types of behaviour in the attractor as thermodynamic phases is hence legitimized. The respective values of $f(\alpha)$ can be interpreted as generalized dimensions of the sets of points supporting the two phases (hence, the set of all "hyperbolic points" has a larger dimension). The "transition temperature" $k_B T_t = 1/(q_t - 1)$ is about 0.8.   □

[6.7] **Cellular automaton 22.** The entropy spectrum $g(\kappa)$ of the asymptotic spatial configuration of the CA 22 has been estimated from the dimension spectrum $f(\alpha)$ in the space of symbolic sequences S mapped to points $x \in [0, 1]$ according to $x(S) = \sum_i s_i 2^{-i}$. The local entropy $\kappa(S)$ is related to the local dimension $\alpha(S)$ through $\kappa = \alpha \ln 2$. After applying the CA rule 10000 times on a random initial configuration, we analysed the middle $M = 10^6$ symbols using the constant-mass method (Badii & Politi, 1984 and 1985; Badii & Broggi, 1988). The local entropies were computed in $m = 2 \cdot 10^4$ randomly chosen neighbourhoods, each containing a mass $p = k/n$. Histograms of the entropies were constructed using 40 channels for $n \in [10^5, 9.8 \cdot 10^5]$ and $k \in \{20, 40, 60, 80\}$. The curve in Fig. 6.13 refers to $n \approx 8 \cdot 10^5$, $k = 40$, and represents the typical shape of the curves obtained for all pairs (k, n). The curves are mostly reliable in the region around the value $\kappa = K_1 \approx 0.51 \pm 0.01$ (which has been estimated independently).

Although the curves are close to each other, more accurate results are obtained with a finite-size scaling analysis. The topological entropy K_0, in particular, is

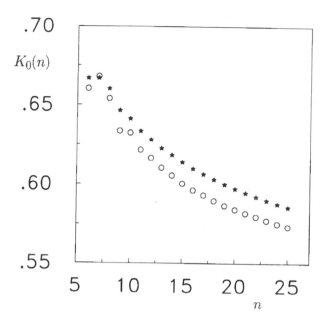

Figure 6.14 Finite-size estimates $K_0(n)$ of the topological entropy K_0 for the asymptotic spatial configuration of the elementary CA rule 22. Stars: logarithm of the largest eigenvalue of the directed graph constructed from all forbidden words of length $\ell \leq n$. Circles: logarithm of the ratio $N(n)/N(n-1)$ between the numbers of legal words of lengths n and $n-1$.

estimated either from the ratio

$$K_0^{(a)}(n) = \ln \frac{N(n)}{N(n-1)} ,$$

where $N(n)$ is the number of sequences of length n, or from

$$K_0^{(b)}(n) = \ln \mu_{\max}(n) ,$$

where $\mu_{\max}(n)$ is the largest eigenvalue of the directed graph that accounts for all prohibitions up to length n. An exponential convergence of the form $K_0^{(i)}(n) \sim K_0 + f_i e^{-\eta_i n}$ is assumed in both cases, indicated by $i = a$ or b, where f_i and η_i are constants. The corresponding curves are displayed in Fig. 6.14. At variance with generic dynamical systems, where the graph provides better approximations (D'Alessandro et al., 1990), the more stringent upper bound to K_0 is here provided by the direct method. The limit value and the convergence rate are estimated to be $K_0 = 0.55 \pm 0.01$ and $\eta \approx 0.08$, respectively, for both curves. Because of the low value of η, a power-law asymptotic behaviour (as conjectured by Fahner and Grassberger, 1987) cannot be completely ruled out. Analogous estimates show a weak dependence of η on q.

The reliability of the $g(\kappa)$ spectrum of Fig. 6.13 is further confirmed by the agreement of the boundaries $\kappa_{\min}$ and $\kappa_{\max}$ of its support with lower and upper bounds that can be obtained in various ways. An upper bound to the minimum local entropy $\kappa_{\min}$ is given by the decay rate $\kappa^{(0)}$ of the probability $P(000\ldots)$ of a sequence composed of n zeros. This is the most frequently observed one (the spatio-temporal pattern, in fact, presents a large number of triangles filled with 0s) and, hence, asymptotically dominates all others. The abundance of such triangles and the existence of a simple recognition algorithm for them permit one to obtain reliable results even for large values of n. Careful simulations indicate that $\kappa^{(0)} = 0.267$ is indeed very close to κ_{min}. An analytic lower bound to the topological entropy can be obtained by realizing that rule 22, when applied four times on an arbitrary concatenation $\mathscr{S}$ of the two words $w_0 = 0000$ and $w_1 = 0001$, simulates the effect of rule 90 on the configuration $\mathscr{S}'$ obtained from $\mathscr{S}$ by substituting each w_0 with 0 and each w_1 with 1 (Wolfram, 1984): in such a simulation, symbol 1 appears only at lattice sites i that are multiples of 4, both at the initial time and every four iterates. Since rule 90 is "linear" (the updated value s_i' at site i is the sum modulo 2 of s_{i-1} and s_{i+1}), it prohibits no sequence and the topological entropy of its limit set is $\ln 2$. Hence, the set $\{w_0, w_1\}^{\mathbf{Z}}$ of all concatenations of w_0 and w_1 in rule 22 contribute to the topological entropy with $\ln 2/|w_0| = \ln 2/|w_1| \approx 0.173$. This value represents a lower bound to K_0 only if all sequences from $\{w_0, w_1\}^{\mathbf{Z}}$ are really contained in the invariant set of rule 22, as is confirmed by numerical simulations.

In the above discussion, it has been implicitly assumed that the irreducible forbidden words were given. Direct inspection of a finite portion of a limit configuration, no matter how long, cannot prove that a sequence is indeed forbidden and not just highly improbable. Because of its relations with formal languages, a procedure to identify forbidden words in elementary CAs is illustrated in Section 7.3 with reference to rule 22, for which we obtain a confirmation of the value 0.55 for K_0. Finally, notice that the approximately linear behaviour of $f(\alpha)$ vs. α for $0.25 < \alpha < 0.5$ is suggestive of a phase transition. The limited accuracy of the estimate and the slow convergence of the thermodynamic functions, however, demand caution. $\qquad\square$

[6.8] **An analytically solvable model.** It is known that the effect of low-pass filtering on chaotic signals may determine the increase of the dimension of the attractor when the cut-off frequency η is decreased (Badii *et al.*, 1988). Consider, without loss of generality, a scalar time series $u = \{u_1, u_2, \ldots\}$ generated by a two-dimensional map $\mathbf{F}$ having Lyapunov exponents $\lambda_1 > 0 > \lambda_2$. The action of the filter on u can be modelled by the linear map

$$z_{n+1} = \gamma z_n + u_n,\tag{6.60}$$

where $\gamma = \exp(-\eta) \in (0, 1)$ and z is the filter "output": hence, filtering is tantamount to adding equation (6.60) to those defining the map itself. Applying the Kaplan–Yorke relation (A4.9) to the new three-dimensional map, the information dimension is found to exhibit three different kinds of dependence on η:

$$D_1(\eta) \sim \begin{cases} 1 + \lambda_1/|\lambda_2|, & \text{for } \eta > |\lambda_2|, \\ 1 + \lambda_1/\eta, & \text{for } \lambda_1 < \eta < |\lambda_2|, \\ 2 + (\lambda_1 - \eta)/|\lambda_2|, & \text{for } 0 < \eta < \lambda_1. \end{cases}\tag{6.61}$$

The discontinuity of the derivative $dD_1(\eta)/d\eta$ is an indication of the occurrence of a phase transition which is better understood by considering the whole dimension function $D_q(\eta)$. The phenomenon is interpretable as a competition between the two contracting directions, characterized by the rates $-\eta$ and λ_2, where the latter fluctuates with the position. Its essence can be captured by a simple hyperbolic model. Let $\mathbf{F}$ be the baker map (3.10), which we rewrite as

$$x_{n+1} = x_n r_0 (r_1/r_0)^{s_{-n}} + (1 - r_1)s_{-n}$$
$$y_{n+1} = y_n p_0 (p_1/p_0)^{s_{-n}} + p_0 s_{-n} \tag{6.62}$$

using the symbolic convention

$$s_{-n} = [1 + \text{sgn}(y_n - p_0)]/2 ,$$

and the "signal" $\{u_n\}$ the superposition $u_n = x_{n+1} + y_{n+1}$. We first take a section of the attractor at $y = 0$ to eliminate the trivial continuous structure along the y-direction. The resulting set of points in the (x, z) plane is studied in the projection onto the z axis (the projection onto the x axis, in fact, yields the ordinary two-scale Cantor set). Each symbolic sequence S of length n identifies an interval of the covering at the nth level of resolution. Lengthy but straighforward calculations (Paoli *et al.*, 1989a) yield, for the length $\varepsilon(s_1 s_2 \ldots s_n) = z(\ldots 000 s_1 s_2 \ldots s_n 111 \ldots) - z(\ldots 000 s_1 s_2 \ldots s_n 000 \ldots)$ of the generic interval, the expression

$$\varepsilon(S) \sim \gamma^n \left[\sum_{l=1}^{n} \left(\frac{r_0}{\gamma} \right)^l \left(\frac{r_1}{r_0} \right)^{\sum_{m=1}^{l} s_{n+1-m}} + O(1) \right] . \tag{6.63}$$

For definiteness, we choose $r_1 > r_0$. The Hamiltonian is most conveniently defined as $\mathcal{H}(S) = -\ln \varepsilon(S)$, since the probabilities $P(S)$ are the same as for the Bernoulli shift $B(p_0, p_1)$ (Ex. 6.1). The sum in Eq. (6.63) lies between its largest addendum

$$L_n(S) = \max_{1 \le l \le n} \left[\left(\frac{r_0}{\gamma} \right)^l \left(\frac{r_1}{r_0} \right)^{\sum_{m=1}^{l} s_{n+1-m}} \right]$$

and $nL_n(S)$, so that the Hamiltonian is an extensive quantity which scales as

$$\mathcal{H}(S) \sim -n \ln \gamma - \max_{1 \le l \le n} \left[l \ln \left(\frac{r_0}{\gamma} \right) + \sum_{m=1}^{l} s_{n+1-m} \ln \left(\frac{r_1}{r_0} \right) \right] \tag{6.64}$$

up to an additive correction of the order of $\ln n$, for $n \to \infty$. Asymptotically, the energy $\mathcal{E}(S)$ for each sequence S is determined by the prevailing term in the argument of the maximum. For $\gamma < r_0$ (weak filtering), the maximum is attained at $l = n$ and the dependence on γ disappears: the usual scaling of the baker map is recovered. Conversely, for $\gamma > r_1$ (strong filtering) the maximum is attained at $l = 1$ and the leading behaviour is trivially $\mathcal{H} \sim -n \ln \gamma$. In the intermediate range, the asymptotic scaling depends on the specific sequence S. Hence, a mixture of two phases may occur. The energy $\mathscr{E}$ is a function $\mathscr{E}(v)$ of the particle density $v = \sum s_i/n$ and scales as

$$\mathscr{E}(v) \sim \begin{cases} -n \ln \gamma \,, & \text{for } v < v_c, \\ -n \left[\ln r_0 + v \ln(r_1/r_0) \right] \,, & \text{for } v > v_c, \end{cases} \tag{6.65}$$

where

$$v_c = \frac{\ln(\gamma/r_0)}{\ln(r_1/r_0)} \,. \tag{6.66}$$

The dependence of the local dimension $\alpha(S) = \ln P(S)/\ln \varepsilon(S)$ on v is accordingly given by

$$\alpha(v) = -n \frac{\ln p_0 + v \ln(p_1/p_0)}{\mathscr{E}(v)} \,. \tag{6.67}$$

The dimension spectrum $f(\alpha)$ is finally obtained by counting the number of sequences that yield a given v, by inverting relation (6.67), and by changing the scaling parameter from n to ε. The transition, illustrated by Eq. (6.65) from the point of view of the energy, induces a cusp at $\alpha_c = \alpha(v_c)$ in the $f(\alpha)$-spectrum. If the resulting curve is not concave, the envelope must be taken and a usual first order transition is observed. In the opposite case, the transition is of the anomalous type mentioned in the previous section. An experimental confirmation of these filter-induced transitions has been reported in Paoli *et al.* (1989b).

The phase diagram in Fig. (6.15) illustrates the two types of transition. The curves $\alpha_1(\gamma)$ and $\alpha_2(\gamma)$ represent, respectively, the lower and upper extremum of the linear portion of the $f(\alpha)$ spectrum. They merge, at a value $\gamma = \gamma_c$, into the curve $\alpha_c(\gamma)$ which corresponds to the cusp in the second type of transition. At the critical point γ_c, the transition is continuous: $\gamma = \gamma_c$ is the value beyond which the information dimension D_1 starts increasing (6.61). This picture remains qualitatively valid in the whole set of the allowed values of the other system parameters.
□

[6.9] **Long-range interactions.** To elucidate the origin of the long-range inter- actions responsible for the occurrence of phase transitions in one-dimensional lattices, consider the simple case in which the Hamiltonian depends only on the sum $\mathscr{V} = \sum s_i$ of the "spins" (more precisely, occupation numbers): $\mathscr{H} = \mathscr{H}(\mathscr{V})$. For a lattice of length n and $s_i \in \{0, 1\}$, $\mathscr{H}$ can take on any of the $n+1$ values $(\mathscr{E}_0, \mathscr{E}_1, \ldots, \mathscr{E}_n)$, corresponding to $\mathscr{V} = (0, 1, \ldots, n)$. The function

$$\chi_m(S) = \sum_{i_1 \neq i_2 \neq \ldots \neq i_m} s_{i_1} s_{i_2} \cdot \ldots \cdot s_{i_m} \prod_{l \notin \{i_1, i_2, \ldots, i_m\}} (1 - s_l) \,,$$

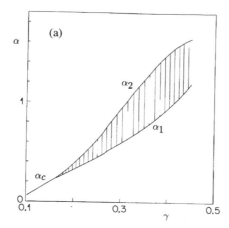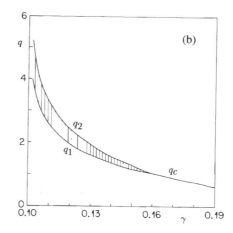

Figure 6.15 Two-dimensional phase diagram of the filtered baker map in the (γ, α) (a) and (γ, q) (b) planes, for the (typical) set of parameters $p_0 = 0.7$, $r_0 = 0.1$, and $r_1 = 0.5$. In the shaded region, delimited by the curves $\alpha_1(\gamma)$ and $\alpha_2(\gamma)$, $q_1(\gamma)$ and $q_2(\gamma)$, two different phases coexist. The curve $\alpha_c(\gamma)$ represents the coordinate of the cusp in the "anomalous" transition. The curve $q_c(\gamma)$ gives the "critical" slope $f'(\alpha)$ in the coexistence region of the "normal" transition.

with $\chi_0 = \prod_{l=1}^{n}(1 - s_l)$ and $1 \leq m \leq n$, is equal to 1 if and only if $\mathscr{V} = m$ for the sequence S. Hence, the Hamiltonian can be written as

$$\mathscr{H} = \sum_{m=0}^{n} \chi_m(S)\mathscr{E}_m .$$

By regrouping all terms containing products of k spins, for all $k \in \{0, 1, \ldots, n\}$, one obtains

$$\mathscr{H} = \sum_{k=0}^{n} J_k \pi_k ,$$

where the symbol J_k denotes the k-body interaction coefficient and π_k the sum of all possible products of k distinct spin-variables (with the convention $\pi_0 = 1$): for example, $\pi_2 = s_1 s_2 + s_1 s_3 + \ldots + s_1 s_n + s_2 s_3 + s_2 s_4 + \ldots + s_{n-1} s_n$. Simple combinatorial analysis (Hardy & Wright, 1938) yields

$$J_k = \sum_{i=0}^{k} \binom{k}{i}(-1)^{i+k}\mathscr{E}_i .$$

Notice, first of all, that the *same* coefficient J_k multiplies all products of k spins, irrespective of the spin positions on the lattice. Hence, the interactions are so long-ranged that they do not decay at all with the interparticle distance. Secondly, many-body interactions are present at all orders $k = 2, 3, \ldots, n$. Moreover, for an arbitrary function $\mathscr{H}(\mathscr{V})$, the coefficients J_k are not expected to decrease with k. The presence of alternate signs in the expression of J_k, however, provides a compensation of energy terms which prevents the coefficients from diverging, as long as the starting Hamiltonian is well-behaved. For example,

$$J_4 = \mathscr{E}_0 - 4\mathscr{E}_1 + 6\mathscr{E}_2 - 4\mathscr{E}_3 + \mathscr{E}_4 \,.$$

In particular, for the Bernoulli shift, $\mathscr{E}_i = a + bi$ (with a and b constants) and $J_k = 0$ for $k \geq 2$. The filtered baker map also yields, for large n, a Hamiltonian that depends on the volume $\mathscr{V}$ only (see Eq. (6.65)). The values $\mathscr{E}_i$ are, in that case, constant for $i \ll nv_c$ and linear in i for $i \gg nv_c$, for each fixed n. Hence, the relevant interactions are those involving more than nv_c bodies! This was to be expected since the transition is ruled by the number $n_o = nv$ of occupied sites. The predominance of a phase over the other is determined by counting n_o. The system can be therefore interpreted as a particle-counting machine. $\square$

Further reading

Baxter (1982), Callen (1985), Feigenbaum (1987), Frisch & Parisi (1985), Fujisaka (1991), Griffiths (1972), Mori *et al.* (1989), Ruelle (1974), Simon (1993), Uzunov (1993).

Part 3

Formal characterization of complexity

Chapter 7

Physical and computational analysis of symbolic signals

The emphasis in the previous part of the book was on mathematical tools that permit the classification of symbolic signals in a general and accurate way by means of a few numbers or functions. The study of complexity, however, cannot be restricted to the evaluation of indicators alone, since each of them presupposes a specific model which may or may not adhere to the system. For example, power spectra hint at a superposition of periodic components, fractal dimensions to a self-similar concentration of the measure, the thermodynamic approach to extensive Hamiltonian functions. It seems, hence, necessary to seek procedures for the identification of appropriate models before discussing definite complexity measures. Stated in such a general form, the project would be far too ambitious (tantamount to finding "meta-rules" that select physical theories). The symbolic encoding illustrated in the previous chapters provides a homogeneous environment which makes the general modelling question more amenable to an effective formalization.

Independent efforts in the study of neuron nets, compiler programs, mathematical logic, and natural languages have led to the construction of finite discrete models (automata) which produce symbolic sequences with different levels of complexity by performing elementary operations. The corresponding levels of computational power are well expressed by the *Chomsky hierarchy*, a list of four basic families of automata culminating in the celebrated Turing machine. This is a conceptually simple mathematical object with the ability of emulating any general-purpose computer. Each automaton defines a formal language as the set $\mathscr{L}$ of all finite strings it can produce. Higher-order families in the hierarchy properly include the lower-order ones. Regular languages, introduced in Chapter 4, belong to the lowermost class.

The importance of Chomsky's hierarchy, largely due to its elegant inclusivity property, is however limited by the nonexhaustive character of the list: in fact, a bewildering multitude of languages having only partial intersection with one of the basic four families has been defined and is currently being investigated in computer science.

It must be remarked that all these models deal exclusively with topological aspects of symbolic sequences, disregarding any metric feature. In particular, they do not include any real parameter. For example, directed graphs reproduce the legal strings of Markov processes but not their conditional probabilities (unless specific extensions are introduced as shown in Chapter 9).

In Section 7.1, we review the Chomsky hierarchy and discuss a few theoretical implications of Turing's work. Formal languages are studied from a physical point of view in Section 7.2. Conversely, the languages generated by a selection of physical systems that currently represent a challenge for research are analysed in the context of automata theory in Section 7.3.

7.1   Formal languages, grammars, and automata

As we have seen in Chapter 4, the set A^* of all finite strings over a given alphabet A can be conveniently represented in the lexicographic order ($A^* = \{\epsilon, 0, 1, 00, 01, 10, 11, 000, \ldots\}$ in the binary case). Accordingly, any formal language $\mathscr{L} \subseteq A^*$ can be represented as an infinite sequence of symbols $\tilde{\mathscr{L}} = l_1 l_2 \ldots$, where l_i is either "1" or "0", depending on whether the ith word in A^* belongs to $\mathscr{L}$ or not. Therefore, the problem of finding the language associated with a given symbolic signal $\mathscr{S} = s_1 s_2 \ldots$ may appear pointless at a first sight, since there is no difference between the object of the study and its description: they are both infinite strings. This correspondence, however, is by no means absolute: the sequence $\tilde{\mathscr{L}}$ defining the language $\mathscr{L}$ is, in fact, a special string, with a peculiar structure, and not an element of a statistical ensemble like the signal $\mathscr{S}$ (the distinction between single items and ensembles will be discussed in detail in the next two chapters, since it characterizes two different families of complexity measures). On the other hand, it is precisely this mapping, from languages to infinite symbolic sequences, that permits a rigorous study of general questions concerning the structure of languages. In particular, it will be shown that certain propositions cannot be proved to be either right or wrong because of their "undecidability".

The Chomsky hierarchy provides a simple ordering of the main language classes. In the following table, we list them in increasing order of generality[1].

Type	Language	Automaton
1	Regular	Finite
2	Context-free	Pushdown
3	Context-sensitive	Linear bounded
4	Unrestricted	Turing

Each automaton is a machine that "accepts" or generates all words in the corresponding language. It can also be viewed as a suitable model to be used to reproduce a given symbolic signal. This classification of formal languages, when these are represented as symbol sequences, is reminiscent of the organization of real numbers into rationals, algebraic irrationals, transcendentals, and so on.

7.1.1   Regular languages

Languages in this class are formally described by symbolic-agebraic formulas, called *regular expressions*, such as

$$R = (r + s)(t + uv)^* , \qquad (7.1)$$

where each of the symbols r, s, t, u, and v is a specific finite word (including the empty word ϵ), the operator $+$ is to be read as "or", "products" like uv denote concatenations, and the superscript $*$ represents an arbitrary number of concatenations of the argument with itself. The regular language $\mathscr{L}(R)$ contains all words compatible with the expression R. The roof map (4.4) at $a = (3 - \sqrt{3})/4$, for example, produces a symbolic signal described by the expression $R = (1 + 0(010)^*1)^*$ or, equivalently, by the set $\mathscr{F} = \{000, 0011\}$ of irreducible forbidden words.

Every regular expression R can be immediately translated into a finite *directed graph*, a collection $\{q_0, q_1, \ldots, q_N\}$ of nodes (vertices) connected by ordered arcs (e.g., see Fig. 4.5). The corresponding mathematical object is called a *finite automaton* (FA). Each symbol in R is assigned to an arc, so as to establish a correspondence between sequences in the language and paths on the graph. In this way, an arc is indicated by a_{ij}^s, where i and j are the indices of the start and end nodes, respectively, and $s \in A$ is the symbol. The "addenda" in each sum correspond to arcs starting from the same node and each expression iterated with the $*$ operator yields a closed path. In order to satisfy these rules, it may be necessary to introduce unlabelled arcs (i.e., "moves" carrying the empty string ϵ)

---

1. We adopt an unconventional enumeration, reversed with respect to the usual one in which unrestricted languages are called "type-0" and regular languages "type-3".

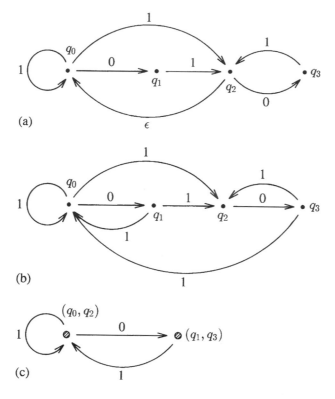

Figure 7.1 (a) Nondeterministic graph associated with the regular expression $R = (1 + (01 + 1)(01)^*)^*$; (b) equivalent graph obtained after removing the unlabelled arc; (c) equivalent deterministic graph.

and *nondeterminism*, in the form of two or more identically labelled arcs leaving the same node.

Example:

[7.1] All of the above features are present in the expression

$$R = (1 + (01 + 1)(01)^*)^* ,$$

corresponding to the automaton illustrated in Fig. 7.1(a). Notice that two arcs with label 1 leave node q_0 and that the transition from q_2 to q_0 is "blank". This can be eliminated by redirecting to q_0 each of the arcs that point to q_2. The result is shown in Fig. 7.1(b).

This is still a nondeterministic graph which can be finally turned into a deterministic one as follows. When several arcs with the same symbol leave one node, all their arrival nodes are regrouped into a single meta-node. The image of a meta-node via a given symbol consists of the images of all its components, which are also regrouped into a possibly new meta-node. The final result, after deletion of an irrelevant transient part, is displayed in Fig. 7.1(c): the graph clearly corresponds to the regular expression $R' = (1 + 01)^*$, i.e., to the language in which two consecutive zeros are forbidden. Different regular expressions (such as R and R') do not necessarily yield different languages. □

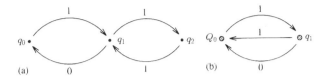

Figure 7.2 (a) Graph representing the regular expression $R = (01 + 11)^*$; (b) minimal equivalent graph.

The reducibility of a regular expression to a deterministic finite automaton has general validity. For a proof, we refer to Hopcroft & Ullman (1979). Vice versa, every FA is equivalent to a regular expression which can be formed by inverting the simple rules given above.

A finite automaton can be seen both as a device to generate strings and as a language recognizer. In the former case, symbols are emitted while following an admissible path on the graph. In the latter, starting on the *initial state* q_0, the next arc to be followed is determined by the current symbol s in the input signal. The language $\mathscr{L}$ is *accepted* by the FA if there is a path for each string $S \in \mathscr{L}$.

As seen in the example, different regular expressions may represent the same language. This reflects in the associated graphs. In fact, it may happen that the sets of sequences originating from two distinct nodes q, q', are identical. This redundance, which would hinder a meaningful comparison between different languages, can be eliminated by finding for each language the graph with the minimum number of nodes. Its existence and unicity is guaranteed (apart from an irrelevant renaming of the nodes) by the *Myhill–Nerode theorem* (Nerode, 1958; Carrol & Long, 1989). Here, we briefly illustrate a recursive procedure for the minimization of a redundant graph, based on the progressive identification of pairs of nonequivalent nodes. All pairs of nodes having different sets of outgoing arcs constitute the initial list L. For each pair (p,q) in L, all pairs (p',q') leading to (p,q) with the same symbol s are included in the list. By iterating this procedure for all pairs and all symbols $s \in A$, until no new entries are found, only the pairs of equivalent nodes are left. The minimal graph is obtained by merging these nodes together according to the equivalence relations.

The language produced by the roof map (4.4) at $a = 2/3$ (see Section 4.3) is represented by the directed graph shown in Fig. 7.2(a). It is immediately seen that the only pairs of nonequivalent nodes are (q_0, q_1) and (q_1, q_2). By merging q_0 and q_2 into the single node Q_0, one obtains the minimal graph of Fig. 7.2(b). As discussed in Chapter 4, the system yielding the language in Fig. 7.2 is strictly sofic. Subshifts of finite type and strictly sofic languages are regular.

A useful tool for proving the nonregularity of certain languages is provided by the *pumping lemma* (Hopcroft & Ullman, 1979), which establishes a general recurrence property of regular languages:

Lemma 7.1 *For any regular language $\mathscr{L}$, there is a finite n such that every word $S \in \mathscr{L}$ of length $|S| \geq n$ may be written as $S = uvw$ with $|uv| \leq n$, v nonempty, in such a way that $uv^i w \in \mathscr{L}$ for all $i \geq 0$.*

The meaning of this lemma is apparent: paths corresponding to sequences S longer than the number of nodes must contain at least one loop. Therefore, all sequences obtained by repeating the loop an arbitrary number of times still belong to the same regular language.

7.1.2   Context-free languages

Regular languages are generated by a *sequential* mechanism: a symbol is *appended* to the sequence each time a transition between two nodes is performed. A qualitatively different generation method is represented by a *parallel* writing system, in which parts of a sequence are *replaced* by words chosen from a given list. The set of *production* rules is usually referred to as a *grammar*. Representative examples in this class are substitutions, encountered in Chapter 4, and *context-free* grammars (CFGs), which belong to the second level of the Chomsky hierarchy and have important string-processing applications.

A CFG operates on strings composed of two classes of symbols: *terminals* and *variables*. The former belong to the alphabet A: concatenations of them eventually constitute the output. Variables are temporary entities to be replaced according to the following rules: each variable V_k is transformed into one of the symbolic expressions $Z_1; Z_2; \ldots; Z_k$, where the semi-colons separate the alternative choices. The strings Z_i, in general composed of terminals and variables, are called *sentential forms*. The *context-free language* (CFL) $\mathscr{L}(\mathscr{G})$ generated by a CFG $\mathscr{G}$ is the set of all words consisting exclusively of terminals which can be produced starting from an initial variable W. Since the substitution of any variable V depends on no symbol other than V itself, the productions can be carried out in any order. The independence from V's neighbours (its "context") motivates the name of these grammars. The action of the rules is effectively illustrated by the following example.

Example:

[7.2]   Consider the productions

$$W \rightarrow \oplus V$$
$$V \rightarrow \oplus VV \; ; \; \ominus, \tag{7.2}$$

where $\oplus$ and $\ominus$ are terminals which can be respectively interpreted as right and left moves on a one-dimensional discrete lattice. Starting with W, the first three production steps yield the following 5 different sequences: $\oplus\oplus\oplus VV\oplus VV$, $\oplus\oplus\ominus VV$, $\oplus\oplus\oplus VV\ominus$, $\oplus\oplus\ominus\ominus$, and $\oplus\ominus$. As can be easily verified, the language $\mathscr{L}_\pm$ generated by (7.2) describes all paths on the positive semi-lattice that eventually return to the origin without ever entering the region to its left: in other words, random walks with an absorbing barrier.

This language can be proved not to be regular by means of the pumping lemma 7.1. For every integer n, choose as a test word $S = uvw = \oplus^n\ominus^n$. No matter how long u is, v will always contain the symbol "$\oplus$" but not "$\ominus$"($|uv| \leq n$). Hence, uv^2w cannot be in $\mathscr{L}_\pm$, since it contains more $\oplus$s than $\ominus$s. Such a sequence is *unbalanced*: this language, in fact, gives all sequences of balanced parentheses, by substituting "(" for $\oplus$ and ")" for $\ominus$. It may be recalled that CFLs satisfy a pumping lemma of their own (Hopcroft & Ullman, 1979). □

All concatenations of terminals can be generated by the grammar (7.2). The unbalanced ones, however, can only appear as subsequences of legal words and do not belong to the language $\mathscr{L}_\pm$ which, therefore, is not factorial (see Section 4.4). Moreover, concatenations such as $\oplus\ominus\oplus\ominus$ of language members are also not in $\mathscr{L}_\pm$ (they also exist as subwords only). Hence, $\mathscr{L}_\pm$ is not even extendible. Thus, it cannot arise from a shift dynamical system. This is not surprising since CFGs have been introduced to model computer operations such as the compilation of syntactically correct expressions, a task that may require a memory allocation comprising the whole string under consideration.

To illustrate the superiority of context-free languages over regular languages, we show how the latter can be generated by CFGs. The nodes $\{q_0, q_1, \ldots\}$ of a given finite automaton are set into a one-to-one correspondence with the variables $\{V_0, V_1, \ldots\}$ of the CFG. The ith production takes the form

$$V_i \rightarrow s_{ij}V_j \, ; \, s_{ik}V_k \, ; \ldots ; \, \epsilon \, ,$$

where s_{ij} is the terminal symbol associated with the transition from node q_i to node q_j. The variables V_j, V_k, ..., correspond to all nodes reachable from q_i. The CFG thus writes the symbols sequentially while keeping at the rightmost end the variable corresponding to the current node on the graph. Notice the simplicity of these grammars, the productions of which never involve more than one variable each.

As soon as multiple-variable productions are allowed, the complete specification of the language generated by the CFG is, in general, a formidable task. Even assessing the nonregularity of the language may be exceedingly hard. In order to simplify the analysis, avoid possible redundancy, and exclude trivial grammars, it is useful to recast each CFG into a *normal form*. In particular, *Greibach*'s normal form is defined by productions of the type

$$V \rightarrow tU \, ,$$

where t is a terminal and U is a (possibly empty) string of variables. Every CFL

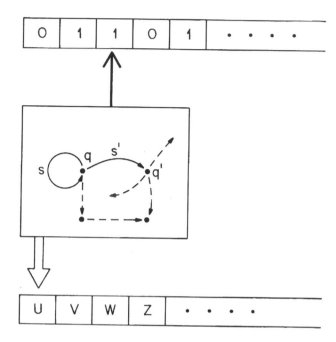

Figure 7.3 Schematic illustration of a pushdown automaton. From top to bottom, one recognizes the tape containing the input, the finite automaton, and the stack. The two arrows indicate the read-only head (top), sliding along the tape, and the read/write head (bottom), acting at the leftmost position of the stack.

which does not contain ϵ can be generated by a Greibach grammar (Hopcroft & Ullman, 1979).

Languages generated by CFGs are recognized by *pushdown automata* (PDAs), nondeterministic devices composed of an input tape T and a finite set of internal states, just as finite automata, with the addition of a *stack register* (see Fig. 7.3). Initially, the sequence to be tested is contained in T and the stack is empty. At each step, the automaton performs the following operations:

1. reads a (possibly empty) symbol s at the current position along T;
2. changes its state from q to $q' = q'(q, s, z)$, where q' is chosen from a *list L* of possible states and z is the leftmost symbol in the stack: L depends on q, s, and z;
3. replaces z with the (possibly empty) finite string U attached to q';
4. if $s \neq \epsilon$, it advances the reading position along T.

The sequence is accepted if its complete reading leads only to legal moves which end with an empty stack.

To elucidate the relationship between CFGs and PDAs, we assume, without loss of generality, that the grammar is in Greibach's form. In this case, it is possible to construct a PDA having a single node q. Each production of the form $V \rightarrow tU$ yields the transformation

$$(q, VZ) \xrightarrow{t} (q, UZ),$$

where the pair (q, S) represents the configuration of the PDA (node, stack content) and $t \in A \cup \{\epsilon\}$ is the current symbol in the input tape T. The variable V is replaced by the string U at the leftmost position of the stack (Z being the rest of the content of the stack). When $U = \epsilon$, V is erased and only the string Z is left. The action of the PDA is understood by referring to legal sentential forms of the type $t_1 t_2 \ldots t_m V_1 V_2 \ldots V_n$, which have all terminals to the left of the variables[2]. It can be easily verified that, once the PDA has read t_1 through t_m, the stack contains $V_1 V_2 \ldots V_n$. Hence, since all words $t_1 t_2 \ldots t_m$ in the language correspond to $V_1 V_2 \ldots V_n = \epsilon$ (i.e., to $n = 0$), they are accepted by the PDA with empty stack.

Pushdown automata in which there is at most one destination node $q'(q, s, z)$ for each triple (q, s, z) are called *deterministic* (DPDA). In contrast with regular languages, which are accepted by both deterministic and nondeterministic finite automata, the languages generated by DPDAs are a proper subset of the CFLs (Hopcroft & Ullman, 1979). The superior capability of nondeterministic PDAs, however, is obtained at a loss of an immediate identifiability of the words in the language. In fact, for any allowed input word, there are many legal operations of the PDA ending with a rejection. In order to avoid this problem, the automaton should be able to explore all possible derivation paths. The search could be rendered automatic by adding a stack where the information about all tested paths is processed (by allowing read/write operations). The resulting automaton, however, would be equivalent to a (much more powerful) Turing machine with a set of *ad hoc* constraints. Rather than resorting to such a hybrid solution, it is preferable to keep a neat separation among computational models by demanding just the recognizability "in principle" of legal words.

Pushdown automata constructed from Greibach's form have an architecture in which the information necessary to process the input is entirely contained in the stack. This is the opposite situation with respect to PDAs that emulate finite automata, where the symbol in the stack coincides with the label of the current node and is, therefore, redundant. For a nonregular CFL, a transfer of information-processing capability from the stack to the nodes is possible, provided that the number of nodes is increased.

Example:

[7.3]   Consider the set of all even palindrome strings SS^r over $\{0, 1\}$, where S^r is obtained from S by reversing the order of the symbols. They constitute a

---

2. Such strings can be obtained by applying the production rules to the leftmost variable only, since the grammar is in Greibach's form.

CFL generated by the production rules

$$W_0 \to 0U_0W_1 \; ; \; 1V_0W_1 \; ; \; \epsilon$$
$$W_1 \to \epsilon$$
$$U_0 \to 0U_0U_1 \; ; \; 1V_0U_1 \; ; \; 0$$
$$U_1 \to 0 \qquad\qquad\qquad\qquad\qquad\qquad\qquad (7.3)$$
$$V_0 \to 0U_0V_1 \; ; \; 1V_0V_1 \; ; \; 1$$
$$V_1 \to 1 \; .$$

Because of its Greibach normal form, the grammar can be associated with a single-node PDA. Since the transitions are nearly a rewriting of the above substitutions, we list only the first three:

$$(q, W_0) \overset{0}{\to} (q, U_0W_1) \, , \; (q, W_0) \overset{1}{\to} (q, V_0W_1) \, , \; (q, W_0) \overset{\epsilon}{\to} (q, \epsilon) \, .$$

The reader can verify that the same language is also generated by the following two-node PDA:

$$(q_1, W) \overset{0}{\to} (q_1, UW) \, , \; (q_1, W) \overset{1}{\to} (q_1, VW) \, , \; (q_1, W) \overset{\epsilon}{\to} (q_2, \epsilon)$$
$$(q_1, U) \overset{0}{\to} \{(q_1, UU); (q_2, \epsilon)\} \, , \; (q_1, U) \overset{1}{\to} (q_1, VU)$$
$$(q_1, V) \overset{0}{\to} (q_1, UV) \, , \; (q_1, V) \overset{1}{\to} \{(q_1, VV); (q_2, \epsilon)\}$$
$$(q_2, W) \overset{\epsilon}{\to} (q_2, \epsilon) \, , \; (q_2, U) \overset{0}{\to} (q_2, \epsilon) \, , \; (q_2, V) \overset{1}{\to} (q_2, \epsilon) \, .$$

Only three variables are involved. $\qquad\qquad\qquad\qquad\qquad\qquad\qquad$ $\square$

By increasing the number of nodes and neglecting the stack altogether, one obtains a sequence of finite automata that approximate the CFL better and better. Asymptotically, an infinite automaton without stack is obtained. One may ask whether there is a general algorithm for the "minimization" of pushdown automata (recall that the minimum automaton can be effectively constructed for every RL). A precise formulation of this question requires a suitable encoding of the whole device (nodes, variables, and transitions) as a symbol sequence, the length of which is the size one is interested in (coding techniques are treated in Chapter 8). Unfortunately, as the following *ad absurdo* argument shows, the answer is negative. If such a procedure existed, it should be applicable also to PDAs generating the language $\mathscr{L} = A^*$. The result would hence be the trivial automaton composed of a single node q and b arcs connecting q to itself. As proved in the theory of computation, however, it is *undecidable* whether a generic CFL is A^* or not. The concept of undecidability will be discussed in detail in Section 7.1.4.

7.1.3   Context-sensitive languages

A straightforward extension of context-free grammars is obtained by making the replacement of each variable V depend on symbols adjacent to V. The productions of the resulting *context-sensitive grammar* (CSG) take the form

$$SVS' \; \rightarrow \; SZ_1S' \; ; \; SZ_2S' \; ; \; \ldots \; , \; \text{with } Z_1, Z_2, \ldots \neq \epsilon \, , \tag{7.4}$$

where S, S', Z_1, Z_2, ... are sentential forms. The substitution of V is permitted only in the "context" $S \cdots S'$. The variability induced by the constraints gives context-sensitive grammars the power of generating a broader class of languages (CSLs) than context-free grammars can do. Productions like those in (7.4) are a special case of substitutions $Z \rightarrow Z'$, where the generic sentential forms Z and Z' satisfy the "noncontractivity" condition $|Z'| \geq |Z|$. This broader set of productions, however, can be proved (Hopcroft & Ullman, 1979) to yield exactly the same class of languages as (7.4). Let us illustrate this equivalence for the general, albeit simple, substitution

$$UV \; \rightarrow \; XYZ,$$

where U and V are variables, and X, Y, and Z are nonempty sentential forms. This production is tantamount to the following sequence of operations

$$UV \rightarrow UV'$$
$$UV' \rightarrow U'V'$$
$$U'V' \rightarrow U'YZ$$
$$U'YZ \rightarrow XYZ \, ,$$

each being of type (7.4). Notice that the dummy variables U' and V' have been introduced in order to avoid illegal productions with substitutions of variables belonging to the context.

Example:

[7.4]   Consider the set $\mathscr{L}$ of all one-dimensional random walks that are confined between the origin O and the first maximum position u_{max} and that, eventually, come back to O. Indicating with $s_i = +1$ the displacement at time i, the walker's position at time n is $u_n = \sum_{i=1}^n s_i$. Setting $s_1 = +1$, the first maximum is defined by $u_{max} = \max\{j : u_j = j\}$. A path of arbitrary finite length n is in $\mathscr{L}$ if $0 \leq u_i \leq u_{max}$, $\forall \, i$, and $u_n = 0$. The set $\mathscr{L}$ is easily recognized as the language generated by the following CSG, in which the terminals $\oplus$ and $\ominus$ correspond to the displacements $+1$ and -1, respectively:

$$W \rightarrow |\oplus V|$$
$$V \rightarrow \oplus V \; ; \; UR_-$$
$$R_-| \rightarrow L_-\ominus|$$

$$R_+| \rightarrow L_+ \oplus |$$
$$R_\pm| \rightarrow L_- \ominus | \; ; \; L_+ \oplus |$$
$$| \oplus U L_- \rightarrow | U \oplus R_+ \; ; \; || \oplus E$$
$$\oplus \oplus U L_- \rightarrow \oplus U \oplus R_\pm$$
$$U L_+ \oplus \ominus \rightarrow \oplus U \ominus R_-$$
$$U L_+ \oplus \oplus \rightarrow \oplus U \oplus R_\pm$$
$$E| \rightarrow ||$$
$$\oplus L_\alpha \rightarrow L_\alpha \oplus \quad , \quad \ominus L_\alpha \rightarrow L_\alpha \ominus \quad (\alpha = +, -)$$
$$R_\alpha \oplus \rightarrow \oplus R_\alpha \quad , \quad R_\alpha \ominus \rightarrow \ominus R_\alpha \quad (\alpha = +, -, \pm)$$
$$E \oplus \rightarrow \oplus E \quad , \quad E \ominus \rightarrow \ominus E \; .$$

In order to satisfy the noncontractivity condition for all productions, we introduced the special symbol | which delimits all sentential forms. This artifact is by no means essential: it could be, however, eliminated only at the price of a considerable complication of the grammar. The first two rules produce the initial arbitrarily long string of consecutive $\oplus$ symbols. The action of the other substitutions is better understood with reference to the following sample sequence:

$$| \oplus \oplus \oplus U \oplus \oplus \ominus \oplus \ominus R_\pm \ominus \oplus \ominus | \; .$$

The variable U indicates the current position of the random walker between the origin (left "|") and u_{max} (leftmost "$\ominus$"). The right- and left-moving variables R_α and L_α, respectively, drop a $\oplus$ or a $\ominus$ at the right end of the current string and carry the corresponding information back to U. The position of U is then accordingly updated. The process can be terminated by the "eraser" E each time the walker returns to the origin.

□

Context-sensitive languages are recognized by *linear bounded* automata (LBAs). Analogously with pushdown automata, they consist of a finite set of internal states and an input tape containing the word w to be analysed. Rather than manipulating just the leftmost symbol on a stack, an LBA can perform read/write operations at any position on a work tape by means of a head H. More precisely, depending on the (optional) reading from the input tape, on the current node, and on the symbol z at the position of H, the LBA changes node, replaces z by another symbol, and moves the head either left or right by one step. The input word w is recognized if the automaton ends in an accepting node. The region of the work tape accessible to H has an extension $l(|w|)$ proportional to the length $|w|$ of w (this explains the origin of the name). Notice that the same limitation is implicit in PDAs: in fact, since a PDA in Greibach's form can delete no more than one stack symbol after each reading from the tape, clearing the stack needs, in the worst case, $|w| - 1$ steps. LBAs have larger freedom in retrieving the information stored on the tape. In order to access the nth symbol on the stack, in fact, a PDA must erase the first $n-1$ entries. To avoid this loss of information, the PDA should store it in suitable internal states: strings with arbitrary length n could not be handled. The LBA, instead, achieves this by "shift" operations analogous to the last three sets of productions in Ex. 7.4.

In its most general implementation, the LBA writes on the work tape the sentential form Z resulting from a sequence of productions until Z either exceeds the allowed portion of tape, or it reproduces the input word w exactly. In the latter case, the word is accepted. The fundamental difference with respect to CFGs is that now the final result depends on the order in which the productions are performed. Hence, assessing the membership of a word w in a given language requires the exploration of all possible paths. This exhaustive search can be rendered automatic (i.e. deterministic) only by ordering all paths and letting the head write on the work tape the code corresponding to the last tested path. This procedure requires access to a tape-region of size $l(|w|) = O(|w|^2)$, which is forbidden by the definition of an LBA.

The grammar in Ex. 7.4 is particularly simple since, at each step, there is only one replaceable sentential form. Hence, the corresponding LBA is deterministic: each move is univocally identified by the current node, reading on the input, and work-tape symbol. Nevertheless, the grammar is strictly context-sensitive since two arbitrary integers (the distances of the walker from the origin and from the maximum) must be continually updated on the tape. A PDA could store both numbers on the stack but could not access the rightmost one without destroying the other. The characteristic of the language in the example is essentially the ability to compute the arithmetic difference between two arbitrarily large numbers. Indeed, all arithmetic operations on integers (including multiplication, exponentiation, etc.) can be carried out within the class of CSGs which thus turns out to be very powerful. Finally, we recall that it is presently unknown whether the language classes accepted by deterministic and nondeterministic LBAs are equivalent.

7.1.4 Unrestricted languages

When the only constraint in the production rules of CSGs is released, by permitting substitutions of the type $Z \to Z'$ with $|Z'| < |Z|$, one obtains the highest class of grammars in the Chomsky hierarchy, which are called *unrestricted*. The relevant novelty introduced by this further generalization is the lack of an upper bound to the number of iterations needed to produce a generic finite word w. At variance with lower-order grammars, in fact, unrestricted grammars (UGs) do not steadily increase the length of a sentential form upon repeated application, but may also reduce it.

A simple but illuminating example of an unrestricted grammar is given by the productions (Hofstadter, 1979)[3]

$$Iz \rightarrow IUz$$
$$v\chi z \rightarrow v\chi\chi z$$
$$III \rightarrow U$$
$$UU \rightarrow \epsilon \qquad\qquad (7.5)$$
$$U \rightarrow u$$
$$I \rightarrow i$$

where I and U are variables, i, u, v, and z terminals, and χ is a generic sentential form. The starting sequence is assumed to be vIz. Apparently, the second production in Eq. (7.5) is not legal, since the structure of χ is not specified and has an arbitrary length. It can be easily seen, however, that the set of all words of the type $\chi\chi$, with a generic χ, is a CSL (Hopcroft & Ullman, 1979). Therefore, that substitution is not a more general kind of production, but just a compact representation of a finite set of legal productions (reminiscent of a "subroutine").

The presence of contracting rules in Eq. (7.5) makes it difficult to test whether even such a short word as "vuz" is in the language or not. We shall see that there is no general procedure (*decision algorithm*) to answer questions of this kind for all unrestricted grammars and intuition is often the only guide to the required derivation path. In this specific example, however, such an algorithm has been recently discovered by L. Swanson (McEliece *et al.*, 1989). The word vuz is not in the language: this consists, in fact, of all words of the type $v\chi z$ where χ is any sequence of i's and u's for which the number of i's is not a multiple of 3 (0 included).

The *Turing machine* (TM) is the automaton associated with an unrestricted grammar. The linear-bounded automaton discussed in the previous subsection is a TM with a limited access to the tape. When this constraint is released, one obtains the basic design of a Turing machine (see Fig. 7.4) which consists of a finite set of internal states, a read/write head, and a semi-infinite tape. The input is stored in the n leftmost cells, whereas the remaining infinity of cells is initially blank. Upon reading one symbol at the head position, and depending on the internal state, the TM changes state, replaces the symbol on the tape and moves either left or right. A given word w is accepted if the TM eventually reaches a preassigned accepting configuration.

This architecture has been proved to be equivalent to many modified versions which apparently extend it by admitting, for example, several heads and (possibly bi-infinite) tapes, or operations on a multidimensional infinite array rather than

3. Unessential modifications have been made to the original grammar in order to conform to the notation adopted in the present book.

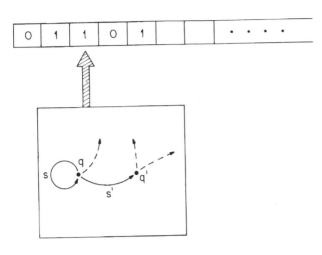

Figure 7.4 Schematic illustration of a Turing machine showing the semi-infinite input/output tape, the read/write head, and the finite automaton.

on a tape. Inclusion of nondeterminism does not increase the computing capability of Turing machines either.

The action of a Turing machine can be interpreted as the computation of a function f from integers to integers. In this view, the content of the tape when the machine halts on input w represents a suitable symbolic encoding of $f(n)$, where n is the integer corresponding to w with the same encoding. For this reason, the theory of TMs is sometimes referred to as the theory of *computability*. Turing machines are believed to constitute the most general class of finite automata capable of performing computations. This is precisely the content of the *Church–Turing hypothesis* which states that any computable function can be implemented on a suitable TM. While there is a widespread belief that the conjecture is true, there is no way to prove it, as long as the concept of "computability" remains informally defined (Hopcroft & Ullman, 1979).

The simultaneous presence of contracting and expanding rules in UGs reflects in the famous *halting problem* of a TM. While finitely many steps are obviously required to accept a legal word w, the TM need not halt if w is not in the language. This phenomenon can be illustrated by interpreting the TM as a discrete-time dynamical system acting on the integer n defined by the symbolic string S on the tape (or, equivalently, on the finite-precision real number $x = \sum_i s_i b^{-i}$, where $S = s_1 s_2 \ldots$). The first type of endless computation is represented by cycles which the TM may enter, in the same way as a dynamical system may do (either by attraction or because of the finiteness of the precision in a numerical simulation). This kind of uncertainty, however, is readily eliminated by supplying the TM with a "subroutine" that detects the onset of periodic behaviour (it is sufficient to store all intermediate results of the computation). Further examples of infinite computations are those involving only expanding substitutions ($|S| \rightarrow \infty$). Also this case can be automatically recognized since the number of substitutions in the process is finite.

Finally, maximal uncertainty occurs when arbitrary sequences of expansions and contractions are possible. The precision of the variable x varies in time and no definite prediction on the outcome of the iteration can be made in general. By pushing the analogy with dynamical systems further, one may imagine the existence of two distinct limit sets Ω_1 and Ω_2 for a dynamical system $(\mathbf{F}, X)$, corresponding to the accepting and the rejecting configurations of the TM, respectively. If one of them, say Ω_2, is a strange repeller, there may be no general algorithm to decide in a finite time whether an arbitrary bounded set $X_0 \subset X$ of initial conditions (encoded by the input word S_0) is entirely mapped to Ω_1 or not. In fact, a response is obtained in a finite time only if X_0 does not intersect any preimage of Ω_2.

The languages generated by UGs (i.e., accepted by TMs) are called *recursively enumerable* (REL). This name originates from the possibility of enumerating all words in any language of this class. In fact, while not necessarily halting, a TM may operate as follows. It carries out (at most) n computation steps, starting with all input words that can in principle be recognized (produced) in no more than n steps. Their number is obviously finite, so that the TM can repeat the procedure for increasing values of n until, eventually, all the words in the language have been generated.

Since TMs are composed of a finite number of parts (internal states), each described by a finite number of properties (the possible moves), the set of all TMs and, in turn, the set of all RELs are numerable. Therefore, although RELs are the broadest class of languages that can be produced by finite automata (according to the Church–Turing hypothesis), they represent only an infinitesimal fraction of all possible languages, which we know are in a one-to-one correspondence with real numbers. One could imagine RELs to be akin to rational numbers. This analogy, however, is not appropriate: there are "many more" RELs than rationals. In fact, each real number admitting a finite specification corresponds to a REL (see, e.g., all irrationals characterized by a periodic continued-fraction expansion). More properly, RELs correspond to computable numbers.

Although the vast majority of languages is nonREL, it is not easy to produce examples of these since they are not algorithmically definable. Such a difficulty can be circumvented by considering the set A^* of all finite words over the alphabet A in the lexicographic order, where w_i is the ith word, and some ordering of the infinite list of all TMs. Let us further indicate with $\mathscr{L}_j = \mathscr{L}(M_j)$ the language generated by the jth TM M_j and define a matrix $\mathbf{M}$ whose entry m_{ij} is 0 if $w_i \notin \mathscr{L}_j$ and 1 otherwise. The language

$$\mathscr{L}_C = \{w_i : m_{ii} = 0, \ i = 1, 2, \dots, \infty\} \tag{7.6}$$

cannot be recursively enumerable. If it were, in fact, it would be accepted by some Turing machine M_k: this is, however, impossible, since $\mathscr{L}_C$ differs from each $\mathscr{L}_k$ by one word, by construction, for all $k = 1, 2, \ldots, \infty$. The argument used in the proof is a modification of the famous diagonal method introduced by Cantor to show that there are more than countably many real numbers. The reason why definition (7.6) cannot be directly turned into an algorithm to construct $\mathscr{L}_C$ is that the procedure to establish whether $w_i \in \mathscr{L}_i$ may require an infinite number of steps if the TM M_i does not halt.

As a possible dynamical-system analogue of nonRELs, one may consider systems with riddled basins (Section 3.3): in fact, no algorithm presumably exists which is able to assess in a finite number of steps the membership of a set X_0 of initial conditions to one of the basins, no matter how small X_0 is. For a more detailed and rigorous discussion of computability (in the context of Julia sets), see Penrose (1989) and Blum and Smale (1993).

An important subclass of RELs is represented by *recursive* languages, i.e., those that are accepted by at least one TM that halts on all inputs. The words of a recursive language can be generated in order of increasing size. It is easy to verify that the complement of a recursive language is also recursive. Moreover, if $\mathscr{L}$ is recursively enumerable but not recursive, its complement $\overline{\mathscr{L}}$ is nonREL.

In order to show that recursive languages represent a proper subset of RELs (i.e., that TMs may indeed fail to halt on some input word), it is first necessary to introduce the concept of a *universal* Turing machine. A universal TM (UTM) is like an ordinary computer with an infinite storage capacity; in fact, it is defined as a machine with the ability of simulating any other TM, that is, to execute any program. As seen before, any language $\mathscr{L}$ can be represented as an infinite binary sequence $\tilde{\mathscr{L}}$. Because of the equivalence between languages and automata, it is clear that any automaton M can also be encoded as a symbol string $C(M)$ (the "program" specifying the internal states and the set of all possible moves of M). The existence of the encoding procedure $C(M)$ permits the construction of a UTM U.

The fundamental component of a UTM is its tape which can be considered as split into three parts: T_1, which contains the code $C(M)$, i.e., the input program for the simulation of M; T_2, where the current internal state of M is stored; T_3, which represents the tape of M. The UTM U, after reading a symbol in T_3, performs the following operations:

1. moves to T_2 where it reads the internal state of M;

2. goes to T_1 to read the next move of M;

3. returns to T_2 and T_3, to store the occurred changes and read the next input symbol.

The three possible actions of U differ from the elementary moves allowed by the definition of a basic TM. It is, however, possible to show that they can be reduced to a list of legal moves. This long and technical exercise is left to the strongly motivated reader. It is important to realize that, although the size of M can be arbitrarily large, the UTM is finite. In fact, the information about the internal state and the program of M is stored on the tape of U (and not in its nodes) which, by the definition of a TM, can be arbitrarily long.

The formal representation of a TM as a string which can be used, in turn, to feed up the machine itself, shows that TMs are sufficiently powerful to answer general questions concerning their own properties. Indeed, the encoding procedure C from the set $\mathcal{M}$ of TMs to A^* is equivalent to the transformation of logic statements into arithmetic expressions that constitutes the core of Gödel's famous proof of the existence of undecidable formal statements (Herken, 1994). While referring to the specialized literature for a rigorous discussion, we recall a few basic results. Let us indicate with $\mathcal{M}$ the set of all TMs and consider the "universal" language

$$\mathcal{L}_u = \{(C(M), w) : M \in \mathcal{M} \text{ and } w \in \mathcal{L}(M)\}$$

which consists of all TM codes $C(M)$, each paired with all words w accepted by the corresponding machine M. The language $\mathcal{L}_u$ is recursively enumerable since it can be accepted by a universal TM U_1. It can be shown, however, that it is not recursive. If U_1 were able to halt on each input, in fact, it would be possible to construct a TM U_2 accepting the language L_C (7.6) which is nonREL. For any word $w \in A^*$, U_2 would proceed as follows:

1. determine the integer n such that $w = w_n$ (in the lexicographic ordering of A^*),
2. find the TM M_n having $n = C(M_n)$ as a code,
3. pass the control to U_1 which verifies whether $(C(M_n), w_n)$ is in $\mathcal{L}_u$ or not.

This algorithm would accept the nth word w_n in the lexicographic order only if w_n is recognized by the nth TM, for all n: i.e., it would construct the complement $\overline{\mathcal{L}_C}$ of $\mathcal{L}_C$. Since $\mathcal{L}_C$ is nonREL, such an algorithm does not exist. Hence, neither $\overline{\mathcal{L}_C}$ nor its equivalent $\mathcal{L}_u$ are recursive. The theorem proves the existence of endless computations. The reason for this unavoidable indeterminacy is that there are countably many computation procedures, whereas the set of all possible statements has the power of the continuum.

Infinite computations can be interpreted as *undecidable* questions: given a word w and a TM M, one may never know whether w is in the language $\mathcal{L}(M)$ or not. The reader might have the impression that, after all, this is no fundamental problem. A little thinking, however, shows that the membership question is equivalent to assessing the truth of a theorem in a formal system. In fact, the initial string(s) from which an unrestricted grammar can generate all words in its language can be seen as the set of axioms of some formal system, while the production rules correspond to all logic implications of the initial statements. The repeated application of the substitutions automatically generates all correct theorems. As shown by the halting problem, it may be undecidable whether a given theorem is true or not!

Finally, we show that CSLs are strictly contained within recursive languages by adapting the argument used to construct $\mathscr{L}_C$ (7.6). Consider the language

$$\mathscr{L}_R = \{w_i \in A^* : w_i \notin \mathscr{L}_i, i = 1, 2, \ldots, \infty, \text{ where } \mathscr{L}_i \text{ is a CSL}\} \qquad (7.7)$$

in which A^* is ordered lexicographically and the CSLs $\{\mathscr{L}_i\}$ arbitrarily. Since $\mathscr{L}_R$ contains w_i if and only if w_i is not in the ith CSL, $\mathscr{L}_R$ is obviously not a CSL. It is, however, recursive since a UTM simulating $\mathscr{L}_R$ (that is, all CSLs) halts on each input. Notice that, instead of giving an explicit example of a recursive nonCSL language, we have constructed an abstract and presumably uninteresting model, the reason being that it is not easy (or even possible?) to prove whether a given recursive language can be reduced to a CSL. We further mention that recursive languages correspond to *total recursive functions*, integer functions that are defined for all integer inputs: examples of them are multiplication, $n!$, $\lceil \log_2 n \rceil$, 2^n. Linear bounded automata, on the other hand, also seem to be sufficiently powerful to reproduce this type of operations. Finally, functions that are computed by generic Turing machines (i.e., not necessarily halting) are called *partial recursive functions*.

7.1.5 Other languages

The language ordering provided by the Chomsky hierarchy is essentially based on the memory requirements of the corresponding finite automata. In fact, starting with RLs, which do not need any memory, we move to CFLs, having just a stack, to CSLs with a storage capacity proportional to the input word length and we eventually arrive at RELs with their unlimited memory. From the point of view of complexity, the above classification is not entirely satisfying as there can exist other relevant differences among languages. For this reason, intermediate levels have been introduced in the original ordering and some classes have been subdivided into disjoint subsets, thus obtaining a tree-like structure. For example, the so-called indexed languages are placed between CFLs and CSLs: they contain the former and are included in the latter. A set of auxiliary variables (the *indices*) is added to the usual ones with the function of representing the context of some productions; eventually, they are deleted (Hopcroft & Ullman, 1979). This family of languages contains, besides the CFLs, the class of substitutions (see Chapter 4) which constitute a side branch of the languages' hierarchy. Substitutions are formally called D0L languages, where the letter "D" stands for "deterministic" and "L" acknowledges A. Lindenmayer (1968) who studied them in connection with biological problems. The number "0" indicates that the substitution of a symbol s with a given word $w = \psi(s)$ occurs independently of the context: substitutions determined by N neighbouring symbols are called DNL. Recall that each symbol s has only one image $\psi(s)$ and that all symbols are updated simultaneously, at variance with CFGs. The classes of CFLs and D0Ls are disjoint: in general, one cannot obtain a given CFL from a D0L grammar nor a D0L language from a CFG. In fact, although D0Ls appear "simpler" than CFLs, the synchronization of their productions requires a computational power that is provided by CSLs but not by CFLs.

The definition of new language classes has proliferated considerably without, however, providing a scheme in which grammars with comparable power (diversity of the corresponding languages) are assigned to the same level. Therefore, we hold to the original Chomsky classification and consider other languages only when they occur in a physical context.

7.2   Physical characterization of formal languages

In this section, we study physical properties of shift dynamical systems associated with the previously introduced languages. In particular, we consider power spectra, ergodicity and mixing, entropy, and thermodynamic functions in a few basic examples. The possible generalization of the results obtained in each case is then briefly discussed. In order to study metric properties of the symbolic signals produced by the various grammars, it is necessary to choose a shift-invariant measure m on the sets of the n-cylinders, for all n, and possibly to assign a numerical value to each symbol $s \in A$ (which, for simplicity, we again indicate with s). Unless otherwise stated, the binary case $A = \{0, 1\}$ is considered.

7.2.1   Regular languages

A shift dynamical system is associated with a regular language $\mathscr{L}$ through the orbit closure of the signal produced by following all legal paths on $\mathscr{L}$'s graph. In general, finite automata may have *transient* states (i.e., such that no return to them is possible from some other state) and disjoint invariant components in which the trajectory may be trapped. By eliminating transients and considering just one component, the resulting symbolic systems are completely equivalent to irreducible Markov shifts and are, therefore, stationary and ergodic.

The canonical way to associate different weights to the various sequences is to attribute a conditional probability $\tilde{P}(j|i)$ to the arc connecting node q_i to node q_j: of course, $\tilde{P}(j|i)$ is just the entry M_{ij} of the graph transition matrix $\mathbf{M}$ (see

Chapter 5). In order to relate these node probabilities to the usual transition probabilities for symbolic strings, notice that each node q in a graph can be unambiguously identified by its "past history": i.e., by a set of suffixes common to all sequences that lead to q. For example, the nodes $(q_1, \ldots, q_4)$ of the graph in Fig. 4.5 correspond to the four mutually exclusive sets $(\{11, 101\}, 10, 100, 1001)$: this list is easily seen to exhaust all possible cases since the sequence 000 is forbidden. Therefore, the value $P(2|1)$, e.g., is the probability of observing 0 after either 11 or 101. The dependence of the label lengths on the node, a general feature of finite automata, shows that the dynamics on the graph is equivalent to a Markov process with "variable memory". In the case of strictly sofic systems, the memory is even infinite at some nodes, as shown by the example in Fig. 7.2: node Q_0 is characterized by the labels $\{0, 1^{2n} : n \in \mathbb{N}\}$ and node q_1 by $\{1^{2n+1} : n \in \mathbb{N}\}$.

The correlation function of an ergodic symbolic signal $\mathscr{S} = s_1 s_2 \ldots$ can be written as the ensemble average

$$C(n) = \sum_{s,s'} \pi(s) P^{(n)}(s'|s) s s' - \left(\sum_s \pi(s) s \right)^2 , \qquad (7.8)$$

where s and s' are the numerical values of the symbols recorded at times 0 and n, respectively, and $\pi(s)$ is the probability of symbol s. The conditional probability $P^{(n)}(s'|s)$ of observing s' n time steps after s can be easily evaluated when the dynamics is Markovian. The node-to-node transitions are indeed of this kind: knowledge of the current node may, however, be insufficient to identify the last symbol emitted by the automaton. This happens whenever some node is reached by arcs carrying different symbols. This problem can be overcome by choosing the arcs themselves as the appropriate internal variables. In fact, indicating with (i, i') the arc pointing from node i to node i' and with $s(i, i')$ the associated symbol, Eq. (7.8) can be rewritten as

$$C(n) = \sum_{(i,i'),(j,j')} \pi_a(i, i') P_a^{(n)}(j, j'|i, i') s(i, i') s(j, j') - \langle s \rangle^2 , \qquad (7.9)$$

where $\pi_a(i, i')$ and $P_a^{(n)}(j, j'|i, i')$ denote arc probabilities. In terms of the node probabilities $\tilde{\pi}(i)$ and $\tilde{P}^{(n)}(j|i)$, they read

$$\pi_a(i, i') = \tilde{\pi}(i) M_{ii'} ,$$

and

$$P_a^{(n)}(j, j'|i, i') = \tilde{P}^{(n-1)}(j|i') M_{jj'} . \qquad (7.10)$$

By the Markov property, $\tilde{P}^{(n)}(j|i)$ is the entry M_{ij}^n of the nth power of the transition matrix $\mathbf{M}$. Since $\mathbf{M}^n$ can be computed by diagonalizing $\mathbf{M}$, the time dependence of $C(n)$ is essentially given by the nth power of the eigenvalues of $\mathbf{M}$: the term containing the largest one, which is equal to 1, cancels the constant

$\langle s \rangle^2$, whereas each of the other eigenvalues yields an exponential decay, possibly accompanied by a periodic oscillation. Accordingly, the power spectrum is a superposition of Lorenzians.

For the evaluation of the generalized entropies K_q (Eq. 6.11), it is useful to introduce the quantity

$$Z_q^{(ij)}(n) = \sum_{\substack{S \\ i \xrightarrow{} j \\ n}} P^q(S), \qquad (7.11)$$

where the sum is taken over the sequences S that are associated with all possible paths of length n from node i to node j. In order to recover the sum in Eq. (6.11), a further sum over all nodes j must be performed, having chosen some initial condition $i = i_0$:

$$\sum_j Z_q^{(i_0 j)}(n) \sim e^{-n(1-q)K_q} .$$

A recursive expression can be written for the matrix $Z_q^{(ij)}(n)$ as follows:

$$Z_q^{(ij)}(n+1) = \sum_{k:k \rightarrow j} Z_q^{(ik)}(n) M_{kj}(q) ,$$

where $M_{kj}(q)$ are the entries of the transition matrix $\mathbf{M}$ raised to the power q^4. Hence, the sum (7.11) reduces to the nth power of the matrix $\mathbf{M}(q)$ with elements $M_{ij}(q)$ and the generalized entropy K_q is obtained from the logarithm of the largest eigenvalue $\mu_{max}(q)$ of $\mathbf{M}(q)$:

$$K_q = \frac{\ln \mu_{max}(q)}{1 - q} . \qquad (7.12)$$

If $\mathbf{M}(q)$, in addition to having nonnegative entries, is primitive (see Chapter 5), the Frobenius–Perron theorem (Seneta, 1981) states that there is a strictly positive eigenvalue μ_{max} such that $\mu_{max} > |\mu|$ for any other eigenvalue μ of $\mathbf{M}(q)$. For $q \rightarrow 0$, one recovers the topological transition matrix discussed in Chapter 4. For $q \rightarrow 1$, $\ln \mu_{max}(q)$ vanishes by the conservation of the total probability and K_1 is computed as the derivative of $\ln \mu_{max}(q)$ with respect to q at $q = 1$. The generalized entropy K_q is positive as long as there is at least a bifurcating node and the associated probabilities are nonzero. Since the calculation of the entropy requires the zeros of some polynomial to be found, its values are effectively computable.

For what concerns thermodynamic properties, a phase transition is observed whenever the "free energy" K_q presents nonanalytic behaviour at some $q = q_c$. Since this may occur only if there is a crossing of the two largest eigenvalues, which is here forbidden, no such phenomenon is possible in the case of Markov

---

4. Whenever there is more than one arc connecting a certain pair of nodes, the matrix entry is $\sum_i p_i^q$, where p_i is the probability associated with the ith arc.

processes. Thus, we recover the known result of the absence of phase transitions in one dimension for short-range interactions.

Let us now consider, as an example of an RL with a nonprimitive transition matrix $\mathbf{M}(q)$, the sofic system represented in Fig. 7.2(b). Every second symbol emitted is a 1 and the system jumps continually back and forth between the two nodes: this residual "periodicity" implies that not all states are accessible at any time n. The eigenvalues of the 2×2 matrix $\mathbf{M}(q)$ are

$$\mu_{\pm} = \pm \sqrt{p^q + (1-p)^q} \,,$$

where p is the probability of emitting symbol 0 when leaving node q_1. Notice that the eigenvalues are degenerate in modulus. No phase transition, however, occurs.

In this section, probabilities have been assigned directly to the arcs in the graph so that the memory is just given by the past history of each node. When, instead, the conditional probabilities depend on a larger number of past symbols, they are accounted for either by adding new nodes or by changing computational class.

Finally, notice that finite automata, although in the lowest computational class, are already able to produce signals with maximal entropy. This shows once more that complexity and randomness are indeed distinct notions.

7.2.2 Context-free languages

Many of the CFGs discussed in textbooks generate rather "unphysical" languages such as, e.g., $\mathscr{L} = \{0^n 1^n\}$. Since we are mainly interested in factorial extendible languages, we ignore them here. Moreover, in order to introduce weights for the different sequences, we assign a probability to each production whenever multiple choices are possible (thus obtaining a so-called *stochastic* CFG).

A simple case is provided by the CFG described in Ex. 7.2. If the production probabilities for V are set to $1/2$, the walker performs a diffusive motion in which the average distance from the barrier grows as $\sqrt{n}$ for n much larger than 1 (but still much smaller than the length of the word under consideration). Hence, the grammatical constraints corresponding to the barrier become less and less effective far from the extrema of the walk. The CFL structure, relevant at the extrema, reduces to a trivial Bernoulli shift in the intermediate region. This crossover is a sign of the nonstationarity of the symbolic sequences which survives even if the walker is allowed to "restart" at the end of a path, e.g., by reflection at the barrier (this modification does not change the computational class). If, in the latter case, the probability p_- of the negative displacements ($\ominus$) is larger than $1/2$, the average distance from the barrier remains bounded and stationarity is recovered.

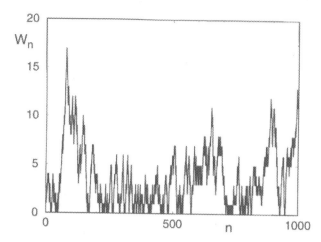

Figure 7.5 Time evolution of a random walker with a reflecting barrier at $x = 0$. The position w_n at time n satisfies $w_n = w_{n-1} + r_n$, where the displacement r_n equals 1 if $w_n = 0$ and ± 1, with respective probabilities p_+ and p_-, if $w_n > 0$. Here, $p_- = 1 - p_+ = 0.52$.

The position of the random walker with a reflecting barrier is displayed in Fig. 7.5 as a function of time, for $p_- = 0.52$. The power spectrum $S(\omega)$[5] of the sequence of the increments (with the symbols $\oplus$ and $\ominus$ mapped to the values $+1$ and -1, respectively) presents a δ-peak at frequency $\omega = \pi$, corresponding to successions of consecutive rebounds on the barrier. The overall shape of $S(\omega)$ is approximately of the form $\omega^2/(\gamma^2 + \omega^2)$. Indeed, the motion of the walker far from the barrier is purely diffusive (thus yielding a constant spectrum for the increments), whereas the probabilistic attraction (governed by p_-) towards the barrier gives a Lorenzian component for the walker's spectrum: its width γ increases with p_-.

The topological entropy K_0 of a CFG can be evaluated by means of the following procedure. For each variable V_i with productions (in Greibach form)

$$V_i \to t_{i1} U_{i1}, t_{i2} U_{i2}, \ldots, t_{ik_i} U_{ik_i},\tag{7.13}$$

we construct the formal algebraic expression

$$V_i = \sum_{j=1}^{k_i} t_{ij} U_{ij}.\tag{7.14}$$

By substituting the analogous expressions of all the variables contained in the U_{ij} and obeying the distributive property of the sum with respect to the multiplication, one constructs a polynomial in the U_{ij}. Iteration of the process eventually yields a so-called *formal power series* which represents the set of all the words that are derivable from the variable V_i. These words appear once each if the grammar is *nonambiguous* (Hopcroft & Ullmann, 1979), as we assume. By

---

5. Notice that the asymptotic statistical properties are not affected if the walker is not required to be exactly at the barrier at both ends of the path.

further replacing every terminal t_{ij} with the auxiliary variable z, one obtains the generating functions

$$V_i = V_i(z) = \sum_{n=1}^{\infty} N_i(n) z^n \,,$$

where $N_i(n)$ is the number of words of length n descending from V_i. Indicating with $N(n) \sim e^{nK_0}$ the largest of the $N_i(n)$ over all i, we see that the series converges if $z < R = e^{-K_0}$. That is, the topological entropy is given by the radius of convergence R as

$$K_0 = -\ln R \,.$$

The functions $V_i(z)$ are computed by replacing all terminals with z in Eq. (7.14) and solving the system for the V_i (Kuich, 1970). In the special case of an RL, the equations are linear and the V_i are rational functions having, in the denominator, the characteristic polynomial $\det \|\mathbf{M} - \mu \mathbf{I}\|$ of the transition matrix $\mathbf{M}$ with $z = \mu^{-1}$. The smallest zero z_1 of the denominator is the radius of convergence R. The same kind of solution occurs for the slightly broader class of the *pseudolinear* languages in which no variable V can ever yield a sentential form U containing V itself twice. Although some of the equations may be nonlinear, they can be solved iteratively (Kuich, 1970).

In the general case, polynomials of any degree are expected, so that no analytic solution is available. A numerical estimate of R can be obtained as follows. The V_i are first computed for a trial value of z. Accordingly, one evaluates the determinant $J = \det \|\mathbf{V} - \mathbf{I}\|$, where $\mathbf{V}$ is the matrix of the partial derivatives $\partial V_i / \partial V_j$ at the given z and $V_i(z)$. The modulus of the z for which $J = 0$ is the sought radius R.

The same method can be adapted to study the whole generalized entropy function K_q: each term in the sum in Eq. (7.14) is weighted by the probability p_{ij}^q of the corresponding production in Eq. (7.13), raised to the power q. The result of the iteration, after substitution of the terminals with z, is then

$$V_i^{(q)}(z) = \sum_{n=1}^{\infty} z^n \sum_{w_i : |w_i| = n} P^q(w_i) \,, \tag{7.15}$$

where the second sum is the canonical partition function (see Eq. (6.11)) for the language of all words w_i of length n that are produced by the variable V_i. The radius of convergence $R(q)$, obtained with the above illustrated procedure, yields K_q as

$$K_q = \frac{1}{q-1} \ln R(q) \,.$$

The functions $V_i(z)$ are the solutions of a system of coupled equations, analogous to Eq. (7.14), in which the generic term U_{ij} is multiplied by its weight p_{ij}^q. An expression of the type (7.15) is called a *zeta-function* and usually appears in the grand-canonical formalism of thermodynamics (Ruelle, 1978).

Concerning phase transitions, it appears that this computational class is powerful enough to yield them. Consider again the random walker of Fig. 7.5 with $p_- > 1/2$. By definition, all length-n sequences of displacements ($\ominus$, $\oplus$) which end below the barrier are forbidden. This implies that all local entropies $\kappa(S)$ smaller than the critical entropy $\kappa_c = \ln \sqrt{p_- p_+}$ (corresponding to paths ending at the barrier) do not contribute to the entropy spectrum $g(\kappa)$. The number of all paths leading to a position above the barrier at time n, on the other hand, is indeed smaller than for an unconstrained walker, because of the exclusion of paths that have temporarily visited the forbidden region, but only by an irrelevant prefactor. The exponential scaling rate is unchanged. Hence, the height of the $g(\kappa)$ curve for $\kappa > \kappa_c$ is the same as for a Bernoulli shift. This heuristic argument shows that a sharp cut occurs in $g(\kappa)$ at $\kappa = \kappa_c$ (for detailed examples, see Radons, 1995).

Finally notice that this example presents analogies with the filtered baker map of Ex. 6.8. In that case, the contribution of the paths depend only on the position $x_n = \sum_{i=1}^{n} s_i$ at time n. If x_n lies below the barrier $n v_c$ (see Eq. (6.65)), the walker is not discarded but contributes as if it were at the barrier.

7.2.3   D0L languages

Limit words of D0L-substitutions exhibit all types of power spectra: pure-point (*pp*), absolutely continuous (*ac*), and singular continuous (*sc*). Different components may be simultaneously present. In order to understand this behaviour, let us consider the spectrum $S_N(\omega)$ as defined in Eq. (6.47) for a discrete-time signal $S = \{s_1, s_2, \ldots\}$ of length N. The perfectly self-similar structure of the substitution permits the writing of recursive relations for the Fourier amplitudes and other interesting quantities. We illustrate the procedure in the binary case $0 \to \psi(0)$, $1 \to \psi(1)$. The words $w_{k+1}^{(0)} = \psi^{k+1}(0)$ and $w_{k+1}^{(1)} = \psi^{k+1}(1)$ satisfy the equations

$$w_{k+1}^{(0)} = \Psi^{(0)}(w_k^{(0)}, w_k^{(1)})$$
$$w_{k+1}^{(1)} = \Psi^{(1)}(w_k^{(0)}, w_k^{(1)}),$$

$$(7.16)$$

where the "formal functions" $\Psi^{(s)}$ yield concatenations of the words $w_k^{(0)}$ and $w_k^{(1)}$ in the same sequence as the single 0s and 1s appear in $\psi(s)$ for $s = 0$ and 1, respectively. In the case of the Fibonacci substitution (Ex. 4.6), for example,

$$w_{k+1}^{(0)} = w_k^{(1)}$$
$$w_{k+1}^{(1)} = w_k^{(1)} w_k^{(0)},$$

so that $\Psi^{(0)}(v, w) = w$ and $\Psi^{(1)}(v, w) = wv$. Hence, the lengths $N_k^{(s)}$ of the words $w_k^{(s)}$ satisfy

$$N_{k+1}^{(0)} = N_k^{(1)}$$
$$N_{k+1}^{(1)} = N_k^{(1)} + N_k^{(0)} .$$

The numbers $N_k^{(s)}$ increase as powers of the eigenvalues of the *growth matrix* $\mathbf{M}_G$ whose generic entry $M_{ss'}$ represents the number of symbols s' in $\psi(s)$. The Fourier amplitude

$$X_k^{(0)}(\omega) = \sum_{n=1}^{N_k^{(0)}} s_n e^{-i\omega n}$$

of the word $w_k^{(0)}$ (s_n being the nth symbol in $w_k^{(0)}$) and the analogous one for $s = 1$ obey the recursion

$$X_{k+1}^{(0)} = X_k^{(1)}$$
$$X_{k+1}^{(1)} = X_k^{(1)} + X_k^{(0)} e^{-i\omega N_k^{(1)}} , \qquad (7.17)$$

where the phase factor accounts for the starting position of the second word $w_k^{(0)}$ inside $w_{k+1}^{(1)}$. In general, if a formal function $\Psi^{(s)}$ is a concatenation $w_a w_b w_b w_a w_b \ldots$ of two words w_a, w_b, the r.h.s. of the equation for the corresponding Fourier amplitude $X_k^{(s)}$ contains as many terms as words in $\Psi^{(s)}$, each with the appropriate phase factor. These recursive equations are very important since they provide efficient schemes for numerical algorithms. An order-k approximation $S^{(k)}(\omega)$ of the power spectrum, evaluated at equally spaced values $\omega = \omega_n$ between $2\pi/N_k$ and π, is obtained by iterating Eqs. (7.17) up to some order k' such that $N_{k'} \gg N_k$ and by summing all contributions within each frequency interval $(n/N_k, (n+1)/N_k]$, for $n \in [0, N_k/2 - 1]$. The initial conditions are $X_0^{(s)}(\omega) = A_s \exp(-i\omega)$, for $s = 0$ and 1, where A_s is the numerical value associated with s.

The singularity spectrum $f(\alpha)$ is estimated by normalizing $S(\omega)$ to unit area, after neglecting the zero-frequency component, and by letting

$$P_i(\varepsilon) = \int_{\omega_i}^{\omega_i + \varepsilon} S(\omega) d\omega .$$

The "free energy" $\tau(q)$ (6.26, 6.28) is then evaluated for given q values and Legendre-transformed as in Eq. (6.31).

An alternative method for the singularity analysis of substitutions is deduced by the recursive relation (7.17) which yields the components $X_k^{(s)}$ of a b-dimensional vector at time k by applying a product $\prod_{i=1}^{k} \mathbf{M}_i$ of matrices $\mathbf{M}_i$ to a suitable initial vector[6]. The exponential growth rate $\Lambda(\omega)$ of the Fourier

---

6. When $\omega = 0$, $\mathbf{M}_i$ is the growth matrix $\mathbf{M}_G$.

components $X_k^{(s)}(\omega)$ for increasing k is related to the spectral exponent $\gamma(\omega)$ (6.48) by

$$\gamma(\omega) = \frac{\Lambda(\omega)}{\Lambda(0)} ,$$

where $\Lambda(0)$ is the growth rate of N_k with k. Accordingly, Eq. (6.50) allows estimation of the local dimension α from the Lyapunov exponent $\Lambda(\omega)$ of the matrix product. The relationship with dynamical system theory is strengthened by observing that the dependence of the ith matrix on the hierarchical level i is uniquely contained in the phase exponent $\omega N_k^{(s)}$ (modulo 2π) and that $x_k^{(s)} \equiv \omega N_k^{(s)}/2\pi$ can be interpreted as the sth coordinate of a point $\mathbf{x}_k$ evolving with k under a *phase map* $\mathbf{F}$ of the unit b-cube. Evidently, the real variable $x_k^{(s)}$ and the integer $N_k^{(s)}$ satisfy the same equations (apart from the modulo operation).

The piecewise linear map $\mathbf{F}$ is chaotic: in fact, the exponential growth rate $\Lambda(0) > 0$ of the word length coincides with the largest Lyapunov exponent λ_1 of $\mathbf{F}$. In the case of the Fibonacci substitution, the dynamics is conservative, since the second Lyapunov exponent λ_2 equals $-\lambda_1$. In general, the phase map $\mathbf{F}$ may have an attractor with dimension lower than b or, on the contrary, be exact (see Section 5.5).

Admissible values of α can be computed from the Lyapunov exponents of the periodic orbits of $\mathbf{F}$: a given $\alpha(\omega)$ is in fact observed when the initial condition $x_0^{(s)} = \omega/2\pi$, $\forall\ s$, is eventually mapped to a periodic orbit with $\Lambda(\omega) = (1 - \alpha(\omega)/2)\Lambda(0)$. In particular, the smallest and the largest α usually correspond to short-period orbits. Notice that although $x_k^{(s)}$ and $X_k^{(s)}$ separately satisfy two linear equations, an interesting Lyapunov spectrum arises because of the nonlinearity of the mutual coupling.

Examples:

[7.5]   The approximation $S^{(16)}(\omega)$ to the power spectrum of the Fibonacci sequence $\psi_F^\infty(0)$ is displayed in Fig. 7.6(a) as a function of $f = \omega/2\pi$. It has been estimated by summing the contributions obtained from 31 iterations of map (7.17). The spectrum consists of δ-peaks with amplitude (Godrèche and Luck, 1990)

$$S(\omega(p,q)) \propto \frac{\sin^2(q\pi\omega'_{gm})}{(q\pi)^2}$$

located at frequencies $\omega(p,q) = 2\pi(p + q\omega'_{gm})$ where $\omega'_{gm} = 1/\omega_{gm} = (1 + \sqrt{5})/2$ and $p, q \in \mathbb{Z}$.

The spectral measure, although concentrated on a countable set of points, presents scaling exponents α different from 0. Consider the peaks centred at the frequencies $\omega(p_k, q_k) = 2\pi|p_k - q_k\omega'_{gm}|$ mod π, where p_k/q_k is the kth diophantine approximant to ω'_{gm}, so that $\omega(p_k, q_k) \to 0$. Choosing an interval

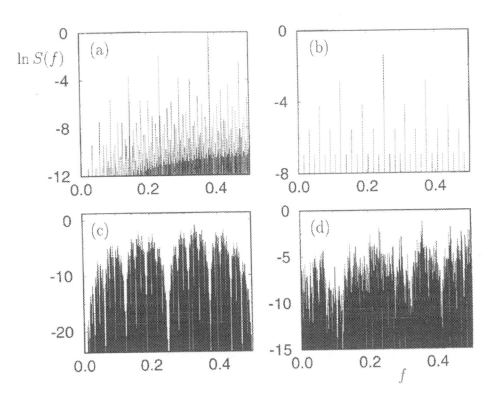

Figure 7.6 Power spectra $S(f)$ for the limit words of the Fibonacci (a), Period-doubling (b), Morse–Thue (c), and ternary (d) substitutions illustrated in Ex. [7.5–7.8]. Natural logarithms of $S(f)$ have been taken after rescaling the spectra so that $\max_f S(f) = 1$.

of size $\varepsilon_k \approx \omega(p_k, q_k)/2$ around $\omega(p_k, q_k)$, for $q_k \gg 1$, the mass $m(\varepsilon_k)$ inside it is proportional to the integral of $S(\omega(p_k, q_k))$ in dq between $q_k - n_1$ and $q_k + n_2$, where $n_i \ll q_k$ are finite integers. Since the result scales as q_k^{-2} and

$$|p_k - q_k \omega'_{gm}| \sim q_k^{-1} \tag{7.18}$$

(see Baker, 1990), $m(\varepsilon_k) \sim \varepsilon_k^2$ and $\alpha = 2$ for large q_k (i.e., close to $\omega = 0$). The same exponent is found on a set of Lebesgue measure 1. In fact, ω'_{gm} is the irrational number with the slowest convergence of the diophantine approximants. On the other hand, almost all numbers are *trascendental* (i.e., nonalgebraic) and any of them, say r, admits a rational approximation satisfying $|r - p_j/q_j| < q_j^{-n_j}$ with $n_j \to \infty$ for $j \to \infty$: that is, a very fast convergence. Because of this property, they are nearly as "hard" to approach by the peaks of $S(\omega)$ (dominated by ω'_{gm}) as are the rationals. Hence, any sequence of peak-frequencies $\omega(p_j, q_j)$ exhibits the same behaviour around any of these numbers as it does close to $\omega = 0$. On the contrary, the exponent $\alpha = 0$ is measured around the peaks which carry most of the spectral mass (at any finite resolution) and is, therefore, typical with respect to $S(\omega)$. The corresponding spectrum of

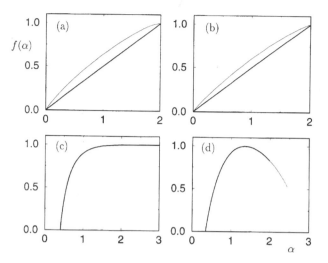

Figure 7.7 Singularity spectra $f(\alpha)$ for the Fibonacci sequence at order $k = 17$ (a), and for the same sequences (Period-doubling (b), Morse–Thue (c), and ternary (d)) as in Fig. 7.6. Solid lines: theoretical result; thin lines: numerical evaluation.

singularities $f(\alpha)$ thus consists of a peak with height 1 at $\alpha = 2$ and of the point $(0,0)$. The two are connected by a segment, so that the generalized dimension D_q is

$$D_q = \begin{cases} \dfrac{1 - 2q}{1 - q}\,, & \text{for } q \leq 1/2, \\ 0\,, & \text{for } q > 1/2. \end{cases} \tag{7.19}$$

The information dimension $D_1 = 0$ confirms the relevance of the atomic part of the spectrum: if a point ω is chosen at random from the *spectral* measure, its index α is almost certainly 0. If, instead, the choice is made with respect to the *Lebesgue* measure, $\alpha = 2$. A direct estimate of $f(\alpha)$ for $k = 17$, reported in Fig. 7.7(a), agrees with the theoretical expectation up to a visible lack of convergence, especially around $\alpha = 2$, which is caused by the phase transition.

All periodic orbits of the phase map **F** except the fixed point $x_0 = (0,0)$ yield $\Lambda = 0$ for the matrix product, so that $\alpha = 2$ as already discussed. The exponent $\alpha = 0$ (pure-point component) is instead found for all frequencies ω corresponding to trajectories of **F** that approach x_0. This analysis exhausts the relevant features of the spectrum in the Fibonacci case. □

[7.6] The recursive equations for the Fourier amplitudes of the period-doubling substitution ψ_{pd} (Ex. 4.7) read

$$X_{k+1}^{(0)} = \left(1 + e^{-i\omega N_k^{(1)}}\right) X_k^{(1)}$$

$$X_{k+1}^{(1)} = X_k^{(0)} e^{-i\omega N_k^{(1)}} + X_k^{(1)}\,. \tag{7.20}$$

By the properties of the power spectrum, already described in Ex. 5.14 and displayed in Fig. 7.6(b), similar arguments to those presented in the previous example show that the $f(\alpha)$ curve (Fig. 7.7(b)) has an atomic component at

$\alpha = 0$, an atypical one at $\alpha = 2$, and a range of "unsubstantial" α-values. There is no definite frequency ω around which an exponent $0 < \alpha < 2$ is actually observable in the infinite resolution limit, in analogy with the scaling properties of the invariant measure of the logistic map (Ex. 6.3).   □

[7.7]   The power spectrum for the Morse–Thue substitution ψ_{MT} (Ex. 4.7), shown in Fig. 7.6(c), is singular continuous, at variance with the two previous ones. Moreover, there is no atomic component and the support of $f(\alpha)$ has infinite extension. The Jacobian of the phase map $\mathbf{F}$ is 0 and the attracting set is the line $x^{(0)} = x^{(1)}$, along which the dynamics is given by the Bernoulli map. The smallest value of α, $\alpha_{min} = \ln(4/3)/\ln 2$, is associated with the period-2 orbit $1/3 \to 2/3 \to 1/3$. All αs between α_{min} and 2 are effectively observable in the power spectrum and correspond to periodic orbits of $\mathbf{F}$. The unboundedness of the support of $f(\alpha)$ is related to the trajectories that asymptotically converge to the fixed point $(0,0)$. These are expected to exhibit the fastest growth rate $\Lambda(\omega) = \Lambda(0)$ for the amplitudes $X_k^{(s)}$, as in the case of the Fibonacci substitution. Here, however, $X_k^{(s)} \sim ae^{k\Lambda} + 2^{-k^2}$ with the prefactor a equal to 0 (Aubry et al., 1988), so that the power spectrum vanishes faster than exponentially in k at the corresponding frequencies and $\alpha = \infty$ (Godrèche & Luck, 1990).   □

[7.8]   The ternary substitution (Godrèche & Luck, 1990)

$$\psi_{GL}(\{0,1,2\}) = \{2,0,101\}$$

also gives rise to a singular continuous power spectrum, which is displayed in Fig. 7.6(d). The smallest local dimension α_{min} is close to 0.3. The convergence of the right part of the $f(\alpha)$ curve is very slow: comparison of different-order approximations suggests that the spectum does not extend beyond $\alpha_{max} = 2$. The phase map $\mathbf{F}$ has three positive Lyapunov exponents: that is, there is no stable manifold (the map is exact). Accordingly, the scaling exponent α_p of a given periodic orbit p can be observed only if the orbit is reached exactly in a finite number of steps. Even though the number of the preimages grows exponentially with their order, this event has probability zero, being equivalent to the intersection of a large but finite set of points with the line $x^{(0)} = x^{(1)} = x^{(2)}$ of the initial conditions. The two fixed points $(0,0,0)$ and $(1/2,1/2,1/2)$ are conjectured to be the only reachable periodic orbits. The former is irrelevant since it corresponds to the isolated trivial zero-frequency component of the power spectrum. The latter yields $\alpha \approx 0.3481$, close to the observed α_{min}. No periodic orbit is associated with $\alpha > 2$, thus reinforcing the reliability of the numerical estimates.   □

A sufficient condition for the existence of an atomic spectral component was found by Bombieri and Taylor (1986, 1987) in a study of quasicrystals (see

Section 7.3). It is required that the largest eigenvalue μ_1 of the growth matrix $\mathbf{M}_G$ be a *Pisot–Vijayaraghavan* (PV) number, i.e., the only eigenvalue with modulus greater than 1. A heuristic interpretation of this condition is that the stable manifold W^s of the phase map $\mathbf{F}$ has codimension 1. Its intersections with the line of initial conditions are therefore typical. In particular, this applies to the fixed point at the origin, which is the only one where $\alpha = 0$ exactly.

The systems in the former three examples above belong to this class. While the former two indeed possess a pure-point spectral component ($\alpha_{min} = 0$), the third exhibits a strictly singular continuous spectrum. As already commented upon in the example, the absence of the atomic part is determined by the fortuitous cancellation of a prefactor. The fourth system, instead, does not enjoy the PV property and lacks the atomic component.

Substitutions, although strictly deterministic, can generate absolutely continuous spectra, as shown by the transformation

$$\psi_{RS}(\{A,B,C,D\}) = \{AC, DC, AB, DB\}$$

named after *Rudin* and *Shapiro* (RS). The power spectrum is constant and $\alpha = 1$ everywhere[7]. The Rudin–Shapiro sequence over the two symbols r and s, obtained by the assignement $(A, C \to r;\ B, D \to s)$ in $\psi_{RS}^{(\infty)}$, has the same property (Queffélec, 1987). The power-spectral analysis does not distinguish the deterministic RS-sequence from a purely random one. Notice that the topological entropy is zero. Moreover, since all subsequences of length n appear with the same frequency, $K_q = 0$ for all q. No general results are available about the abundance of absolutely continuous power spectra for substitutions.

For all substitutions obeying the conditions listed in Section 4.3 (independence of the language from the start symbol) the number $N(n)$ of words of length n is linearly upper bounded. We briefly illustrate the theorem (Queffelec, 1987). For any $n \geq 1$, there is an integer k such that

$$\inf_{s \in A} |\psi^{k-1}(s)| \leq n \leq \inf_{s \in A} |\psi^k(s)|. \tag{7.21}$$

Every word of length n is then contained in the kth image of a word w_0 consisting of at most two symbols. In fact, if w_0 contained three symbols, the middle one would yield, in k iterates, a sequence with more than n symbols, by Eq. (7.21). Any pair of symbols generates a word w of maximum length $2 \sup_{s \in A} |\psi^k(s)| \leq 2Cn$, since $|\psi^k(s)| \leq C\theta^k$ for all s, where C is a constant and θ the largest eigenvalue of the growth matrix $\mathbf{M}_G$. Hence, w contains at most $(2C - 1)n$ words of length n, and w itself can be generated from at most b^2 initial pairs. As a result, the number $N(n)$ of distinct n-blocks cannot increase faster than n.

---

7. Depending on the numerical values assigned to the symbols, a trivial atomic component may arise at $\omega = \pi$: in fact, A and D appear at odd sites only, while B and C occupy the even ones, all with the same probability.

Notice that directed graphs for the reproduction of D0L languages diverge. In fact, any finite graph with zero entropy yields a periodic signal since it must have no bifurcating nodes: this is not the case for D0L systems. On the other hand, no D0L can simulate an aperiodic regular language, since the latter has positive entropy. Hence, it is not possible to say which of the two classes D0L and RL is more complex with a criterion based on the entropy: they are just disjoint, except for trivial cases.

Dynamical systems arising from substitutions are not strongly mixing; some, however, are weakly mixing, as the one generated by the Dekking–Keane transformation ψ_{DK} (Ex. 5.21). A necessary condition for this property to hold is that the substituted words have different lengths (Martin, 1973).

7.2.4 Context-sensitive and recursively enumerable languages

We have seen that shifts arising from CFGs can display phase transitions in the associated thermodynamic formalism, provided that weights are included in the grammar, while it seems that no striking topological properties can be expected in that class of languages. We show in this section that CSLs yield new and interesting phenomena already at the topological level, by discussing self-avoiding random walks (SAWs): stochastic processes in which a "particle" wanders in discrete steps on a d-dimensional lattice in such a way that returns to an already visited site are forbidden (Madras and Slade, 1993). Among the fields in which SAWs occur, we mention polymer physics (de Gennes, 1979).

On a two-dimensional square lattice, the four possible moves are encoded as u (up), d (down), l (left), and r (right). The language we are interested in consists of all sequences of these four symbols and refers, therefore, to the "derivative" of the walk. The prohibition of self-intersections causes the exclusion of an infinite list of words from the language (lr, ud, $uldr$, etc.). Since a detailed description of the grammar would be exceedingly tedious, we sketch the main ideas only. During the generation of admissible symbols, the current coordinates of the walker are saved on the tape. After a new symbol is emitted, the head moves back along the tape to check whether any of the previous positions coincides with the last visited site. Whenever this is the case, the word is rejected. Since none of the above logical operations involves a reduction of variables or of symbols (except for a finite number of irrelevant final productions where the "working" variables must be deleted), the corresponding grammar is certainly context-sensitive[8].

The number $N(n)$ of different paths of length n is conjectured to scale as

8. While this approach is appropriate for addressing correctly the membership of SAWs to CSGs, we do not recommend the reader to follow the same lines for an actual numerical simulation: much better performance can be achieved by suitably exploiting memory, i.e., by simulating a SAW with an unrestricted grammar!

$N(n) \sim An^{\gamma-1}\mu^n$, where the constant μ depends on the dimensionality of the problem and on the lattice geometry. The walker performs a diffusive motion in which the average $D(n) = \sqrt{\langle\delta^2(n)\rangle}$ of the displacement $\delta(n)$ from the origin attained in a run of length n grows as $n^{2\nu}$. The exponent $\gamma \geq 1$ is conjectured to be 43/32 in two dimensions, irrespective of the lattice type and possible restrictions on the allowed paths (Madras and Slade, 1993). A similar "universal" character is shared by the exponent ν which is estimated to be 3/4 (Nienhuis, 1982). For $d \geq 5$, the exact values are $\gamma = 1$, $\nu = 1/2$.

The topological entropy is given by $K_0 \approx \ln\mu$, where the currently best estimate for a two-dimensional lattice yields $\mu \approx 2.638$ (Masand et al., 1992). The power-law correction to $N(n)$ indicates a slow convergence of the finite-n estimates for K_0. Recall that an analogous scaling behaviour has been numerically observed in the elementary cellular automaton 22. The implications of this phenomenon for the enumeration of the forbidden words in the language will be discussed in Chapter 9.

It is proved that no general algorithm exists for the computation of the topological entropy of CSLs (Kaminger, 1970). Furthermore, there is no procedure to decide whether the language of any CSG is finite or not (Landweber, 1964).

The additional freedom in the productions of unrestricted grammars is expected to allow for any sort of behaviour. It is therefore unclear which new features are introduced with respect to CSGs from the point of view of a physical characterization. Moreover, the fraction of all unrestricted grammars (with given bounds on the number of their terminals, variables, productions, and symbols) that yield "meaningful" languages is obviously uncomputable.

7.3   Computational characterization of physical systems and mathematical models

Even apparently simple physical systems may turn out to correspond to high computational classes. A few representative cases are pointed out in this section.

7.3.1   Dynamics at the borderline with chaos

Dynamical systems at or below the onset of chaos may produce a rich variety of D0L languages, a few examples of which have already been discussed. Therefore, such systems also provide a sequential generation mechanism for the limit words.

The simplest class of substitutional dynamical systems consists of quasiperiodic motions modelled by the bare circle map (3.13) with a parameter α having periodic continuous-fraction expansion (Eq. (3.16)). This characterizes the *quadratic irrationals*, i.e., the solutions of second-order algebraic equations with integer coefficients. For example, the symbolic dynamics generated from $\alpha = [0,\overline{1,2}] = \sqrt{3} - 1$ is obtainable by iterating the substitution $\psi_{12} : \{0,1\} \rightarrow \{011,0111\}$. In all such cases, the determinant of the growth matrix is ± 1 and the largest eigenvalue is a PV number, as expected, since the dynamics is quasiperiodic by construction. Given the parameter α, the corresponding substitution can be found with standard procedures (Luck *et al.*, 1993). A slight generalization of quasiperiodic systems is represented by Sturmian systems, introduced in Ex. 5.19, which are also associated with substitutions.

Strange nonchaotic attractors represent a further class of systems that can be described by D0L languages. In particular, the dynamics of model (6.52), recorded by the symbols $s_t = (1 + x_t/|x_t|)/2$, admits a description based on three substitutions when ω is the inverse golden mean (Feudel *et al.*, 1996). To what extent generic strange nonchaotic attractors are reducible to substitutions is still unclear.

Smooth bimodal maps of the interval having two increasing branches (in $[a, x_-)$ and $(x_+, b]$) and a decreasing one (in (x_-, x_+)) present two separate period-doubling cascades around the critical points x_- and x_+ and one which corresponds to the bifurcations of a period-two orbit having its points close to x_- and x_+. The former two follow the usual pattern (with metric properties depending on the polynomial order of the extrema), topologically described by the substitution ψ_{pd} (Ex. 4.7). The latter cascade requires a three-symbol description: 0 for $x \in [a, x_-)$, 1 for $x \in (x_-, x_+)$, and 2 for $x \in (x_+, b]$ (the critical points having special symbols c_- and c_+). The resulting symbolic dynamics, extensively studied by Mackay and Tresser (1988), has been recently shown to have the computational complexity of context-sensitive languages (Lakdawala, 1995).

Finally, we mention the modified Lorenz system

$$\dot{x} = \sigma(y - x) + \sigma dy(z - r)$$
$$\dot{y} = -y + rx - xz$$
$$\dot{z} = -bz + xy \, ,$$

introduced by Lyubimov and Zaks (1983) to model the thermal convection in a fluid layer subjected to transversal high-frequency gravity oscillations, which yields the Morse–Thue sequence at $r \approx 15.8237$ and $d \approx 0.052634$ (with $\sigma = 10$ and $b = 8/3$).

One should notice that all of the above dynamical systems exhibit a renormalizable behaviour at special parameter values (corresponding either to particular frequencies, or to the onset of chaos).

7.3.2 Quasicrystals

The observation of a crystal atomic structure (the metallic alloy Al_4Mn) with a five-fold symmetry (Shechtman *et al.*, 1984) raised considerable new interest into the subject of quasicrystals (Steinhardt & Ostlund, 1987) which had previously developed as a branch of mathematical modelling. Further experimental investigations have revealed that many alloys exhibit icosahedral phases: a kind of symmetry that is forbidden in periodic patterns. Crystals, in fact, possess long-range translational and orientational order, the latter being confined to a subgroup of rotations characterized by five two-dimensional and fourteen three-dimensional Bravais lattices (Ashcroft & Mermin, 1976). Quasicrystals share the same two types of order but in connection with *forbidden* crystallographic symmetry[9]. This, in turn, generates incommensurate length scales: linear combinations of the orientation vectors of a pentagonal cell, e.g., yield collinear vectors with length ratios which involve the golden mean ω_{gm}. Quasiperiodicity alone, however, is not sufficient to characterize a quasicrystal. In fact, there also exist quasiperiodic structures, called incommensurate crystals, with crystallographically *allowed* orientational symmetry. They may be produced by superposing periodic lattices with incommensurate lattice spacing or with rotation angles incommensurate with respect to the symmetry of the unit cell. While their length (angle) ratios can vary continuously (with the intersection points coming arbitrarily close to one another), this is not the case in quasicrystals because of the geometric constraints imposed by the orientational symmetry. Quasicrystals are a fundamentally new class of ordered atomic structures.

A special model for quasicrystalline geometries is the *quasilattice*, an array of polygons (polyhedra), chosen from a finite set of shapes (at least two) and having perfectly matching edges (faces), which fill the available space *necessarily* in an aperiodic way.

The most famous example in two dimensions is the Penrose tiling (1974), illustrated in Fig. 7.8, which is composed of wide and narrow rhombic unit cells. The pattern can be generated by means of an inflationary self-similar transformation, sketched in the lower part of the figure. From each basic cell, one obtains two smaller cells of each type, scaled by a factor ω_{gm}. The procedure is then iterated inside each new piece (Levine & Steinhardt, 1986). Each edge in the tiling is normal to a symmetry axis of a pentagon, so that the pattern has pentagonal point symmetry. Its quasiperiodicity has been demonstrated by Amman (Grünbaum & Shephard, 1987). The existence of a self-similar construction method immediately suggests that quasilattices are computationally akin to D0L substitutions. While this is indeed true, as will be shown in the following, it must be stressed that not every quasicrystal can be

9. Glasses, the only other pure solid form known before the discovery of quasicrystals, do not exhibit any long-range correlations (neither orientational, nor translational).

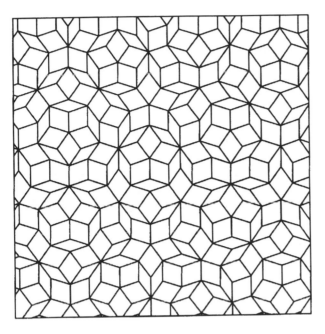

Figure 7.8 A sample of a Penrose tiling (top) generated with the two types of rhombic cells (wide and narrow) that are displayed at the bottom with thick edges. The eight cells obtained in one step of the inflationary self-similar transformation are also shown (thin lines, bottom).

generated by placing atoms on a quasilattice. Hence, D0L languages describe just a subset of all quasicrystals.

The Penrose tiles can be decorated with line segments in such a way that the latter join to form sets of parallel lines superimposed on the quasilattice. Each set is parallel to a symmetry axis of a pentagon. The position x_n of the nth line in one of these sets satisfies the relation $x_{n+1} = x_n + \omega'_{gm} + s_n$, where $s_n \in \{0, 1\}$ is the nth symbol in the Fibonacci sequence $\psi_F^\infty(0)$. This is a special case of the more general equation

$$x_{n+1} = x_n + \ell_0 + (\ell_1 - \ell_0)s_n , \qquad (7.22)$$

where ℓ_0 and ℓ_1 are mutually incommensurate real numbers and s_n is the nth symbol of a Sturmian sequence with parameters α and β (see Ex. 5.19). Indeed, Eq. (7.22) defines a one-dimensional lattice in which the interparticle distance is either ℓ_0 or ℓ_1, depending on the information provided by an external source. Taking $x_0 = 0$ yields $x_n = n\ell_0 + (\ell_1 - \ell_0)\sum_{i=1}^n s_i$, so that the mean interparticle

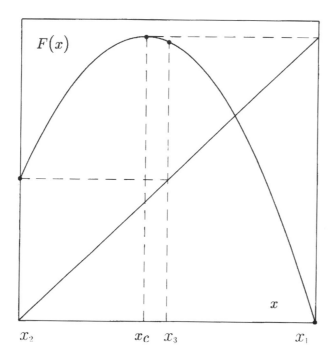

$F(x)$

x

x_2 x_c x_3 x_1

Figure 7.9 A prototypical unimodal map. The first three iterates of the critical point x_c are indicated.

distance is $d = \lim_{n \to \infty} x_n/n = \ell_0 + (\ell_1 - \ell_0)(1 - \beta)$, since $\langle s \rangle = 1 - \beta$. The fluctuation of the position x_n with respect to the nth mean-lattice point nd is therefore proportional to $\delta_n = \sum_{i=1}^{n} s_i - n\langle s \rangle$. In the limit $n \to \infty$, this quantity remains bounded if and only if $1 - \beta = j\alpha \mod 1$ for some integer j (Kesten, 1966). In such a case, the corresponding structure is an ordinary incommensurate lattice; otherwise, one has a quasilattice. Hence, the structure of some one-dimensional quasicrystals can be traced back to D0L substitutions, since these are known to describe the symbolic dynamics of Sturmian maps for certain choices of the parameters α and β. When the first two eigenvalues of the growth matrix satisfy $\lambda_1 > \lambda_2 > 1$ (i.e., in the non-PV case), the divergence of δ_n is of the type $\delta_n \sim (\ln n)^\mu n^\nu g(n)$, where $\mu \in \mathbb{R}$, $\nu = \lambda_2/\lambda_1$, and the continuous, nowhere differentiable function $g(n)$ satisfies $g(\lambda_1 x) = g(x)$ (Dumont, 1990).

7.3.3   Chaotic maps

Generic chaotic systems are not reducible to Markov chains. This is especially clear for unimodal maps F of the type of the logistic map (3.11). They admit a finite Markov partition if and only if the kth image $x_k = F^k(x_c)$ of the critical point x_c belongs to a periodic orbit for some finite k, as illustrated by the following argument. The domain of the transformation F is delimited by the first two iterates x_1 and x_2 of x_c which are necessarily border points of the partition, together with x_c itself (see Fig. 7.9). The intervals $I_0 = [x_2, x_c)$ and

$I_1 = [x_c, x_1]$ define the generating partition. Unless it coincides with x_c (or x_2), x_3 is not a border point: therefore, we store it in a list for later reference. The procedure is iterated until $x_k = x_j$, for some $k \geq j$: when this happens, the points in the list constitute the borders of the Markov partition. In general, however, the orbit of x_c is aperiodic and this condition is not satisfied.

In order to see how an infinity of forbidden words arises when a Markov partition does not exist, consider first the interval I_1. Since its image is the whole phase space $I_0 \cup I_1$, no prohibitions occur at the first iterate. They may descend only from points in I_0. If x_c is contained in the image $I = (x_3, x_1)$ of I_0, $I = (x_3, x_c) \cup I_1$ and still no prohibition is found. If, instead, $I \subset I_1$, the sequence 00 (i.e., the transition $I_0 \rightarrow I_0$) is forbidden. In both cases, future prohibitions are detected by following the image J of a single interval (at this stage, either (x_3, x_c) or I itself, depending on whether $x_c \in I$ or not). At a generic step of the procedure, one checks whether the current interval J intersects both I_0 and I_1 or not: in the former case, one retains the subinterval delimited by x_c and its currently known highest-order image (the "future" of the other subinterval being already accounted for); in the latter, J itself is kept. The next J is finally the image of the chosen interval. A forbidden word occurs each time no intersection is found and consists of the past symbolic sequence of J concatenated with the symbol of the missed element of the generating partition. Notice that if some x_i coincides with a previous x_j, there exists a Markov partition, but the procedure may continue identifying new forbidden sequences. This is the case of a strictly sofic system, whose finite graph yields an infinity of forbidden sequences. Analogous arguments hold for maps of the interval having a finite number of monotonic branches. Extensions of the above constructions show that Markov partitions are even more exceptional in higher dimensional maps (D'Alessandro & Politi, 1990). Hence, the languages produced by chaotic systems are in general nonregular. Finally, Friedman (1991) has proved that the sets Δ_r and Δ_u of parameter values at which the language is, respectively, regular and nonREL (i.e., uncomputable) have positive measure for certain families of unimodal and circle maps in one dimension[10]. The remaining set, for which computable nonregular dynamics is found, has zero measure.

7.3.4   Cellular automata

In the global study of a cellular automaton (Section 3.4), one follows the time evolution of all doubly-infinite symbol sequences. The topological properties of the limit set $\Omega^{(\infty)}$ (3.22) can be elucidated with a complete listing of the forbidden words arising from the repeated application of the rule.

10. The former result is well-known (Collet & Eckmann, 1980) and concerns, mostly, periodic behaviour; the latter is a restatement of the ubiquity (measure 1) of irrational numbers.

In the previous subsection, we have illustrated a finite procedure which allows, at least in principle, identification of all forbidden sequences in unimodal maps. Although a similar approach is conceivable for CAs as well, fundamental limitations reduce here its efficacy. The basic scheme is the following: given a CA rule, all possible preimages of a test sequence S of length n are computed up to order k (their length being $n + 2kr$ in elementary automata with range r). Then, S is forbidden if an empty set is found for a finite k. This procedure, however, is not guaranteed to halt: there is no bound, in principle, to the number k of backward iterates that must be considered in order to assess the legality of S. The detection of forbidden words, on the other hand, is not too hard since knowledge of the previously identified ones may be used to speed up the computation. Usually, a few iterates are sufficient to identify the first prohibitions.

A more fundamental difficulty is represented by sequences that are "asymptotically" forbidden. Let $P(S, n; k)$ denote the occurrence probability of a sequence S of length n in the k-th image of a uniformly random initial configuration. Then,

$$P(S, n; k) = N_p(S, n; k) b^{-(n + 2rk)} , \qquad (7.23)$$

where $N_p(S, n; k)$ is the number of k-th order preimages of S and the factor $b^{-(n+2rk)}$ represents the Lebesgue measure of the preimage (a sequence of length $n + 2rk$). A sequence S of length n must be considered asymptotically forbidden if $P(S, n; k) \to 0$ for $k \to \infty$.

The existence of such sequences is easily verified for the elementary rule 22. The sequence $S' = 10101$, e.g., has only one preimage of order k, consisting of $5 + 2k$ alternating 0s and 1s. Accordingly, the probability of observing 10101 after k iterations of the rule is $P(10101, 5; k) = 2^{-5-2k}$, which vanishes exponentially for large k. Further asymptotically forbidden words are 010110001 and 100110011: their probabilities $P(S, n; k)$ are plotted in Fig. 7.10 versus k.

The difficulty of identifying the forbidden words is better understood by noticing that the generation of all the preimages of the test word S can be described by a context-sensitive grammar. The following sequence of substitutions represents a simplified CSG simulating the procedure for rule 22:

$$B0 \to B00'0' , \; B01'1' , \; B10'1' , \; B11'0' , \; B11'1'$$
$$B1 \to B10'0' , \; B01'0' , \; B00'1'$$
$$0'0'0 \to 00'0'$$
$$0'0'1 \to 00'1'$$
$$0'1'0 \to 01'1'$$
$$0'1'1 \to 01'0'$$
$$1'0'0 \to 10'1'$$

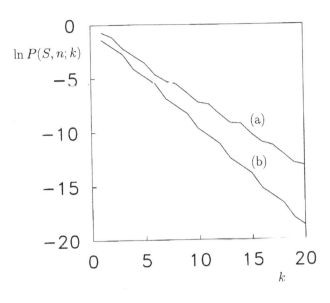

Figure 7.10 Evolution of the probabilities $P(S, n; k)$ of the n-blocks 010110001 (a) and 100110011 (b), where $n = 9$ and k is the time index, under repeated application of the elementary CA rule 22 with a uniform random initial condition. Natural logarithms have been used.

$$1'0'1 \rightarrow 10'0'$$

$$1'1'0 \rightarrow 11'1' \; ; \; 11'0'$$

$$0'0'E \rightarrow 00E$$

$$0'1'E \rightarrow 01E$$

$$1'0'E \rightarrow 10E$$

$$1'1'E \rightarrow 11E \; .$$

The grammar is supposed to act on the input string BSE, where B and E are variables marking the borders of the sentential forms and S is the tested sequence consisting of a concatenation of the variables 0 and 1. Productions involving B "guess" the possible prefix of the preimage S' from the first symbol in S. The rest of S' is constructed via productions involving neither B nor E and completed by one of the last four productions. Primed 0s and 1s are working variables which eventually disappear[11]. Since there is no general algorithm to compute the entropy of a CSL whatsoever (Kaminger, 1970), there cannot be one that is able to determine the forbidden words of a generic CA.

   If a finite number of iterations is performed on an initial condition corresponding to a subshift of finite type (regular language), the spatial configuration of an elementary cellular automaton remains in the same class (Wolfram, 1984; Levine, 1992). In the special case of a single iteration, the directed graph can be constructed as follows. Let us start with a graph having 8 nodes labelled by the possible binary triplets and arcs representing the transitions allowed by the left shift operation (as, e.g., 010 → 100; 101). Each arc is then labelled by the

---

11. A formal definition of the grammar should include productions that collapse 0s and 1s onto true terminals.

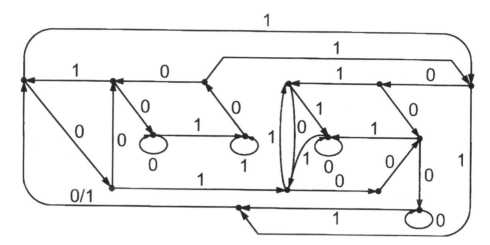

Figure 7.11 Directed graph for the language generated by the elementary CA rule 22 in one step. The symbol 0/1 attached to the bottom left arrow means "either 0 or 1".

image symbol of the arrival triplet under the CA rule, so that the whole image configuration is generated while scouring all paths. If the two arrows departing from a node carry the same symbol, both a prohibition and nondeterministic behaviour occur. The final directed graph is obtained by removing the nondeterministic parts (see Section 7.1) and minimizing the resulting automaton. The graph for a one-step iteration of CA 22 is reproduced in Fig. 7.11.

The size of the graphs grows rapidly with the number of iterations. One expects the nature of the asymptotic limit language to be qualitatively different. One-dimensional automata over $b = 6$ symbols and ranges $r = 2$ and 4 are able to reproduce the simplest context-free and context-sensitive languages (Hurd, 1990a). The complement of the language associated with the limit set is recursively enumerable (Hurd, 1990b). The topological entropy K_0 of a CA is always finite but uncomputable: there is no algorithm that can compute it for any CA (Hurd et al., 1992).

Examples have even been found of CAs having the same power as a universal Turing machine. In one dimension, it has been possible to reduce the number of necessary symbols to 7, with a range $r = 1$, and to 4, with $r = 2$ (Lindgren & Nordahl, 1990). On a two-dimensional square lattice, two symbols suffice. This is the case of the celebrated *Game of Life* (Berlekamp et al., 1982) which uses a nearest-neighbour majority rule. The equivalence between a given CA and a UTM is shown by encoding the initial data of a generic program by means of either stationary or periodic structures admitted by the CA and by identifying the machine operations with the modifications occurring to the pattern when these objects collide with other structures which propagate in space with different speed. Rules with a small number of symbols and range may admit very large,

Table **7.1**. *Substitution rules for Moore's generalized shift. The letters from A to H label the 8 phase-space partition elements for the corresponding two-dimensional map.*

A	$0\hat{0}0 \rightarrow \hat{0}11$
B	$0\hat{0}1 \rightarrow 10\hat{1}$
C	$0\hat{1}0 \rightarrow 11\hat{1}$
D	$0\hat{1}1 \rightarrow \hat{0}00$
E	$1\hat{0}0 \rightarrow 00\hat{1}$
F	$1\hat{0}1 \rightarrow \hat{0}10$
G	$1\hat{1}0 \rightarrow 01\hat{1}$
H	$1\hat{1}1 \rightarrow \hat{0}01$

composite propagating forms which are hard to discover. Moreover, it may be rather difficult to associate their interactions with the usual computational operations of a UTM. A Turing machine with b symbols and n states can be simulated by a one-dimensional automaton with range 1 and $b+n+2$ states per cell (Lindgren & Nordahl, 1990).

7.3.5 Relationship between Turing machines and dynamical systems

A new class of dynamical systems, called *generalized shifts* (GS), has been recently introduced by Moore (1990, 1991) with the purpose of relating the motion of low-dimensional dynamical systems with the action of a Turing machine. Given an infinite sequence $\mathscr{S} = \ldots s_{i-1}\hat{s}_i s_{i+1} \ldots$, with a distinct position (marked by the hat), the updating rule replaces the word w of length $2r+1$ centred at the reference symbol and displaces the pointer by one position. Both the substituted word and the direction of the shift are entirely determined by w. The $2r+1$ sites involved in the substitution are called the *domain of dependence* (DOD). This mechanism is best illustrated by the model introduced by Moore (1990) and described in the table, which refers to a rule with range $r = 1$. This mechanism is reminiscent of a cellular automaton, the difference being the restriction of the GS-substitution to a single movable finite domain. The analogy is strengthened by the resemblance of the motion of the pointer with that of defects (gliders) in CAs.

Any GS can also be represented as a piecewise linear map. Indeed, the two semi-infinite sequences at the sides of the pointer can be interpreted as the binary expansions of a pair (x, y) of real numbers in the unit interval, whereby x is determined by the left one and y by the right one which, by convention, includes the symbol under the pointer. Hence, a right shift of the pointer yields a contraction along the x-direction and an expansion along the y-direction,

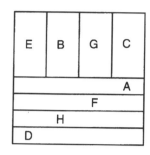

Figure 7.12 Illustration of the action of the GS-map on the unit square. An eight-element partition (left) and its image (right) are indicated. The letters label the domains corresponding to the eight DODs, as coded in the table.

both by a factor 2. The opposite occurs for the left shift. Furthermore, the substitutions occuring in the DOD correspond to translations of rectangles in the unit square. The action of the map corresponding to the above GS is sketched in Fig. 7.12. The dynamics is evidently area-preserving. It is also nonhyperbolic, since there is no clear distinction of expanding and contracting directions in the whole phase space.

The behaviour of the generalized shift can be profitably studied by following the movement of the pointer. As shown by the table, the pointer finds itself on a 0 (1) after a left (right) move. The shift dynamics produced by the pointer movement (left=0, right=1) is a full representation of the GS since it corresponds to the symbolic dynamics of the GS-map of Fig. 7.12 with the binary partition defined by $y = 1/2$ (see the picture on the right). In fact, if $y < 1/2$ ($> 1/2$), the pointer is positioned on a 0 (1) in the image triple, after being moved to the left (right). Moreover, the partition is generating because each symbolic triple on the left column in the table univocally identifies one of the rectangles of Fig. 7.12. With this representation, we estimated the topological entropy to be $K_0 \approx 0.14$ by counting all legal sequences up to length $n = 80$[12]. The finite-size estimates $K_0(n)$ are shown in Fig. 7.13.

The Lyapunov exponent can be evaluated by noticing that the distance $d(n)$ between nearby points grows as $d(n) \sim 2^{|i(n)|}$, where $i(n)$ is the pointer's position at time n and $i(0) = 0$. The absolute value of $i(n)$ is needed because a net expansion occurs independently of whether the pointer drifts to the left or to the right. The Lyapunov exponent is, hence, $\lambda = (\ln 2) \lim_{n\to\infty} i(n)/n$, which is proportional to the pointer's velocity v. Numerical simulations (Moore, 1990) indicate that the movement is diffusive ($i(n) \sim n^{1/2}$), so that $\lambda = 0$. Therefore, this model can be interpreted as a further example of a system at the "edge of chaos". The positivity of the topological entropy, however, is a remarkable feature.

The association between GSs and Turing machines is intuitive, since the substitutions in the DOD are assimilable to a change in the machine's internal state and the shift of the pointer to a movement of the tape head (Moore, 1991).

12. It has been possible to achieve such unusually large sequence length by directly exploiting the GS-rules.

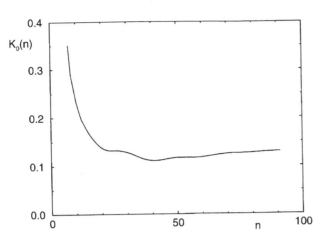

Figure 7.13 Finite-size estimate $K_0(n)$ of the topological entropy of the generalized shift described in the table versus the sequence length n.

Therefore, the long-time behaviour of a GS is unpredictable, not much because of sensitivity to the initial conditions, but for the undecidability of any general question concerning the state of the machine. For example, no algorithm can establish whether the pointer will ever reach any preassigned position for generic initial conditions. Of course, this remark holds only if it is indeed impossible to reproduce the specific GS with a machine belonging to a lower class in the Chomsky hierarchy.

The analogy with a Turing machine does not render the GS "more interesting" than a shift dynamical system. In fact, as shown before, the GS does correspond to an ordinary shift dynamics. Its possible complexity depends on the specific structure of the language.

7.3.6   Nucleotide sequences

Nucleic acids represent the most stimulating challenge for the application and development of formal language theories. In fact, sequences of the four bases A, C, G, and T contain the instructions for the generation of a single individual within a given species. Thus, it is tempting to interpret a DNA sequence as a computer-program code written in an unknown language to be deciphered. Although appealing, this assertion is probably too strong since DNA sequences are believed to act also as *input* data carrying information for the regulation of certain chemical processes which develop in a cell. Very likely, both functions of DNA are simultaneously active and so finely interlaced as to make the deciphering a formidable task.

Although a general approach has not been devised yet, the recent availability of large sets of experimental data provides a marvellous opportunity for investigating at least some specific problems. As usual, Markov chains, i.e. regular languages, have been widely employed to model DNA structure for their simple and general character. These languages, however, do not appear very appropriate for capturing the long-range correlations that are present in DNA sequences (see Ex. 5.16 for the correlation function analysis). One possible source of long-range couplings is the existence of proteins which, synthetized by decoding a certain tract of a DNA sequence, intervene in the regulation of the transcription process at some other location along the same sequence. A further contribution stems from the so-called "secondary structure", i.e. from the folding properties of the double-helix in space, which is related to the ordering of the bases themselves.

Here, we focus our attention on the relatively simple context of tRNA, one of the three types of RNA so far discovered, which plays the major role in translating the 4-base sequence contained in the mRNA (see Section 2.6) into the corresponding sequence of amino acids. tRNA sequences can be interpreted as the entries of a dictionary containing the codon table reported in Section 2.6. The most important ingredients of each tRNA sequence are the (anti)codon region, which is the only fragment that binds itself to the mRNA codon to be translated, and the corresponding amino acid, carried at one tRNA end. Only if codon and anticodon are mutually complementary as in the two DNA strands (uracil U replacing thymine T), tRNA attaches to mRNA thus supplying the proper amino acid.

Except for the presence of an amino acid, each tRNA is made of the same four bases as any other RNA fragment, so that its peculiarity must be hidden in the structure of the symbolic sequence. In particular, it has been realized that tRNA generally contains relatively long palindromic sequences (see Section 2.6) which, by favouring a local folding, contribute to determine tRNA's secondary structure. Obviously, the spatial folding is a key element to understand why only the tRNA anticodon region can bind to mRNA.

The presence of palindromic sequences immediately suggests contex-free grammars as the simplest scheme to model tRNA and account for the large variability experimentally observed. Stochastic CFGs have indeed been success-fully employed to recognize tRNA sequences and to describe their secondary structure. Sakakibara *et al.* (1994) have introduced the following productions: $V \rightarrow sV\bar{s}$, where s is any of the four bases and $\bar{s}$ its complement ($A \leftrightarrow U$, $C \leftrightarrow G$), $V \rightarrow sV$, $V \rightarrow s$, and $V \rightarrow VV$. In fact, any variable V can generate pairs of palindromes through the first production: they correspond to possible folding regions of tRNA molecules. The second production, by increasing the number of variables, originates secondary branching structures, while the last two rules give rise to unpaired bases. An example of a sequence generated by this grammar and of its corresponding secondary structure is reported in Fig. 7.14. The above grammar generates all possible sequences: its usefulness emerges if different productions are attributed different probabilities. This can be done either *a priori*, on the basis of general knowledge of tRNA sequences,

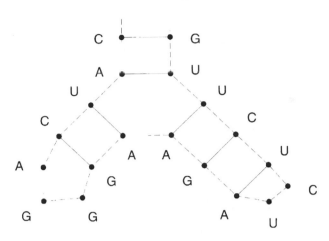

Figure 7.14 Secondary structure of a tRNA tract. The sequence is to be read along the dashed line. The solid segments join the elements of the base pairs.

or, more effectively, by maximizing the probability of a few training sequences (Lari & Young, 1990). A generic sequence is then recognized as a member of the "tRNA language" if its probability turns out to be larger than some preassigned threshold[13]. Successful recognitions have been reported in nearly 100% of the cases for mitochondrial and cytoplasmic tRNA and around 80% for sequences with "unusual" secondary structure. Whether the employment of formal language theory in classifying "short" tRNA sequences (less than 100 bases) will extend to include long DNA sequences is, however, still an open problem.

7.3.7   Discussion

In Chapters 5, 6, and 7 we have reviewed some traditional and some more recent mathematical tools for the characterization of one-dimensional symbolic patterns. On the basis of physical motivations, we have further specialized the object of study to stationary processes equivalent to shift dynamical systems. Before setting out to examine topical quantities which are more explicitly addressed to complexity, it must be realized that guidelines for a meaningful general classification can already be traced with the so-far available instruments.

To begin with, the researcher may ask which properties must be exhibited by a system in order to legitimate a sharper study of its complexity. According to a power-spectral analysis, periodic and quasiperiodic signals and, more generally, all those with pure point spectrum are simple. A step beyond, one finds systems with singular continuous spectrum (such as, e.g., quasicrystalline structures), in some cases with singularity exponents α which fill a whole interval. The correlation functions slowly decay with sporadic "resurgences" to finite values.

13. A proper normalization is required to account for different sequence lengths (Sakakibara *et al.*, 1994).

At the opposite extremity, there are absolutely continuous spectra with little or no structure at all, typical of short- or zero-memory processes which, therefore, appear as simple.

The ergodic classification generally agrees with the above in that, on the one hand, periodic and quasiperiodic signals lie at the bottom of the scale and, on the other, short memory signals are strongly mixing or even exact. The intermediate class of weakly mixing systems includes several families of substitutional (D0L) systems and others which are obtained from them through a single transformation. The intermediate regions in both classifications are, however, to a large extent *terra incognita*.

Within information theory, entropy is the basic measure of disorder. Once again, extreme situations agree with the conclusions of the previous schemes, a minor difference being that *all* systems from periodic to "quasicrystalline" (singular–continuous spectrum) are recognized as equally simple (zero entropy). The entropy is maximal for completely random systems and decreases for increasing memory.

A more refined way to quantify the strength of memory and correlations in the signal is provided by the thermodynamic formalism. In fact, the inhomogeneity of the pattern is traced back to the range of the microscopic interactions in the Hamiltonian. Short-ranged interactions are typical, in one dimension, of structureless systems in which no emergence of order is observed (except, perhaps, at zero temperature). Sufficiently slowly decaying interactions, instead, can lead to coexistence of different phases and, in particular, to patterns which possess elements of order and disorder simultaneously.

The Chomsky hierarchy is a qualitatively different ordering. Regular languages, for example, although in the lowermost class, include both periodic and Markov processes. More exotic types of behaviour are encountered when climbing up the hierarchy.

None of these schemes appears sufficient to give a clear-cut definition of the domain in which the observer is entitled to speak about complexity. Moreover, there is only a partial correspondence among them. Indeed, there are zero-entropy strongly mixing systems, substitutional limit words with white power spectrum, minimal systems with arbitrary topological entropy. This incompleteness and lack of concordance demands that the analysis be performed with all available tools, with the objective of achieving a consistent picture. The occurrence of serious discrepances is a clue to the actual existence of complexity.

Further reading

Aho *et al.* (1974), Culik *et al.* (1990), Howie (1991), Kolář *et al.* (1993), Lind (1984), Luck *et al.* (1990), Rozenberg and Salomaa (1990).

Chapter 8

Algorithmic and grammatical complexities

The core of the problem when dealing with a complex system is the difficulty in discerning elements of order in its structure. If the object of the investigation is a symbolic pattern, one usually examines finite samples of it. The extent to which these can be considered regular, however, depends both on the observer's demand and on their size. If strict periodicity is required, this might possibly be observed only in very small patches. A weaker notion of regularity permits the identification of larger "elementary" domains. This intrinsic indefiniteness, shared alike by concepts such as order and organization, seems to prevent us from attaining a definition of complexity altogether. This impasse can be overcome by noticing that the discovery of the inner rules of the system gives a clue as to how its description can be shortened. Intuitively, systems admitting a concise description are simple. More precisely, one tries to infer a model which constitutes a *compressed* representation of the system. The model can then be used to reproduce already observed patterns, as a verification, or even to "extend" the whole object beyond its original boundaries, thus making a *prediction* about its possible continuation in space or time.

As we shall see, a crucial distinction must be made at this point. Either one is interested in studying a *single item* produced by an unknown source, or in modelling the source itself through the properties of the *ensemble* of all its possible outputs. In both views, however, complexity is naturally related to randomness through the concepts of compressibility or predictability of the data in so far as genuinely random objects are most incompressible and unpredictable.

The former approach has been developed in greater depth and generality in the fields of information theory (Kolmogorov, 1965) and computer science (Solomonoff, 1964; Chaitin, 1966 and 1990ab). Because of the statistical nature of complexity problems, the second view is the more appropriate one in a physical context (Grassberger, 1986).

Accordingly, two classes of definitions can be distinguished. We call the first algorithmic or grammatical complexities, to remark that the accent is on the computational skills that a model (algorithm or grammar) needs to have in order to reproduce a *given input object*. We designate the second class hierarchical scaling complexities, since an infinite or infinitely refinable object is considered as a particular combination of finite parts chosen from an ensemble which can be hierarchically ordered. The associated measures will be the subject of Chapter 9.

This chapter treats the most widely diffused definition of complexity (Solomonoff, 1964; Kolmogorov, 1965; Chaitin, 1966) and some representative alternatives. In order to appreciate the meaning of this approach in detail, it is necessary to discuss first the fundamentals of coding theory and model inference (Sections 8.1 and 8.2) which, dealing with the problem of information compression, represent the main motivation for algorithmic complexity theory (Section 8.3). A few variants of the original approach, which have been proposed to improve some of its unsatisfactory aspects, are reviewed in the final sections.

8.1   Coding and data compression

A *code* is the assignment of symbol sequences to the elements of a set. The Morse code, for example, translates letters and punctuation from natural languages to concatenations of dots (".") dashes ("-") and blanks (" "): letter A is represented by ".-", letter B by "-...", and so on. Accordingly, a text is transmitted as a sequence of the three basic symbols ".", "-" and " ", and later decoded by the receiver.

More formally, a code is a mapping ϕ from a set Ξ of *source symbols* (such as typographical characters or articles exhibited on a shelf) to A^*, the set of all finite words over an alphabet A with cardinality b, whereby $\phi(s) \in A^*$ is the codeword assigned to $s \in \Xi$ and $\ell(s) = |\phi(s)|$ is its length. Let the source be a discrete stationary stochastic process $\{s_i\}$. Usually, the purpose of introducing a code is to transmit data as economically as possible. Hence, it is desirable to associate the most frequent outcomes of the source with short codewords. The achieved compression is expressed by the following quantity.

Definition: The *expected length* $L(\phi)$ of a code ϕ for a random variable $s \in \Xi$ with probability distribution $P(s)$ is given by

$$L(\phi) = \sum_{s \in \Xi} \ell(s)P(s) . \tag{8.1}$$

In the Morse code, letter E (the most frequent in English) corresponds to just a dot. Notice that the source symbols may be rather general objects, not necessarily one-character strings.

Example:

[8.1]   Consider a random variable taking the values s_1, s_2, and s_3, with probabilities $p_1 = 1/2$, $p_2 = 1/3$, and $p_3 = 1/6$, respectively. The choice $\phi(s_1) = 0$, $\phi(s_2) = 10$, and $\phi(s_3) = 11$ yields $L(\phi) = 3/2$. If the objects s_i are labelled with a *fixed-length* code, such as $\phi'(s_1) = 00$, $\phi'(s_2) = 01$, $\phi'(s_3) = 11$, the length of the encoded binary message will be larger than with the code ϕ: in fact, $L(\phi') = 2$. One says that ϕ is more *efficient* than ϕ'. □

The advantage of codes consisting of variable-length words, as in Ex. 8.1, is that the same information can be represented with fewer symbols, on average, than with fixed-length codes. This is especially true if the source symbols have a strongly nonuniform distribution. The construction of an efficient code, hence, requires knowledge of the statistics of the signal.

Efficiency is not the only property that a good code must satisfy. It is essential for the received message to have a unique interpretation. A code with this characteristic is called *uniquely decodable.* If $\phi(a) = 0$, $\phi(b) = 11$ and $\phi(c) = 011$, the message 0011 is ambiguous since it could represent either *aab* or *ac*. To clarify this point, it is useful to consider the *n*th *extension* of a code ϕ: namely, the code that labels all possible concatenations $s_1 s_2 \ldots s_n$ of n source symbols and satisfies $\phi(s_1 s_2 \ldots) = \phi(s_1)\phi(s_2)\ldots$, where the expression on the right indicates the concatenation of the individual codewords. For unique decodability, it is necessary that no two encoded concatenations coincide, even for extensions of different order n.

A final fundamental property is needed to obtain an effective code: the immediate recognizability of each codeword as soon as its reception is completed. Without it, in fact, it may be necessary to scan the whole incoming message before even the first word has been identified.

Example:

[8.2]   Consider the code $\phi(s_1) = 0$, $\phi(s_2) = 01$, $\phi(s_3) = 011$, $\phi(s_4) = 111$ (Hamming, 1986). It can be verified that it is uniquely decodable. When the string $0111\ldots 1111$ is received, however, it is not possible to decide whether the first word corresponds to s_1, s_2, or s_3 until the number of 1s in the whole message has been determined.

□

This difficulty arises when one or more codewords are prefixes of other ones. For a code to be immediately recognizable or *instantaneous*, it is necessary and sufficient that none of its codewords is the prefix of another one (Hamming, 1986). For this reason, such codes are also called *prefix-free*: they form a subset of the uniquely decodable codes. The construction is straightforward. A five-symbol source, for example, could be encoded by reserving 1 for the first symbol, 01 for the second, 001 for the third, 0000 and 0001 for the last two. Alternatively, one may start with pairs of code symbols, such as 00, 01 and 10 for the first three source symbols, and complete the code with 110 and 111. This procedure is tantamount to a successive pruning of the branches of a tree, starting at some level (1 and 2, respectively, in the examples above) and stopping at a lower level with two (in general, with b) end leaves. In this way, no codeword is the ancestor of any other label on the tree. The choice of the codeword lengths for a prefix-free code is limited by the Kraft inequality (Hamming, 1986):

Theorem 8.1 *A necessary and sufficient condition for the existence of an instantaneous code* $\phi : \Xi \to A^*$ *having codeword lengths* $\{\ell_i : i = 1, \ldots, c\}$ *is*

$$\sum_{i=1}^{c} b^{-\ell_i} \leq 1, \tag{8.2}$$

where c and b are the cardinalities of the source and code alphabets Ξ and A, respectively.

This result can be extended to the case of a countable infinity of codewords (Cover & Thomas, 1991). Notice that these codes are not unique: in particular, one may interchange the symbols at any stage of the construction and still obtain a meaningful code.

The list $\tilde{b} = (b^{-\ell_1}, b^{-\ell_2}, \ldots)$ can be interpreted as a probability vector when the equality in (8.2) holds. Indeed, by arranging the probabilities $(p_1, p_2, \ldots)$ and the components of $\tilde{b}$ in the same (e.g., increasing) order, the indices must be in a one-to-one correspondence if the code is efficient, otherwise its average length L could be reduced by exchanging some entries. This observation can be rendered quantitative by finding the optimality condition for a code, which is defined by a set of lengths $\{\ell_i\}$ satisfying Eq. (8.2) and having minimal average length $L(\phi)$ among all instantaneus codes. Minimizing $L(\phi) = \sum p_i \ell_i$ with the simultaneous constraint (8.2) yields the lengths

$$\ell_i^* \geq -\frac{\ln p_i}{\ln b} \tag{8.3}$$

of the optimal code ϕ^*. Evidently, the equality holds in the ideal case which, however, is not usually attained since in practice the lengths must be integers. The expected codeword length

$$L(\phi^*) = -\sum_i p_i \ln p_i / \ln b \qquad (8.4)$$

in this event would equal the entropy H_b of the distribution of the encoded random variable, measured in units of $\ln b$. The expected length of any other instantaneous code is greater than or equal to this lower bound.

By defining $\tilde{\ell}_i = \lceil \ell_i^* \rceil$ as the smallest integer larger than or equal to the optimal length ℓ_i^*, for every i, it is possible to determine the upper bound as well. In fact, multiplying the obvious relation

$$-\log_b p_i \le \tilde{\ell}_i < -\log_b p_i + 1 \qquad (8.5)$$

by p_i and summing over i, one obtains $H_b \le \tilde{L} < H_b + 1$, where $\tilde{L}$ is the average code length for the set $\{\tilde{\ell}_i\}$. Exponentiating (8.5) and summing again over i shows that the set $\{\tilde{\ell}_i\}$ satisfies the Kraft inequality. Finally, recalling that $H_b \le L(\phi^*) \le \tilde{L}$, the optimal expected length is proved not to exceed $H_b + 1$. This means that, in general, the compression is not maximal but lies one unit of information (when measured to the base b) from the ideal one. This waste of information is caused by the restriction of the coding procedure to the *individual* source symbols. The distance between the lower and upper bounds can be arbitrarily reduced by allowing for more variability in the set of objects to be encoded.

Let us consider all words S of length n emitted by the source. Rather than taking the nth extension of a code for the source symbols, one can directly design a new code $\phi^{(n)}$ for the n-blocks using knowledge of their probabilities $P(S)$ and define the expected codeword length per input symbol as

$$L_n = \frac{1}{n} \sum_{S:|S|=n} |\phi^{(n)}(S)| P(S) = \frac{1}{n} L(\phi^{(n)}) . \qquad (8.6)$$

Using the known bounds on $L(\phi^{(n)})$, we obtain

$$\frac{H_n}{n} \le L_n \ln b < \frac{H_n}{n} + \frac{\ln b}{n} , \qquad (8.7)$$

where H_n is the n-block entropy (5.36). Hence, in the limit $n \to \infty$, the minimum expected codeword length per symbol equals the metric entropy K_1 (5.44). This illustrates further the meaning of the entropy as the expected number of characters (in base b) necessary to describe the process.

Example:

[8.3] The improvement in compactness achieved by a word encoding over the elementary character encoding can be seen by considering a binary source with $P(s_1) = 3/4$ and $P(s_2) = 1/4$. Setting $\phi(s_1) = 0$ and $\phi(s_2) = 1$ yields $L_1 \equiv L(\phi) = 1$. Supposing that $P(s_i s_j) = P(s_i)P(s_j)$, for simplicity, the four 2-blocks can be encoded as $\phi^{(2)}(s_1 s_1) = 1$, $\phi^{(2)}(s_1 s_2) = 01$, $\phi^{(2)}(s_2 s_1) = 000$ and $\phi^{(2)}(s_2 s_2) = 001$, yielding $L_2 = 27/32$. Hence, further compression has been achieved with the second code. The assumed factorization of probabilities is irrelevant. Highly nonuniform distributions, for any n, are the most easily compressible ones.

□

The code defined by the set $\{\tilde{\ell}_i\}$ (Eq. (8.5)) is named after Shannon and Fano. Although not optimal in general, it has the advantage of defining each length from the corresponding word probability only. Optimal codes, named after Huffman (1952), are obtained with an iterative procedure involving a reduction and a splitting process. The source objects are listed in decreasing order of probability. The lowermost two (in the binary case) are regrouped into a metasymbol, which is repositioned according to its total probability. This procedure is repeated until two metasymbols remain, which are hence labelled by 0 and 1. The groups are then split by following the inverse path, whereby the symbols 0 and 1 are attached to the right of the current group labels each time a splitting occurs (Hamming, 1986; Cover & Thomas, 1991). The price to pay for optimality is that the codeword lengths depend on all probabilities, which must be estimated *a priori*.

It is instructive to see what happens when the estimated distribution $Q = (q_1, q_2, \ldots)$ differs from the unknown true distribution. Considering again the Shannon–Fano code $\{\tilde{\ell}_i\}$, the length assignment now reads

$$\tilde{\ell}_i = \lceil -\log_b q_i \rceil . \tag{8.8}$$

The expected code length $L_{p|q}$ satisfies

$$-\sum_i p_i \log_b q_i \leq L_{p|q} = \sum_i p_i \tilde{\ell}_i < \sum_i p_i (1 - \log_b q_i) \tag{8.9}$$

which can be rewritten as

$$H(P) + H(P\|Q) \leq L_{p|q} \ln b < H(P) + H(P\|Q) + \ln b , \tag{8.10}$$

where $H(P)$ is the entropy of the true distribution P and $H(P\|Q)$ is the relative entropy (5.48). Relation (8.10) should be compared with Eq. (8.7) for $n = 1$. Evidently, the average code length is burdened with the penalty $H(P\|Q) \geq 0$ which can hence be interpreted as the increase in descriptive complexity due to incorrect information.

It is worth mentioning that other coding procedures may be preferable to Huffman's even for the purpose of achieving signal compression. Some of these are discussed in Hamming (1986) and in Cover and Thomas (1991). Specifically, Huffman's method is impractical when long blocks of source symbols have to be coded, because this requires evaluation of the probabilities of a large number of words (of the order of b^n for n-block encoding) and construction of a tree for their labelling.

Finally, notice that the complexity of the system, if properly defined, ought not to be affected by a finite number of coding steps. It can be shown, on the contrary, that the metric entropy increases upon compression of the signal (for an optimal code, which yields maximal compression, it is maximized). This noninvariance under finite coding once more indicates that entropy is not a good measure of complexity.

8.2   Model inference

A more interesting and challenging task is the construction of a global model which is able to reproduce topological and metric features of the signal $\mathscr{S}$ with a prescribed accuracy. The field that deals with algorithms for the solution of such problems is called *inductive inference*. A complete, operative theory is still missing. Only approximate methods and approaches with a limited range of applicability have been developed. The progressive discovery of the rules that underlie the formation of the symbolic pattern is akin to a *learning* process. This can be achieved in various ways, listed by Dietterich *et al.* (1982) under the following names: by heart, from a teacher, from examples, and by analogy. While the former two imply no real understanding, the latter two may represent the basis for a theory of "automatic learning". Because of the abstractness of the concept of analogy (suggestive of mathematical thought), only learning by examples has been precisely defined and implemented in practice.

The inference method is formalized as follows. Upon receiving a sequence $\{i_1, i_2, \ldots i_n\}$ of information units, the observer guesses a language $\mathscr{L}_n$ out of a fixed class $\mathscr{C}$, by means of some algorithm. A language is *identified in the limit* if all guesses are correct after a finite training time $n = n_0$. The information is usually collected either from a "text" (e.g., the list of all legal words) or from an "informant" (who provides a list of words in A^*, specifying which of them are legal). In both cases, the list is arbitrarily ordered.

The former presentation of data permits recognition of only finite languages since the observer cannot decide whether a missing word is indeed forbidden or simply not yet transmitted. With the second, more powerful method, even context-sensitive languages can be recognized (Gold, 1967). The identification occurs by mere enumeration of all languages in the class $\mathscr{C}$, ordered according to the size of their grammar. Notice that the observer can never be sure of his success since the right language might still differ from the current guess for words that have not yet been received. This is similar to a child's learning of a natural language: while increasing his knowledge, he may still incur a grammatical mistake and yet make good use of the acquired experience. Unrestricted grammars are not learnable, despite their enumerability, because of the unboundedness of the verification time (halting problem). The most serious drawback of the enumeration method for exact identification is the dramatic divergence of the number of attempts with the grammar size, which excludes any practical implementation, except in trivial cases. Therefore, either *ad hoc* techniques are developed for certain classes of models or approximate recognition schemes must be devised. In the following, we shall explore both possibilities.

With a stronger notion of learnability, defined by the requirement that the observer be aware of the recognition, almost no language can be identified. Then, an effective procedure can be implemented only if more information is made available. For example, one might be told that the language arises from the limit set of an elementary (i.e., range-1) cellular automaton.

In the context of shift dynamical systems, an observer who makes real-time guesses while recording the signal $\mathscr{S}$ operates in the text framework. If, instead, a sufficiently long stretch has been read and processed, stationarity permits the identification of forbidden words with high precision (on the basis of a low-probability threshold), so that the operation approximately takes place within the informant framework.

In order to construct an actual model, the condition of exact recognizability is relaxed in favour of a suitable notion of approximation. This relies on a definition of a metric in the space of languages: object and predicted language ($\mathscr{L}$ and $\mathscr{L}'$) are required to lie within a preassigned distance d from each other. General inference theorems depend on the chosen metric. Wharton (1974) considers the lexicographic ordering of A^* and attributes the weight $p_i \geq 0$ to the ith word w_i, so that $\sum_{i=1}^{\infty} p_i = 1$[1]. The sum of the probabilities of the discordant words of $\mathscr{L}$ and $\mathscr{L}'$ is an appropriate distance. With these prescriptions, Wharton proved that any recursive grammar can be recognized in the limit from a text presentation with an arbitrarily small error d. Under an informant presentation, instead, recursive grammars are identifiable even in finite time, this term meaning that the observer realizes that the solution has been found.

To illustrate the utility of approximation methods, consider the sequence of directed graphs that account for all prohibitions with length up to n found in a language $\mathscr{L}$. The corresponding sequence of languages need not converge in a finite time as, e.g., when $\mathscr{L}$ is sofic. Nevertheless, it is clear that for any given tolerance d an identification in the limit takes place.

Among the many *ad hoc* grammatical inference methods, it is worth men-

---

1. Notice that the normalization of these weights differs form the usual one in which each set of equal-length words individually has probability one. High-order hierarchical levels are here quickly damped to achieve the overall normalization.

tioning an algorithm for stochastic context-free grammars (Section 7.2). Given a test CFG $\mathscr{G}'$ with language $\mathscr{L}'$, the size of its production set is evaluated by means of an entropy-like quantity h which accounts for the probabilities of the substitutions and the lengths of the image words. The distance δ between object- and test-language is then calculated and added to h to form a cost function $f = h + \delta$ which must then be minimized upon change of the trial grammar $\mathscr{G}'$. The joint presence of the distance and the size of the grammar in f reflects the necessity for a good model to be not only accurate but also concise. Rather than searching for the minimum of f via mere enumeration, Cook $et\ al.$ (1976) propose a combination of three transformations in the space of grammars. The procedure starts with the trivial grammar $\mathscr{G}'$ defined by the productions $W \rightarrow w_1; w_2; \ldots ; w_n$, where the set $\{w_i\}$ is the available finite presentation of $\mathscr{L}$. This is an exact but sizable grammar (i.e., with $\delta = 0$ but large h). Its form is then changed by taking the following actions:

1. substitution;

2. disjunction;

3. removal of productions.

In (1), each subword Z_j that appears frequently in the sentential forms is replaced by a variable V_j and the production $V_j \rightarrow Z_j$ is added to the list. In (2), productions of the type $V \rightarrow ZZ'; ZZ''$ are transformed into the pair $V \rightarrow Z U_k, U_k \rightarrow Z'; Z''$. In order to reduce h, it is sometimes convenient to identify V_j and U_k with an already existing variable, although this approximation may increase the discrepancy δ. Finally, at each stage of the process, useless, redundant, and inaccessible productions are removed (3). The major drawback of $ad\ hoc$ methods is that one cannot tell $a\ priori$ whether they are successfully applicable to any new problem.

The limitations of the above models can be overcome at the expense of restricting the class of acceptable models. Markov chains are undoubtedly the most generally applicable approximation schemes in mathematics and physics. In the case of a shift dynamics, the model reconstructs the original signal by drawing the nth symbol s_n at random according to the k-symbol conditional probability (see Eq. (5.11)). In the presence of long-range correlations, large values of k (and storage space) are needed. This sets an upper bound on the value of k in actual implementations. Independently of this practical difficulty, interesting cases are those in which convergence either is slow or does not take place at all. There are, however, situations in which only a small fraction of histories exhibits anomalously long memory. In type-I intermittency (Ex. 6.5), these are sequences of n consecutive 0s, the probabilities of which decay as $n^{-\gamma}$. In such cases, it is useful to single out blocks of symbols and $recode$ them as new symbols s'_i. Accordingly, the signal $\mathscr{S}$ can be viewed as the concatenation

$w_0 w_1 \ldots$ of *basic* or *primitive words* w_i which can be renamed as $s'_i = \phi(w_i)$, using a code ϕ. The resulting signal $\mathscr{S}'$ is again amenable to a Markov analysis, possibly with a smaller order k. In order to maintain a meaningful correspondence between $\mathscr{S}$ and $\mathscr{S}'$, unique decodability is required.

The probability $P'(s')$, with $s' = \phi(s_1 s_2 \ldots s_m)$, is in general different from $P(s_1 s_2 \ldots s_m)$ in a signal with memory. Consider, for example, the code $(00, 01, 1) \rightarrow (a, b, c)$ in a Markov process with conditional probabilities $P(s_{i+1}|s_i)$. The probability $P'(c)$, for example, can be expressed as

$$P'(c) = P'(a)P(1|0) + \left[P'(b) + P'(c) \right] P(1|1)$$

since c consists of a single 1 which either follows the final 0 of a previous a or the final 1 of b or c. Similar equations hold for $P'(a)$ and $P'(b)$. For an independent process (Eq. (5.6)), the previous symbol is irrelevant and the equations reduce to $P'(a) = P(0)^2 = P(00)$, $P'(b) = P(01)$, and $P'(c) = P(1)$. For an order-1 Markov process one obtains, instead,

$$P'(c) = \frac{P(1|0)\left[1 + P(0|0) - P(0|1)\right]}{1 - P(0|0)\left[P(0|0) - P(0|1)\right]}$$

which, in general, differs from $P(1)$. Such considerations are valid independently of the code. In the case of the logistic map (3.11) at $a = 1.85$, letting $(1, 01, 001) \rightarrow (a, b, c)$, one finds $P(1) \approx 0.5896$, $P(01) \approx 0.3236$, and $P(001) \approx 0.0868$, while $P'(a) \approx 0.451$, $P'(b) \approx 0.401$, and $P'(c) \approx 0.148$.

With the new probabilities, the model produces a signal $s'_1 s'_2 \ldots$ which can be coded back into the symbols s by means of ϕ^{-1}. This reconstruction method is therefore the composition of a *sequential* (Markov) process with a (deterministic) *parallel* operation (the substitution $s' \rightarrow \phi^{-1}(s')$). These two mechanisms act in many physical systems and are more or less explicitly incorporated in most mathematical models. In general, however, there is no clear-cut separation between them. The Fibonacci sequence (Ex. 4.6) can be generated either by a D0L substitution (parallel) or by a map (sequential); the limit set of the elementary CA 90 can also be trivially obtained from the Bernoulli map. Depending on whether one wants to reconstruct the signal or to compute quantities which describe the structure of the system, different mixtures of the two ingredients can be included in the model. A hybrid automaton, composed of directed graphs connected with special nodes and arcs at which substitutions take place, has been constructed by Crutchfield and Young (1990) to reproduce the period-doubling sequence.

If the primitive words w have different lengths, not all k'-blocks in $\mathscr{S}'$ correspond to k-blocks in $\mathscr{S}$ with some fixed k. The order k' of the Markov process $\mathscr{S}'$ translates into the order (average Markov time)

$$k = \sum_{S' : |S'| = k'} \ell(S') P'(S') , \tag{8.11}$$

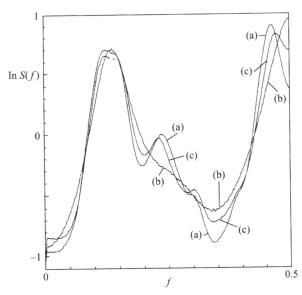

Figure 8.1 Power spectrum $S(f)$ of the symbolic signal of the Hénon map (a), compared with those obtained with an order-5 Markov approximation (b), and with a first order process with variable-length encoding (c). The natural logarithm has been taken after setting $S(0) = 0$ and normalizing to 1 the area under the curve $S(f)$ vs. f.

where $\ell(S')$ denotes the number of symbols s in $\phi^{-1}(S')$ and the sum is performed over all S' composed of k' codewords s'.

A clever choice of the code may accelerate the convergence of the statistical properties of the reconstructed signal, while weakening the storage requirements for the model. Wang (1989a) has been able to estimate thermodynamic properties of an intermittent map using the code $(1, 01, \dots, 0^i 1, \dots)$ which corresponds to a countable Markov partition.

The Hénon map (3.12) at $(a, b) = (1.4, 0.3)$ is a clear example in which recoding is definitely advantageous. In Fig. 8.1, the power spectrum of the actual symbolic signal is compared with those obtained with a conventional order-5 Markov approximation and a with a first order process over the seven primitive words 1, 01, 0011101, 0011111, 00111101, 00011101, and 00011111 (with normalized total probability). The former model involves 21 5-blocks and 33 6-blocks, the probabilities of which yield the transition matrix. The latter has 7 basic words and 48 legal transitions (2-blocks): on average, time advances more than twice as fast as in the conventional model, according to Eq. (8.11).

The variable-length encoding clearly shows better performance with a comparable amount of information. Even the zeroth-order approximation is able to reproduce the main peaks of the spectrum (Badii *et al.*, 1992). This success obviously depends on the choice of the primitive words. It is hence useful to discuss a few criteria that may help in the search. The assumed ergodicity of the signal implies recurrence for all n-cylinders (Chapter 5). This suggests that one should look for sequences with distinguished recurrence properties (this term meaning, for example, an anomalously large occurrence probability or a higher frequency of consecutive repetitions). It is then natural to use these sequences in the construction of a code. The codewords used above to reconstruct the dynamics of the Hénon map are the shortest periodically extendible sequences that do not contain a prefix with the same property. The words 0^n and $(001)^n$ do not exist for all n, since 0000 and 0010 are forbidden, while 1^n and $(01)^n$ do. Hence, 1 and 01 are primitive and the other primitives must begin with at least two 0s (the list is actually infinite: an arbitrary truncation has been made here).

Structural inhomogeneities may arise both from metric and topological features. Regions with weak instability are characterized by larger probabilities (see, e.g., the small-x region in the intermittent map of Ex. 6.5). Similarly, the existence of "close" forbidden sequences may constrain the trajectory along obliged paths. Consider, for instance, the set $\mathscr{F} = \{00, 111, 11011\}$ of irreducible forbidden words. All possible continuations of the word 011 must begin with 0101: the language is given by the regular expression $(01 + 10101)^*$. Using the two basic words 01 and 10101 as a code, one obtains a zero-memory Markov process (forgetting possible correlations generated by slowly decaying conditional probabilities). The ordered components of the language are thereby incorporated in the code and the random feature is singled out. In general, one cannot expect the code to yield a complete decomposition of the dynamics, especially if the language is nonregular.

Further guidance in the search of elements of order in the signal comes from the pumping lemmas for regular and context-free languages (Chapter 7) which state the existence of arbitrary repetitions of certain words. For instance, every sufficiently long word in a regular language $\mathscr{L}$ can be written as uvw in such a way that also $uv^iw \in \mathscr{L}$, for any i. The (shortest) blocks v satisfying this condition are natural candidates for the code. The same argument holds for context-free languages.

In minimal systems, although every point is recurrent, periodic orbits do not exist, so that arbitrary repetitions of words do not occur. Nevertheless, the above exposed ideas can still be fruitfully applied to recode the signal. In the Fibonacci sequence, pairs of 0s do not appear, while both isolated 1s and pairs of 1s can be found, so that the natural choice for the code is $1 \rightarrow a$, $01 \rightarrow b$. The resulting a-b sequence turns out to coincide with the original one upon identifying a with 0 and b with 1^2. In this case, coding does not permit compression of the correlations. The discovered self-similarity, however, readily yields an exact, purely parallel, deterministic model. A slightly different method is required by the Morse–Thue sequence. The single symbols 0 and 1 have a perfectly symmetric behaviour. The first inhomogeneity is found for 2-blocks: while both 01 and 10 may concatenate once with themselves, neither 00 nor 11 can. Moreover, 01 and 10 alone are sufficient to parse the whole signal. The recoding $01 \rightarrow a$, $10 \rightarrow b$ maps the signal onto itself by further setting $a = 0$

2. This applies to the limit-word $\psi_{F'}^\infty(0)$ which is specular and, therefore, equivalent to $\psi_F^\infty(0)$ (see Ex. 4.6).

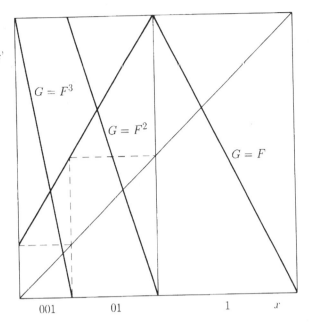

Figure 8.2 Illustration of the recoding procedure for the roof map $x' = F(x)$ (4.4) at $a = (3 - \sqrt{5})/4$. The branches of the renormalized map G correspond to F's iterates of order 3, 2, and 1 in the intervals labelled by 001, 01, and 1, respectively.

and $b = 1$. In general, the above heuristic criteria do not guarantee finding the optimal code, as the reader may verify for the square-free Morse–Thue sequence of Ex. 4.8.

The recoding process is susceptible to a geometric interpretation. Every symbolic dynamical system has a counterpart in a one-dimensional piecewise linear map, although possibly having an infinite number of branches. The renaming of blocks with symbols is tantamount to a modification of the map, so that an appropriate choice of the substitutions "simplifies" its form. Consider the regular expression $(1 + 01 + 001)^*$ and the corresponding roof map F (4.4) at $a = (3 - \sqrt{5})/4$ reproduced in Fig. 8.2. The substitutions $001 \to a$, $01 \to b$, and $1 \to c$ yield a signal $\mathscr{S}'$ which can be seen as the symbolic output of a map G, constructed from F with the following argument. The word 001 is observed at time n if and only if x_n belongs to the interval labelled 001. Hence, in order to emit $a = 001$ with G in a single move and to keep track of the current orbit point x_{n+3}, it is sufficient to replace F with its third iterate F^3 in that interval. The same argument applies to the other two intervals, so that G is the map represented in Fig 8.2. The utility of coding clearly emerges from the figure. Map G is complete: there are no forbidden words. Each of its pieces coincides with a different iterate of F. Of course, such a remarkable simplification is extremely rare.

This operation is called *renormalization*, in analogy with the block transformations of statistical mechanics (Chapter 6). The renormalized Hamiltonian $H'(S') = -\ln P'(S')$ describes interactions among spin blocks of the old configuration $\mathscr{S}$, obtained through the inverse code $S = \phi^{-1}(s')$. An infinity of renormalization steps can be performed on certain families of maps associated with D0L systems, such as the Fibonacci or the Morse–Thue substitutions. A careful choice of the intervals I and of the powers F^i shows that the resulting map G closely approximates, within each I, the original map F (Procaccia *et al.*, 1987). Upon rescaling and iterating the procedure, one arrives at a limit function $G_\infty{}^3$. Hence, a D0L substitution (also called an "inflation") can be seen as the inverse of a renormalization step. Map F is correspondingly trasformed into the so-called *induced* or *tower* map G (Kakutani, 1973; Petersen, 1989).

8.3   Algorithmic information

Following the present terminology (Chaitin, 1990ab), this section has been entitled algorithmic *information*, rather than *complexity*, since the quantity it introduces, Kolmogorov's complexity, is actually closer to a measure of randomness than to complexity. Intuitively, the descriptive effort for a finite string S (the "object") is related to the information content $I(S) = -\ln P(S)$ of S, according to a suitable probability measure. In the algorithmic approach, the length of the shortest computer program that reproduces S plays the role of $I(S)$ in information theory. Since the discussion applies to arbitrarily long sequences, reference to a universal Turing machine is made in order to render the definition essentially computer independent. Notwithstanding the apparent abstractness of the algorithmic setting, its conclusions are surprisingly close to those of information theory, apart for a few noteworthy exceptions. In fact, there is a close relationship between the expected description length of a random variable and its entropy.

Consider a Turing machine operating on three tapes (Chaitin, 1990b): the first contains the program in the form of a finite binary string which is read sequentially from left to right, the second is a work tape, and the third is reserved for the output. A string π, with finite length $\ell(\pi)$, is a *program* if it lets the Turing machine halt after reading all of its $\ell(\pi)$ input symbols. This scheme is closer to the functioning of a real computer than the basic one. Notice that, since the reading of the input is unidirectional, the programs form a prefix-free set: in fact, no halting program can be the prefix of another one. Therefore, the following Kraft inequality holds for any Turing machine U:

$$\sum_{\pi\,:\,U \text{ halts on } \pi} 2^{-\ell(\pi)} \leq 1 , \tag{8.12}$$

where the sum is taken over all programs π on which U halts and base-2 has

---

3. The best known example is the already mentioned function g which represents the fixed point of the period doubling operator (see Ex. 3.3).

been chosen for simplicity. Let U be a universal computer (schematized as above) and $U(\pi)$ its output when instructed by program π.

Definition· The *algorithmic complexity* $K_U(S)$ of a finite string S with respect to a universal computer U is

$$K_U(S) = \min_{\pi:U(\pi)=S} \ell(\pi), \qquad (8.13)$$

where $\ell(\pi)$ is the length of program π and the minimum is taken over all programs π that yield S and halt when processed by U.

Whenever a sequence S can be described by a short code π, which is able to reproduce it unambiguously in a finite computational time, the size of the code provides an upper bound to the Kolmogorov complexity of S. The apparent dependence on the computer U is irrelevant: in fact, for any other computer U', the inequality

$$K_U(S) < K_{U'}(S) + c_{U'}$$

can be proved to hold for all strings S, with the constant $c_{U'}$ not depending on S. The value of $c_{U'}$ can be very large: it represents the length of the program that enables U to emulate U'. For very long sequences S, however, the algorithmic complexity becomes independent of the particular computer since the length $c_{U'}$ of the emulation program can be neglected (hence, the subscript U is usually dropped). This result is known as the "universality of the Kolmogorov complexity". Generally speaking, compression is obtained if $\ell(\pi) < \ell(S)$. For short sequences S, the length of the program π is heavily affected by the constant c: the theory is meaningful only for very long sequences.

The necessity for the *minimal program* is a serious restriction on the practical applicability of algorithmic complexity. The most remarkable results have been obtained in formal language theory. Among them, we recall a new characterization of regular languages, the separation between DCFLs and CFLs and a measure of nonrecursiveness (Li & Vitányi, 1997).

To deepen the discussion, it is useful to consider the algorithmic complexity $K(S|R)$ of a string S relative to the input R supplied to the computer U. Obviously, $K(S) = K(S|\epsilon)$, where ϵ is the empty string. A particularly convenient input string is just the length $|S| = \ell(S)$ of the studied sequence (a theory which accounts for all possible types of input is, in fact, hard to formulate). When $R = \ell(S)$, one obtains the bound

$$K(S|\ell(S)) \le \ell(S) + c,$$

where the constant c is essentially the number of bits necessary to define a print statement for S. Knowledge of $\ell(S)$ obviously renders the machine's task easier: the program is then called *self-delimiting*. When this is not the case, the end of the sequence S must be marked, so that it can be recognized by the computer.

A simple but inefficient method consists of repeating every symbol in the binary expansion of the number $\ell(S)$ twice and by closing it with the string 01. This description requires $2\lceil \log_2 \ell(S) \rceil + 2$ bits, in addition to those of the minimal program for S: hence,

$$K(S) \leq K(S|\ell(S)) + 2\log_b \ell(S) + c$$

for strings written in a b-symbol alphabet. A more stringent bound is (Chaitin, 1990a)

$$K(S) \leq K(S|\ell(S)) + \log_b^* \ell(S) + c \,,$$

where $\log^* n = \log n + \log \log n + \log \log \log n + \ldots$ is the sum of iterated logarithms (to be interrupted after the last positive term).

The Kolmogorov complexity of a string $sss\ldots s$ of n equal symbols is $K(sss\ldots s) = c + \log_b n$. Similarly, the first n digits of $\pi = 3.14159\ldots$ can be calculated using a series expansion with a constant-length program: hence, $K(\lfloor \pi \times 10^n \rfloor | n) = c$. Other simple examples can be found in Cover & Thomas (1991). The complexity of an integer n is defined as in Eq. (8.13), by requiring that $U(\pi) = n$. Because of Eq. (8.12), one has the following theorem:

Theorem 8.2 *There is an infinite number of integers* $I = \sum_{i=0}^{n-1} s_i 2^i$ *such that* $K(I) > \log_2 n$.

In fact, if $K(I)$ were smaller than $\log_2 n$ for $n > n_0$, one would obtain

$$\sum_{n=n_0}^{\infty} 2^{-K(I)} > \sum_{n=n_0}^{\infty} 2^{-\log_2 n} = \infty \,,$$

that is, a contradiction.

One of the main "conceptual" applications of Kolmogorov's complexity is in the definition of randomness and compressibility. Since there are 2^n binary strings of length n but only $2^n - 1$ shorter descriptions, it follows that $K(S) \geq |S|$ for some binary string S with any length $n = |S|$. Such sequences are called *incompressible* or *algorithmically random* (Chaitin, 1990b). Notice that $\pi = 3.14159\ldots$ is not algorithmically random, although its digits satisfy the usual statistical tests for stochastic randomness. If an algorithmically random sequence failed one of these tests, this information would reduce the description length for S, thus yielding a contradiction. Hence, algorithmic randomness is a stronger condition than statistical randomness. In particular, the strong law of large numbers (Feller, 1970) holds for any incompressible string: different symbols occur in the sequence with frequencies that become equal in the infinite-length limit. Substrings of incompressible strings can be compressible (Li & Vitányi, 1997): e.g., every sufficiently long random sequence contains long strings of zeros.

The overwhelming majority of strings (with respect to Lebesgue's measure) display hardly any (computable) regularity and are, therefore, algorithmically random. The probability $P(K(s_1 s_2 \ldots s_n|n) < n - k)$ that a sequence of length n (chosen with respect to the Lebesgue measure b^{-n}) can be compressed by more than k symbols does not exceed b^{-k} (Cover & Thomas, 1991). In spite of this, no particular string, except finitely many, can be proved to be algorithmically random. Usually, the mathematical proofs obtained in the theory of Kolmogorov complexity yield lower bounds to the program size needed for some task and require consideration of all inputs of a given length. Although the lower bounds are shown to hold for some "typical" input, finding such a string may be hard or even impossible. A typical input consists of a Kolmogorov-random string and cannot be exhibited, in general, since no explicit string can be proved to be random (see Li & Vitányi, 1997, for a technical discussion).

Since the Kolmogorov complexity is the length of the shortest program that is able to reproduce the input string S, it is closely related to the information carried by S. Indicating by $S_{\langle n \rangle}$ an n-block contained in the infinite string $\mathscr{S}$, the *algorithmic complexity density* of $\mathscr{S}$ is defined as

$$K(\mathscr{S}) = \limsup_{n \to \infty} \frac{K(S_{\langle n \rangle})}{n} . \tag{8.14}$$

This quantity represents the complexity per symbol in a given string and is the relevant algorithmic indicator for an infinite sequence. The following theorem (Brudno, 1983) holds:

Theorem 8.3 *If m is an ergodic invariant measure for a shift dynamical system $(\Sigma', \hat{\sigma})$, then*

$$K(\mathscr{S}) = K_1(\Sigma', \hat{\sigma}) \tag{8.15}$$

for m-almost all $\mathscr{S} \in \Sigma'$.

Hence, the complexity density coincides with the metric entropy with probability one for an ergodic shift transformation; moreover, K is translationally invariant: $K(\hat{\sigma}^n(\mathscr{S})) = K(\mathscr{S})$. Notice, however, that stationarity of the sequences is not mandatory in the algorithmic context, at variance with dynamical systems theory.

We already commented upon the "special" character of numbers such as $\pi = 3.14159\ldots$, $e = 2.71828\ldots$, or $\sqrt{2}$, the digits of which can be computed with a finite program. The corresponding complexity density is zero. The sequence $0\,1\,0\,0\,01\,10\,11\,000\ldots$ (the list of all binary words in lexicographic order) also has $K = 0$, although its entropy is $\ln 2$. Sturmian trajectories (see Ex. 5.19) have both K and K_1 equal to 0. Finally, we recall a more significant result (Brudno, 1983): minimal systems with positive topological entropy have zero complexity density K, since their symbolic trajectories can be constructed with an explicit procedure (Grillenberger, 1973).

The above discussion shows that algorithmic complexity is a measure of information, almost equivalent to the metric entropy. In fact, its study is now included in the so-called algorithmic information theory (Chaitin, 1990a). With a careful definition of the programs allowed to run on Turing machines (the self-delimiting property is required), differences between Solomonoff's approach and those by Kolmogorov, Chaitin and Martin-Löf (1966) become irrelevant (Chaitin, 1990b). Moreover, algorithmic information theory can be formulated in complete analogy with the usual information theory.

We have seen that, although appealing, the idea that complex strings must be incompressible unavoidably leads to entropy, i.e., to randomness. An even more fundamental limitation to the applicability of algorithmic information to physical problems is that $K(\mathscr{S})$, unless it is zero, is not *effectively computable* for an infinite sequence $\mathscr{S}$ (Grassberger, 1989): i.e., it cannot be ascertained, in general, whether the *shortest* description has actually been found. Furthermore, algorithmic complexity classifies the object (S) by the properties (length) of the process (the shortest program π): it is, therefore, a machine-complexity rather than an object-complexity. The object, however, may be the result of a *long* process taking place on a *small* Turing machine. The time needed to yield S, an important characterization of S, is completely ignored. These objections motivate the two notions of complexity to be treated in Sections 8.4 and 8.5.

8.3.1   P–NP problems

Before concluding this section, we briefly survey a topic related to algorithmic complexity which is central in applied computer science and frequently mentioned in connection with statistical mechanical disordered systems. Rather than asking whether an algorithm exists "in principle" (decidability problem), or what its size is (Kolmogorov complexity), one may be interested in the "run-time" limitations arising in practical implementations. The so-called *computational complexity theory* is concerned with the time and storage space required by actual algorithms to solve problems in a given class.

Usually, the search for the most efficient, exact algorithm that reproduces a given sequence S is of the *exhaustive* type: i.e., all programs of increasing size $n = n_0, n_0 + 1, \ldots$ are tried until one (or several) yield S. The most efficient algorithm is usually the fastest among those of minimum length. This procedure often requires an exponential time in n or, equivalently, in the size $|S|$ of the "problem" S (since the least sufficient program size $n_{min} = K(S) \geq |S|$ for almost all S). Not all decidable questions, however, require such a time-consuming solution scheme. When a solution can be found in a time that grows at most polynomially in n, the problem is called *computationally simple* or *tractable*.

Conversely, a problem for which no polynomial time algorithm exists is called *intractable*.

More precisely, an algorithm $\mathscr{A}$ is said to solve a problem Π if it can be applied to any instance of Π and is guaranteed always to produce a solution for that instance (Garey & Johnson, 1979). A typical problem may consist of verifying whether a given string S is a member of the language generated by a grammar $\mathscr{G}$. An instance is then a finite specification of the grammar, and the number n of symbols needed for it formally represents the size of the problem. The algorithm is required to answer in a finite time T for all grammars in a given family (e.g., the context-free grammars). The theory of computational complexity (Hopcroft & Ullman, 1979) addresses exclusively to *decidable* problems and separates them into "mild" and "hard" on the basis of the *time complexity function* $T(n)$ for the solving algorithm $\mathscr{A}$: the function $T(n)$ represents the *largest* computation time employed by $\mathscr{A}$ to solve an instance of size n. The cause of intractability is not restricted to exponential computation times $T(n)$ but includes cases in which the solution itself is so "vast" that it cannot be described with a number of symbols $S(n)$ having a polynomial upper bound as a function of n. Undecidability corresponds to the extreme case in which no algorithm is guaranteed to halt on any arbitrary computer and input string for a certain problem.

Both *decision* problems (i.e., requiring a *yes* or *no* answer) and *optimization* problems (which ask for the structure that has minimum "cost" in a specific class) can be formally encoded as languages, the elements of which constitute all possible instances (Garey & Johnson, 1979). The yes or no answer to each instance S is provided by the final state (q_Y or q_N) of a Turing machine with input S. The language $\mathscr{L}$ consisting of the finite strings with affirmative answer is associated with the problem Π. This is said to belong to the *class P* if there is a Turing machine that accepts each entry S of $\mathscr{L}$ in a polynomial time in $|S|$.

The second important class of problems (languages) is best understood with reference to a concrete example: "given a finite set $C = \{c_1, c_2, \ldots, c_k\}$ of cities and the distances d_{ij} between them, find the ordering $(c_{p(1)}, c_{p(2)}, \ldots, c_{p(k)})$ of the cities that minimizes the total length of the tour starting and ending at $c_{p(1)}$ after visiting each city in sequence" (*travelling salesman problem*). This typical optimization problem is usually turned into a decision problem by asking whether there is a tour with length less than some bound ℓ_{max}. There is no known polynomial time algorithm for solving this question. If, however, a solution were available for some instance I (set of cities, distances and bound), its truthfulness could be tested by a general "verification algorithm" with time complexity function $T(|I|)$ polynomial in $|I|$. This procedure is formalized by a special type of Turing machine, called *nondeterministic* (NTM), which has a first "guessing" stage, after which the operation is taken over by an ordinary TM. The input string S is accepted if the NTM halts in state q_Y for at least one guess (the NTM may perform an infinite number of guesses for each given S). The time required for accepting S is the *minimum* number of steps over all accepting computations for S. The class of languages that are recognized by a polynomial time NTM-program is called "NP" (Garey & Johnson, 1979). The travelling salesman problem is in NP. This class points out the difference between *finding* a concise proof of a statement and *verifying* it efficiently (an intuitively simpler task).

At variance with class P, class NP is not exclusively defined in terms of practical decision procedures but requires the unrealistic notion of a "guessing" step. The other important difference between the two classes is related to the complementarity of "yes" and "no" answers. If the question "is X true?" belongs to P, also the converse ("is X false?") does. This is not the case for NP: a simple example is the complement of the travelling salesman problem in which one asks whether *no* tour with length less than l_{max} exists, since verifying a "yes" answer to this problem implies examining all possible tours. Clearly, P⊆NP: every problem solvable by a polynomial time deterministic algorithm can be solved by a polynomial time nondeterministic algorithm. The superior capabilities of the latter suggest that P is strictly contained in NP. So far, no polynomial time solution has been found for several problems, including the travelling salesman's. For these reasons, it is believed that indeed P≠NP, even though no proof of this conjecture has been found yet.

Finally, we mention an important subclass of NP: a language $\mathscr{L}$ is defined to be *NP-complete* if $\mathscr{L} \in$ NP and all other languages $\mathscr{L}' \in$ NP are *polynomially transformable* to $\mathscr{L}$: i.e., if there is a function f, computable in polynomial time by a deterministic Turing machine, that maps each $w \in \mathscr{L}'$ to some $f(w) \in \mathscr{L}$ and if, for all $w \in A^*$, $w \in \mathscr{L}'$ if and only if $f(w) \in \mathscr{L}$. NP-complete problems are, in principle, the hardest in NP since solving any one of them in polynomial time would imply that all of them could be solved in that way. The travelling salesman problem is NP-complete. For the closely related notion of NP-hardness, we refer the reader to Garey & Johnson (1979).

Although not directly related to self-generated complexity, NP-completeness often occurs in mathematics and physics. For example, algorithms for the construction of preimages in two- and higher-dimensional cellular automata (Lindgren & Nordahl, 1988) or of phylogenetic trees (see Section 2.6) with a minimum number of mutations (Fitch, 1986) fall in this class.

Summarizing, the algorithmic information $K(S)$ represents a "static" characterization of the string S, whereas time (or space) complexity functions may be viewed as its "dynamic" counterpart. Algorithmic information theory is often useful in proving results about computational classes. The joint usage of the two approaches may lead to a refined classification of symbolic sequences in some cases (e.g., signals with $K = 0$ that are generated by programs with different space-time requirements). It must be stressed, however, that typically algorithmic questions have usually little, if any, relevance from a physical point of view. In a physical investigation, it is very uncommon to search exactly for the nth digit of a number (such as an entropy or the period-doubling rate δ_τ) rather than for bounds on that number or for the convergence properties of the corresponding algorithm. Physics usually deals with approximations and their speed: although this seems to be a "looser" approach to the solution of problems, the answer need not be simpler. There are actually systems for which modelling average, long-term, characteristics is as hard as answering questions about exact properties. Furthermore, statistical fluctuations around some "optimal solution" (such as a spin configuration or a parameter value) are often more important and difficult to study than the solution itself. Physical questions are, in general, not algorithmic and, hence, not directly related to Kolmogorov's complexity. The theory of P–NP, on the other hand, refers to problems that are algorithmic or have been recast in an algorithmic form: it essentially studies how manageable these problems are.

8.4 Lempel–Ziv complexity

Almost all symbolic sequences are algorithmically complex with respect to the Lebesgue measure in the limit of infinite length. The minimal program to reproduce any of them reduces to a mere print statement. The actual value of $K(S)$, however, cannot be determined since there is no certainty, in general, that the shortest possible description has been indeed found.

Coding theory provides a way to overcome this problem. As already discussed, a signal may be transmitted in a condensed form by exploiting the nonuniformity of its word-structure. Moreover, the optimal compression is achievable by a code satisfying Eq. (8.3). An effective estimate of the algorithmic complexity of a sequence can therefore be obtained from the compression ratio $\rho_c = L(N)/N$, where $L(N)$ and N are the lengths of the encoded and the original sequence, respectively.

Since the optimality condition requires prior knowledge of the probability distribution for all subsequences, which is usually unavailable for very long strings or even undefined for nonstationary signals, several coding procedures have been proposed which are always applicable and approach optimality very closely. The most elegant algorithm is due to *Lempel* and *Ziv* (1976 and 1978). The source sequence $S = s_1 s_2 \ldots$ is sequentially scanned and rewritten as a concatenation $w_1 w_2 \ldots$ of words w_k chosen in such a way that $w_1 = s_1$ and w_{k+1} is the shortest word that has not appeared previously. For example, the string 0011101001011011 is parsed as 0.01.1.10.100.101.1011. Once a new word has been identified, the signal is further inspected, with the addition of one more symbol if the current block is already present in the list. In this way, each word

w_{k+1} is the extension of some word w_j in the list, where $0 \leq j \leq k$ and, by convention, w_0 is the empty word ϵ. Since $w_{k+1} = w_j s$, with s the current input symbol, w_{k+1} is encoded as the pair (j, s). The specification of the integer j requires at most a number $\log_b N_w(N)$ of alphabet symbols, where $N_w(N) \leq N$ is the total number of codewords and N is the length of the input sequence. Hence, the total length $L(N)$ of the encoded sequence grows as

$$L(N) \sim N_w(N)\big(\log_b N_w(N) + 1\big) \tag{8.16}$$

for increasing N. As long as N is small, $L(N)$ may well exceed N (as in the example above). When $N \gg 1$, however, the words w_k become so long that they are more efficiently identified by means of their "order number" j (the position of the prefix) in the list and by the suffix s rather than by a direct description (for details on the algorithm, see Bell, Cleary, and Witten (1990)). The number $N_w(N)$ of words in a *distinct parsing* (i.e., consisting of different words) of a sequence $S = s_1 s_2 \ldots s_N$ satisfies (Cover & Thomas, 1991)

$$N_w(N) \sim c\,\frac{N}{\ln_b N}, \tag{8.17}$$

for $N \to \infty$, with $c \leq 1$. The *Lempel–Ziv complexity* C_{LZ}, defined as

$$C_{LZ} = \limsup_{N \to \infty} \frac{L(N)}{N}, \tag{8.18}$$

is hence less than 1, as can be verified by substituting Eqs. (8.16, 8.17) into Eq. (8.18). The following rigorous bound holds (Cover & Thomas, 1991): given a stationary shift dynamical system with metric entropy K_1 and invariant measure m,

$$C_{LZ} \leq K_1 / \ln b \tag{8.19}$$

for m-almost all sequences $S = s_1 s_2 \ldots s_N$. Since the codeword length per symbol L_n satisfies relation (8.7), C_{LZ} is readily seen to achieve optimality (i.e., the bound (8.19) holds as an equality). Once again, a measure of algorithmic complexity (although with a restriction on the coding method) turns out to be a measure of randomness. This result holds as an equality (Ziv & Lempel, 1978) not only for all cases of physical interest but also for sequences like the digits of π, which have zero Kolmogorov's complexity, since Lempel–Ziv's coding is sensitive to the statistics of the source. In general, the efficiency of the code increases with the length of the input sequence: in the first steps, in fact, nearly every word is "new". In sequences with low entropy, many repetitions are present and the codewords are longer, so that fewer of them are found.

At variance with Shannon's information, for the computation of which the probabilities are to be estimated in a preliminary scan of the signal, the Lempel–Ziv complexity takes into account the statistics of the source through the frequency of occurrence of new codewords along the signal: hence, the code automatically incorporates the information on the probability distribution itself. The Lempel–Ziv code is an example of a *universal code*, i.e., a code that can be constructed without previous knowledge of the source statistics. Moreover, it is asymptotically optimal, since it achieves a compression rate equal to the metric entropy of the source.

The Lempel–Ziv procedure has been used to estimate the metric entropy K_1 by computing the average word length (Grassberger, 1989a) $\langle |w| \rangle_N = N/N_w(N)$ as a function of N. In fact, by recalling Eq. (8.19) as an equality, one can fix the proportionality constant c in Eq. (8.17) to $K_1/\ln b$. Then, $\langle |w| \rangle_N \sim \ln N/K_1$. Notice that every new word w is, according to the metric (4.5), the nearest neighbour sequence of some previously encountered codeword, lying at a distance $\varepsilon(w) = b^{-|w|}$. The average $\langle |w| \rangle_N$ can hence be rewritten as

$$\langle -\log_b \varepsilon(w) \rangle_N \sim \frac{\ln N}{K_1} \sim \frac{\log_b N}{D_1} ,$$

where the last relation follows from Eq. (6.27) for the generalized dimension computed with the constant-mass method (Badii & Politi, 1985), in which we have set $P = 1/N$, $q \to 1$, and $\tau \to 0$. This relation, which once more shows that the metric entropy is an information dimension in sequence space, has been turned into an algorithm by Grassberger (1989a).

Finally, we mention that the dependence of the total code length $L(N)$ on N for the Lempel–Ziv procedure and similar ones has been studied in Rissanen (1986 and 1989). For simple stochastic models (finite automata) defined by k conditional probabilities, one has $L(N) \sim NK_1/\ln b + k \ln N/2$.

8.5   Logical depth

As we have seen, algorithmic complexity is an estimate of randomness for almost any sequence (exceptions, like the digits of algebraic irrationals, form a set of Lebesgue measure zero in any finite interval of the real axis): its numerical value essentially coincides with Shannon's information. A complex message, however, is intermediate in entropy between a totally redundant and a random sequence. Bennett (1988) proposed to define the "organization" or complexity of a sequence not through its information content but through the "value" or importance it has for the reader: "*... the value of a message is the amount of mathematical or other work plausibly done by its originator, which its receiver is saved from having to repeat*". Of course, any sequence could be produced by mere coin tossing (a Bernoulli process). The receiver, however, when confronted with a passage of, say, a literary work, would reject this zeroth-order hypothesis on the ground that it entails a large number of *ad hoc* assumptions (or random guesses) and would seek an explanation based on a model with a richer structure. As a consequence, Bennett defines "*the plausible work involved in creating a message as the amount of work required to derive it from a hypothetical cause without making unnecessary* ad hoc *assumptions*".

These remarks were formalized by identifying the most plausible cause of the message S with its minimal algorithmic description and its complexity or *logical depth* with the time required to retrieve it from this minimal description. An object is deemed complex or *deep* if it is implausible except as the result of a long calculation (Bennett, 1988). This definition was inspired by life itself, the most striking example of a "pattern" with great logical depth: in fact, life was originated by a presumably short "program" which has been working for about 10^9 years. This analogy is, however, only approximate: as a more pertinent example, Grassberger (1989b) proposes the signal $S = \{s_0(0), s_0(1), \ldots, s_0(t)\}$, obtained by recording as a function of time t the symbol at site 0 in a spatial configuration that evolves under the elementary cellular automaton 86. When the initial configuration consists of an isolated "1" at site 0 ($s_i = \delta_{i0}$, $i \in \mathbb{Z}$), S appears to have maximal entropy (Wolfram, 1986). Since no other more efficient algorithm than the direct simulation is known for the generation of this sequence (which involves $O(t^2)$ operations), its depth is expected to be large.

Notice that the notion of logical depth mediates between algorithmic information, where the *size* of the *shortest* program is considered, and computational complexity, where the *run time* for the *fastest* program is investigated. The requirement of using the shortest program, in the present case, is tantamount to the need for the unknown underlying mechanism of the observed phenomenon or, in other words, for a compact "theory". A successful but redundant model ought to be discarded as "unlikely" or "artificial", since the probability that it can be assembled by a random cause is too low.

In order to arrive at a more formal definition of depth, it is useful to notice that a concisely coded message looks "more random" than the original one (i.e., it has larger entropy) and is harder to describe. In particular, the minimal encoding of a sequence carries all the necessary information to retrieve the original but appears nearly structureless. A decrease in (program) size corresponds to an increase in (computation) time. An object is deep if there is no *short cut* between its concise description and its actual structure. Accordingly, Bennett defines a finite string S *d-deep* with *k bits significance* or (d, k)-deep if every program to compute S in time $t \leq d$ is compressible by at least k bits. Hence, logical depth is a relation $d(k)$ that accounts for the minimal computational time d achievable by a program of size $k + K(S)$, where $K(S)$ is the algorithmic complexity of S (the length of the minimal program for S). Notice that depth cannot be characterized uniquely by a single pair (d, k): in fact, when the minimal program π for a string S is only a few bits shorter than some much faster program π' with the same output, a tiny change in S may cause a large change in the run time if some program π'', close to π', turns out to be the minimal one after the modification of S. This instability, which is more likely to occur for small k, should be visible in the graph of d versus k. To overcome this difficulty, Bennett proposed to average over all programs π that yield S, the weight being given by the expression

$$p(\pi) = b^{-|\pi|} .$$

Accordingly, the *reciprocal mean reciprocal depth* was defined as

$$d_{\mathrm{rmr}}(S) = P_a(S) \left[\sum_{\pi : U(\pi) = S} \frac{b^{-|\pi|}}{t_\pi(S)} \right]^{-1} , \tag{8.20}$$

where $t_\pi(S)$ is the run time needed by π to compute S and

$$P_a(S) = \sum_{\pi : U(\pi) = S} b^{-|\pi|} \tag{8.21}$$

is the *algorithmic probability* of S with respect to a universal computer U (hence, d_{rmr} is machine-dependent for finite sequences). The use of reciprocals is necessary to yield a converging average dominated by the fast, short programs for S: the various alternatives leading to S are treated like an array of parallel resistors in which fast computations "short-circuit" the slow ones (Bennett, 1988). Depending on S, d_{rmr} may be polynomial or exponential in the length $|S|$.

Analogously with algorithmic complexity, logical depth is not effectively computable: in fact, given a string S and a program π such that $U(\pi) = S$, it cannot be excluded, in general, that some shorter program π' yielding S can be found since its halting time might be arbitrarily large. The uncomputability of depth may be partly circumvented by estimating bounds for it. A lower bound could be obtained by finding a program π that produces S in time t and showing that no shorter program can do the same in less time. Correspondingly, the size $|\pi|$ would provide an upper bound to the algorithmic complexity of S. Logical depth accords with the intuitive notion of complexity, for which both periodic and random strings are simple: in fact, the PRINT program provides a rapid generation of a given random string, whereas any periodic sequence is immediately produced by a very short program. The digit sequences of $\sqrt{2} = 1.41421\dots$ or $e = 2.71828\dots$ would be deeper, although no definite value for their depth can be given.

8.6   Sophistication

The lack of an intrinsic difference between program and data is at the basis of the theory of algorithmic information. Turing machines with separate data and program tapes are fully equivalent to one-tape machines: in the latter, part of the input represents the program which, in turn, can be considered as a data string for an interpreter included in the design of the machine. The simplest application of this equivalence is the Lempel–Ziv coding algorithm.

A seemingly complex symbolic sequence S can be seen as a particular realization of a "purposeful" procedure, such as a program π for the construction of syntactically correct algebraic expressions, whereas different sequences S' result from different inputs w' to π. The whole class of admissible sequences would then be described by the set π of "grammatical" rules alone and the input words would be interpreted as data. The common properties of the objects in this class are encoded in the program and thus manifest its *function*. Every sufficiently long correct expression S carries in its *structure* an amount of information which may allow recovery of part or all of the production rules. Hence, we see that a distinction between the structure of an individual object and the (usually unknown) function of the originating machine suggests that one could distinguish program from data and characterize the complexity of the sequence through the functional properties of its source. This approach has been proposed by Koppel (1987 and 1994) and by Atlan & Koppel (1990)[4]. Consider, for example, the sequences $S = 01011101011$ and $S' = 11101101011$, two different instances produced from the regular expression $R = (01 + 1)^*$. The body of R (consisting of the replacing of each 0 on input by 01) constitutes the program, whereas the strings $D = 0011001$ and $D' = 11101001$ are the input data corresponding to S and S', respectively: they give the sequence of arcs to be followed on the directed graph for R.

Koppel and Atlan (1991) consider a Turing machine U with two unidirectional input tapes (for program and data), a nonerasable unidirectional output tape and a two-way work tape. The program π is self-delimiting: i.e., the first output symbol is printed while the last program symbol is read. Moreover, they define π *total* if the output $U(\pi, D)$ of U is infinite for all infinite data sequences D. Then, (π, D) is a description of a finite or infinite string S if π is a total self-delimiting program and S is the prefix of the output $U(\pi, D)$. While the mimimum of the size $|\pi| + |D|$ over all descriptions of S essentially coincides with the algorithmic information $K(S)$, the program size $|\pi|$ alone measures the "meaningful complexity" or *sophistication* of S (Koppel & Atlan, 1991). A random string has zero sophistication, as appropriate, since it entails no rules at all: the length of its program (a mere PRINT statement) may be conventionally set to zero.

The identification of the minimal description for the whole class of strings that share the same properties may not be achievable by inspecting a single finite sequence S: in fact, the minimal description of S may have *ad hoc* features and be, hence, too concise. In order to allow for longer descriptions which

---

4. The authors call "meaningful structure" of the sequence S what is here called "function" of the program. For a non-technical discussion, see Atlan (1987).

characterize the whole class, Koppel and Atlan introduced the concept of *c-minimal* descriptions, consisting of pairs (π, D) such that $|\pi| + |D| \le K(S) + c \le |\pi'| + |D'| + c$, where $K(S)$ is the algorithmic information of the finite string S, c is the free parameter in the definition, and (π', D') is any other description for S. Accordingly, a program π is called *c*-minimal for S if (π, D) is a *c*-minimal description of S for some D and $|\pi| \le |\pi'|$ for any *c*-minimal description (π', D') of S. The *c-sophistication* of S, $\mathsf{SOPH}_c(S)$, is the length of a *c*-minimal program for S.

As remarked by Koppel & Atlan (1991), a *c*-minimal program should exploit patterns in a given sequence S so as to allow its compression into a noncompressible data string. The role of the parameter c is illustrated by the following example (Atlan, 1990). The first N prime numbers can be generated either by the PRINT program using a list of prime numbers or by a program PRIME that applies a test for primeness to every integer, without any need for data. For N smaller than some N_0, the former description is certainly more concise. For increasing N, however, program PRIME becomes more and more economical. Allowing for descriptions that exceed the minimal one by some constant c, it may be possible to identify a pair (π, D) that remains the shortest one for large values of c, i.e., for long (or even arbitrarily long) strings.

Although appealing, the concept of *c*-sophistication requires the finding of the minimal description not of just a string but of the class of all strings that share certain common (unknown) properties, independently of their lengths, and allowing for some freedom in the choice of the tolerance c and of the distinction between program and data: quite a daunting task indeed! A practical implementation of this idea, hence, appears generally unfeasible. Notice that the definition of sophistication does not rely on the finiteness of the string S: the program size may be finite for infinite strings. In such a case, the constant c becomes irrelevant.

Typical systems that require almost no input data are D0Ls: the limit sequence is obtained by iterating a deterministic substitution starting with some initial symbol. Context-free grammars, instead, involve the nondeterministic choice of alternative derivation paths which thus have to be encoded as data in order to reproduce a given string S from the seed. Although strings produced by a CFG are, in general, algorithmically more complex than D0L strings, they are not so from the point of view of sophistication: their nondeterministic features, in fact, do not belong to the program part of the description. A similar remark applies to regular languages, since their rules form a finite list of allowed transitions and the data account for the succession of graph bifurcations that lead to the required symbol sequence. Hence, they do not have, in general, lower sophistication than context-free grammars, in spite of their lower classification in the Chomsky hierarchy.

Cellular automata have a finite specification, whereas their input data are usually infinite (the initial configuration). Whenever an attractor is reached after a finite transient, independently of the initial condition, the sophistication is very low since the direct simulation can be by-passed. Hence, sophistication is indeed able to distinguish among CA rules, even when the latter's specification lengths are equal. Automata with transients that grow polynomially with the lattice size, in fact, have presumably a different minimal description from those with exponential dependence (in which case it may well be impossible to find a better program than the CA itself).

Like algorithmic information and logical depth, sophistication is generally uncomputable and aims at a topological characterization of symbol sequences: no metric features are examined, since real numbers, such as probabilities, are not taken into account. The introduction of sophistication was inspired by an analysis of DNA (Atlan & Koppel, 1990). "*The widespread metaphor of DNA as a genetic program*" is criticized by Atlan (1990) who rather considers DNA strings as "*data for the cell metabolism machinery which would work like a program*". As remarked by von Neumann (1966), in order for a system to evolve to a higher degree of complexity, it is necessary that it contains its own self-description. The ability of DNA strings to replicate themselves could be interpreted as a sign of their complexity. This is presumably not sufficient, however, to conclude that they are akin to the encoding of some "biological Turing machine" but just that the laws of physics perform complex operations on sufficiently complex objects.

8.7   Regular-language and set complexities

All previous definitions, apart from Lempel–Ziv's, share the undesirable property of not being effectively computable. This is due to the request of minimality (optimality) for the program, associated with that of machine-independence for the complexity measure. An easily computable indicator has been proposed (Wolfram, 1984) for those symbolic signals that are recognized as being described by a regular-language $\mathscr{L}$: given the minimal deterministic finite graph G for $\mathscr{L}$, the *regular-language complexity* (RLC) is defined as

$$C_{\mathrm{RL}} = \log_b N ,$$

(8.22)

where N is the number of nodes of G. This is motivated by the consideration that the information needed to specify one state out of N possible ones, irrespective of its probability, is precisely $\log_b N$ (to the base b). A more refined measure might take into account also the arcs joining the nodes: possibly, not just their number but also their disposition (a graph with a fixed number of out- or in-going arcs per node is simpler than one with forbidden connections). The regular-language complexity is finite even for periodic signals, unless the arcs are labelled by words rather than by single symbols.

Higher-order languages, such as CFLs or D0Ls, cannot be modelled by finite directed graphs and C_{RL} diverges. Such graphs for CFLs have a tree-like shape, whereby the nodes represent internal states of the pushdown automaton with a certain stack content (Lindgren & Nordahl, 1988). The top node corresponds to the empty stack. Other languages give rise to different structures with arbitrary connections among the nodes.

The RLC is infinite for nonregular languages, which are the most interesting ones. Therefore, it has a rather limited domain of applicability and, even there, its definition might be improved. In order to extend it so as to include metric features of the subshift, when this supports an invariant measure m, one may consider the probability distribution induced on the nodes of the graph. Denoting by p_i the probability of visiting the ith node ($i = 1, 2, \ldots, N$) in a scan of the path corresponding to some input signal $\mathscr{S} = s_1 s_2 \ldots$, Grassberger (1986) defines the *set complexity*

$$C_{SC} = -\sum_{i=1}^{N} p_i \log_b p_i \qquad (8.23)$$

as the information of the set of the nodes on the graph. For this to be possible, restriction to deterministic automata is necessary. Furthermore, transient parts in the graph must be eliminated to guarantee the stationarity of the paths.

Obviously, the set complexity (SC) never exceeds the RLC. Moreover, it may be finite in some cases in which the RLC is not. This happens when the node probabilities p_i decrease sufficiently fast with the distance from the start node, as in the case of the (nonfactorial) CFL consisting of all palindromes (Grassberger, 1989b): namely, words of the type $w\overline{w}$, where $\overline{w}$ is w's mirror image. Notice that the "motion" on infinite graphs need not be stationary but may display diffusive behaviour: after a number t of steps nodes at a distance $x(t) \sim \sqrt{t}$ from the origin are visited.

If the grammar is not known but must be inferred from inspection of a symbolic signal $\mathscr{S} = s_1 s_2 \ldots$ with arbitrary length, there is no algorithm that can say whether a node or a finite subsequence is forbidden or just appears very rarely. This can even prevent the application itself of definition (8.22). To see this, consider the sofic language $\mathscr{L}_2$ produced by the roof map (4.4) at $a = 2/3$. The irreducible forbidden sequences form the set $\mathscr{F}_2 = \{01^{2n}0, n \in \mathbb{N}\}$ (Section 4.3). If only the first (00) is known, a one-node minimal graph is obtained. If the first two are taken into account, the graph has four nodes; with three forbidden strings, the nodes are 6 and their number increases when more and more prohibitions are included. This is illustrated in Fig. 8.3, where the labels (0), (1), (2), (∞), refer to the number N_f of forbidden sequences. In spite of the regularity of the language, the RLC of the approximating graphs diverges

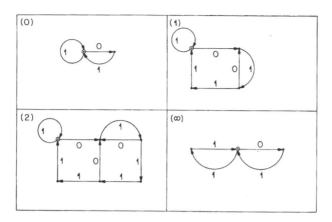

Figure 8.3 Directed graphs for the subshifts of finite type having the sets $\mathscr{F}_n = \{0(11)^i0, \, i = 0, 1, \ldots, n\}$ of forbidden words with $n = 0, 1, 2,$ and ∞, as indicated by the labels.

for $N_f \to \infty$! The graph corresponding to the language (labelled by ∞ in the figure) is indeed finite but it cannot be obtained by this limit procedure. Notice that the minimal exact graph (see Fig. 7.2) contains only two nodes: the one in Fig. 8.3 has the additional property that all arcs leading to a given node carry the same symbol. This problem does not arise if one determines the smallest graph exhibiting *at least* (and not *exclusively*) the detected prohibitions. This may be seen as an application of Occam's razor: among all models guaranteeing the same accurate reproduction of all the words up to the desired length, the smallest one is chosen (assuming complete ignorance about longer words).

It must be remarked that the minimal graph, while yielding the RLC, need not correspond to the minimal SC: there might be a larger graph in which fewer nodes are visited more frequently (Grassberger, 1986). For these reasons, neither the RLC nor the SC are particularly useful quantities in general. The definition of set complexity has been extended by Crutchfield and Young (1989) to a generalized index $C_{SC}(q)$ defined as the limit for $n \to \infty$ of the Renyi information I_q of Eq. (6.13) divided by n, where the probabilities involved are computed at the nodes of the graph as for the SC. Moreover, when the tree representing the language exhibits some form of self-similarity, as for the period-doubling substitution ψ_{pd} (examples 2.3 and 3.7), Crutchfield and Young (1990) identify subtrees with identical link structure (called a "morph") up to some depth ℓ (a similar regrouping was proposed by Huberman and Hogg, 1986) and assign a node in the graph to each class of identical morphs. A transition between two nodes is allowed if there are branches on the tree that connect vertices in the two corresponding classes. In this way, one obtains a "renormalized" picture of the language. The SC, however, still diverges for languages arising from substitutions.

Finally, one might speculate about possible complexity measures referring to the size of the automata characterizing higher computational classes. Since there are no general minimization algorithms for these automata, any comparison among the languages based on such criteria would be unreliable.

The discussion of the RLC and the SC has introduced a new point. Both of them, in fact, refer to the *whole set* of all substrings of an infinite sequence $\mathscr{S}$ and not to a particular finite string S as the algorithmic information. Hence, although closely related to automata and to the algorithmic approach, they anticipate the main feature of the measures to be discussed in the next chapter.

8.8 Grammatical complexity

Nonregular languages give rise to infinite directed graphs and, hence, cannot be characterized by the RLC. One might instead consider the growth rate of the graph size as a function of the the largest word length n. Although this route has not been directly pursued, similar ideas inspire the notion of complexity introduced by Auerbach (1989) and Auerbach & Procaccia (1990). Consider a symbolic dynamical system $(\Sigma', \hat{\sigma})$ having as an invariant set the closure $\Sigma' = \overline{\mathcal{O}(\mathscr{S})}$ of the orbit of an infinite sequence $\mathscr{S}$. Can one find a set of basic words $\mathscr{V} = \{v_1, v_2, \ldots\}$ such that $(\Sigma', \hat{\sigma})$ is a renewal system (see Chapter 4) over these words (i.e., such that $\mathscr{S}$ consists of *all* infinite concatenations of words in $\mathscr{V}$)? The answer is obviously no, unless the number and size of the v_i are allowed to diverge. In such a case, the type of divergence may be employed as an indicator of complexity. It is immediately realized that the notion of basic word must be broadened in order to obtain meaningful results. Given, for example, the regular expression $R = (1 + 01(01)^*11)^*$ (the irreducible forbidden words being 00 and 0110), the language $\mathscr{L}(R)$ should be approximated by $\mathscr{L}'(R) = (v_1 + v_2 + \ldots + v_{n(\ell)})^*$ where $n(\ell)$ is the least number of basic words necessary to yield all legal subsequences of $\mathscr{S}$ up to length ℓ. Hence, one finds the list $\mathscr{V}_3 = \{1, 01\}$ for $\ell = 3$, $\mathscr{V}_4 = \{1, 0111, 010111\}$ for $\ell = 4$, and $L_{n+3} = \{1, 01(01)^i 11, i = 0, 1, \ldots, n\}$ for a generic $\ell = n + 3$. That is, the number of the basic words may diverge even for a regular language! Moreover, these lists do indeed account for all the *forbidden* words up to the length ℓ but they do not reproduce all *legal* strings (such as 0, 01, 10, 11, 010, ...) through concatenations of the v_i's: most allowed sequences appear just as *subsequences* of the words thus formed.

To overcome these problems, Auerbach (1989) and Auerbach & Procaccia (1990) permit the v_i to be *regular expressions* so that they are more properly called *building blocks* rather than basic words. More precisely, given the directed graph that accounts for all substrings of maximum length ℓ, the building blocks are identified as the regular expressions associated with all families of independent paths leading from a suitable start node n_0 back to itself (without earlier returns to n_0). By "family" is meant an expression of the type uv^*w, where word v corresponds to a loop in the graph which may be scanned an arbitrary number of times. The graph may not have transient parts. Since it is desirable to minimize both the number and size of the blocks and to render the procedure independent of the start node for $\ell \to \infty$, expressions containing the operator "+", like $u(v + v')^* w$, must be weighted according to the number of alternatives they allow (Auerbach & Procaccia, 1990): graphically, they correspond to subtrees of the global complete tree of regular expressions that approximates the language. Identical subtrees are counted only once. Notice that, in general, the blocks may change when changing ℓ (their "length" is considered to be the number of symbols they contain, although they may yield infinite sequences or trees).

Once the blocks are known, the *dynamical* (Auerbach, 1989) or *grammatical complexity* (Auerbach & Procaccia, 1990) is defined as

$$C_G = \limsup_{\ell \to \infty} \frac{n(\ell)}{\ell}, \qquad (8.24)$$

where the supremum is taken when the ordinary limit does not exist. By construction, the grammatical complexity is identically zero for all regular languages: it measures how higher-order languages deviate from regularity.

For the Fibonacci sequence, one has two blocks for each ℓ: in fact, $\mathscr{V}_2 = \{1, 01\}$ (00 forbidden), $\mathscr{V}_3 = \{01, 101\}$ (00 and 111 forbidden) and, in general, $\mathscr{V}_{F_n}$ consists of the two Fibonacci words S_{n+1} and S_n. Hence, $C_G = 0$ (as desired, given the simplicity of the language) but only the trivial concatenation $S_{n+1} S_n = S_{n+2}$ is allowed and the legal strings of the language must be searched within S_{n+2}: the rules of the shift dynamics are not recovered by concatenating many (short) building blocks but are merely recorded within the two basic words (i.e., there is a transfer of "complexity" from a possible infinity of building blocks to finitely many but infinitely long basic words). By the minimality of the system, in fact, no periodic subsequences exist: regular expressions, instead, usually contain cycles. In this example, the basic words cannot be considered as low-order data units out of which higher-order structures are formed.

Indeed, the definition of C_G is tailored to subshifts having a number $N(n)$ of periodically extendible substrings that grows exponentially with their length n. As an example, consider the logistic map (3.11) at $a = 1.7103989\ldots$, where the symbolic orbit $S = s_1 s_2 \ldots$ of the critical point $x_c = 0$ (the so-called *kneading sequence*) is aperiodic and can be generated by a D0L substitution (Auerbach & Procaccia, 1990). Defining $t_k = (\sum_{i=1}^k s_i) \bmod 2$ and $\tau = \sum_{k=1}^\infty t_k 2^{-k}$, the largest value of τ is given by the kneading sequence and the smallest one by its first iterate (under the left shift). All other symbolic strings $S' = s_1' s_2' \ldots$ that can be produced by the logistic map have an intermediate value of τ (Collet & Eckmann, 1980). They correspond to all periodic and aperiodic orbits of the map (in case one of the periodic orbits is stable, a strange repeller may exist simultaneously: then, the interesting symbolic dynamics arises from the latter). Hence, although the kneading sequence at $a = 1.7103989\ldots$ is minimal, the map admits an invariant set with infinitely many unstable periodic orbits. Its grammatical complexity (Auerbach & Procaccia, 1990) is $C_G = (\sqrt{5}-1)/2$. This is in accordance with the nonregularity of the language. The rules underlying it, however, are straightforward (in particular, the set $\mathscr{F}$ of irreducible forbidden words is analogous to that of the Fibonacci sequence given in Chapter 4) so that their complexity seems to be overestimated. The building blocks were constructed so as to reproduce all periodically extendible strings explicitly, through concatenations: aperiodic strings appear then as subsequences only. Therefore, the grammatical complexity is referred to the set of all periodic orbits of the map and not to the kneading sequence itself. Evidently, this "nonlinear-dynamical" approach has a limited domain of applicability: in general, one wishes to study a given symbolic sequence, whatever its origin is, and not an infinite set of words related to it through some transformation.

For the special case of one-dimensional maps with positive topological entropy and a binary alphabet, the grammatical complexity is bounded between 0 and 2 (Auerbach & Procaccia, 1990). As will be shown in Section 9.3, the number $N_f(n)$ of irreducible forbidden sequences of length n, which is bounded for one-dimensional maps, in general diverges exponentially with n. As a consequence, the number of building blocks also exhibits exponential divergence since it must be increased each time a new prohibition is found (unless the blocks are chosen as in the paradoxical situation illustrated above for the Fibonacci substitution). Therefore, the grammatical complexity is expected to diverge in most cases.

Finally, we mention that the grammatical complexity C_G can be extended to a function $C_G(q)$ of generalized complexities by introducing a suitable weight for the building blocks and an averaging parameter q (Auerbach & Procaccia, 1990): $C_G = C_G(0)$ is recovered for $q = 0$.

Similarly to the RLC and to the SC, the grammatical complexity refers to the set of all sequences compatible with a shift dynamics and not to a single string with a fixed origin. Moreover, it is suggestive of a hierarchical order since larger-scale structures are obtained by joining smaller ones (although the transfer of "complexity" from the tree representing the language to the building blocks, while yielding a plain complete tree, produces blocks that often cannot be considered elementary at all). Hence, this approach provides a suitable crossover to more explicitly hierarchical complexity measures such as those discussed in the next chapter.

Extended further reading

The following is a list of recent articles on complexity which could not be discussed individually in the present chapter, although the ideas they expose have been partly taken into account.

- *"Algorithmic" approaches:* Agnes and Rasetti (1991) connect deterministic chaos and coding algorithms for geodesics in curved spaces with undecidability and algorithmic complexity; Crisanti *et al.* (1994) apply algorithmic complexity to study maps acting on a discrete phase space and to a fully connected neural network; Uspenskii *et al.* (1990) and Zvonkin and Levin (1970) elaborate on the meaning of randomness in discrete mathematics.

- *Entropy and rational partitions:* Casartelli (1990) introduces the concept of *rational partitions*, constructed from pairs of ordinary partitions in such a way that their inverse under intersections exists (in analogy with the embedding of integers in the rational field). An extended entropy functional, which is sensitive to the antisimilarity between two probabilistic experiments, is then proposed as a measure of complexity.

- *Information fluctuations:* Bates and Shepard (1993), consider the node probabilities p_i of a nondeterministic finite automaton (NFA) and form the quantities $\Gamma_{ij} = \ln(p_i/p_j)$, where i and j are consecutive nodes in a scan of the graph, and $\Gamma_{0j} = \ln(p_0/p_j)$, where node 0 is fixed and j corresponds to the current state. The fluctuations $\langle \Gamma^2 \rangle - \langle \Gamma \rangle^2$ of these information differences, computed as time averages during a long walk on the graph, are proposed as complexity measures and applied to NFA models of elementary cellular automata. Their values increase with the graph size N. "Random" and periodic rules are classified as simple.

- *Integral coarse-grained entropy:* Zhang (1991) and Fogedby (1992) perform a coarse-graining of a symbolic signal of length N by averaging the values s_i of the observable over τ consecutive sites and compute the corresponding 1-block entropy $H_1(\tau)$ (Eq. 5.36). By defining complexity $K(N)$ as the sum of $H_1(\tau)$ over all $\tau = 1, 2, \ldots, N$, they show that this quantity (which diverges for $N \to \infty$) is maximal for a Gaussian signal with $1/f$-type power spectrum.

- *"Physical complexity":* Günther *et al.* (1992) evaluate the probability $P(s, s'; n)$ for a system to be in the state s at time n and in s' at time $n + 1$ and compare it with the test distribution $Q(s, s'; n) = P(s; n)P(s'; n + 1)$ by forming the mutual information $H(P \| Q)$ (Section 5.6), which is taken as the definition of a "physical complexity" $C(n)$. The formula is applied in the limit $n \to \infty$ to a stochastic model for population dynamics with an increasing number of species.

"Renormalized entropy": Saparin *et al.* (1994) investigate the suitability of the renormalized entropy (Klimontovich, 1987) as a complexity measure. Denoting by $\rho(x, a)$ the probability density of the x variable in a dynamical system with control parameter a, they evaluate the entropy $S(a) = -\int \rho(x, a) \ln \rho(x, a) dx$ and propose to compare it with the entropy at a different parameter value a', after the corresponding distribution $\rho(x, a')$ has been modified to a function $\tilde{\rho}(x, a, a')$ which yields the same mean "energy" as $\rho(x, a)$. Consequently, they define an effective Hamiltonian H_e as $H_e(a) = -\ln \rho(x, a)$ and $\tilde{\rho}(x, a, a') = c(T_e) \exp[-H_e/T_e]$, where T_e is an effective temperature which must be tuned in order to satisfy the equal-energy constraint and $c(T_e)$ is a normalization constant. The difference $\Delta S = \int \rho(x, a') \ln[\rho(x, a')/\tilde{\rho}(x, a, a')] dx$ is an estimate of the change of "order" in the transition $a \to a'$. Its value depends on the reference state a. For the logistic map, with a fixed at the first bifurcation value, ΔS is never zero.

Thermodynamic depth: Lloyd and Pagels (1988) identify the complexity of a system in state s_0 at time zero, reached in n discrete time steps through the sequence $S(n) = s_{-n} s_{-n+1} \ldots s_{-1}$ of intermediate states, with the "thermodynamic depth" $D_T(s_0)$, defined as $D_T(s_0) = -k \ln P(s_{-n} \ldots s_{-1}|s_0)$, where $P(s_{-n} \ldots s_{-1}|s_0) = P(s_{-n} \ldots s_{-1} s_0)/P(s_0)$ is the probability that s_0 is attained through the particular path $S(n)$ and k is an arbitrary constant. No clue is given as to how large n should be chosen (see also remarks in Bennett (1990) and Grassberger (1989b)).

Chapter 9

Hierarchical scaling complexities

The three notions of complexity discussed in Sections 8.3–8.5 measure the difficulty of reproducing a given symbolic sequence S *exactly*. As a result, random signals either have maximal complexity (algorithmic or Lempel–Ziv) or, if recognized as simple (logical depth), require an infinite storage (as for the PRINT program). These unsatisfactory characteristics are partly eliminated by the proposals of Sections 8.6–8.8 which attempt to identify and describe rules common to a *set* of sequences "equivalent" to S. In this way, zero complexity is attributed to randomness which corresponds to the absence of rules whatsoever (besides stationarity and a finite number of specifications for the probabilities). In the new approach, however, the problem seems to have become even harder: how can one infer general rules of an unknown equivalence class of objects by simply observing a single element of it? How can one be sure that the "minimal" class has been found, and not one that contains also elements with additional properties? A solution can be found or approached when the objects under investigation are virtually infinite symbolic sequences for which stationarity holds, as we assume. In such a case, in fact, the shift invariance of the probabilities ensures that different subsequences contained in the signal $\mathscr{S}$ share the same properties, provided that they are sufficiently long. Hence, they can be interpreted as distinct members of the equivalence class that is looked for. Strings of length $n_1 \gg n_0 > 1$ contain sufficient information about substrings of length up to n_0, strings of length $n_2 \gg n_1$ about those of maximum length n_1, and so on. Therefore, a hierarchical ordering of the language arising from the subshift defined on $\mathscr{S}$ is the natural framework to study the properties of the system. The rules of the class are learned gradually, from the most elementary to the most involved ones, and the accuracy of the description increases with the level of the approximation. The complexity measures discussed in this section refer explicitly to a hierarchical structure and to the infinite-depth limit of some suitable quantity associated with the corresponding tree.

Not only more practical and physically meaningful, the task of describing the "average" or "global" properties of the object as an element of a class requires new concepts which shed additional light onto the characterization of complexity. In fact, the intuitive belief that such a description is easier to find than an exact one is not correct. There are, e.g., non-random or moderately random systems having a number of rules that increases exponentially with their length: while an exact prediction of the future outcome may be possible with appreciable frequency (because of the low value of the KS-entropy), the listing of all legal outcomes requires inspection of the past history of the current state, possibly for an infinite time. The optimal model may then need to have infinite size. On the other hand, a real understanding of the system is not exhausted by a mere display of its mechanism, not even when this is specified by a few finite rules. Self-generated complexity, in fact, is produced by the infinite application of these rules, and manifests itself through the interplay of parts that one may recognize at different levels of resolution. The study of the scaling behaviour of physical observables from finite-resolution measurements appears, therefore, as an essential instrument for the characterization of complexity.

9.1 Diversity of trees

The tree structure underlying a hierarchical system can be characterized both topologically, i.e., by referring solely to its shape, and metrically, by considering the values of the observables associated with each of its vertices. The former approach is especially relevant when self-similarity does not hold. This appears to be the general case, as can be observed in biological organisms, electric discharges, diffusion-limited aggregates, dynamical systems, and turbulence, to name a few. Huberman and Hogg (1986) proposed to associate complexity with the lack of self-similarity, evaluated by means of the *diversity* of the hierarchy. This is best defined by referring to a tree with a finite depth n and maximal branching ratio b (Fig. 9.1).

The leaves are considered as identical units (i.e., not distinguishable by means of labels as for a language) representing components of a physical system. The tree is coarse-grained into subtrees with a given depth m which are then sorted into isomorphism classes: trees with identical shape (up to depth m) form a single unit (Huberman & Hogg, 1986; Ceccatto & Huberman, 1988). The tree T of Fig. 9.1 is composed of the subtrees T_1, T_2, and T_3 which, in turn, contain subtrees indicated with T_{ij}, and so on. Taking $m = 1$, for simplicity, the T_i can

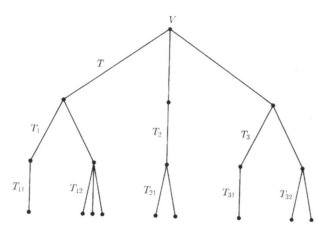

Figure 9.1 Sketch of a tree T having three subtrees T_i with 2, 1, and 2 branches, respectively. The subtrees of the T_i, indicated by T_{ij}, have, in turn, 1, 3, 2, 1, and 2 branches, from left to right.

be grouped into two classes: $\{T_1, T_3\}$, containing two-leafed trees, and $\{T_2\}$, containing one-leafed trees. Clearly, larger values of m discriminate the trees more sharply, so that more classes with fewer elements each are found: the number $k_T = k_T(m)$ of isomorphism classes is a nondecreasing function of m. The diversity $D(T)$, is defined recursively as[1]

$$D(T) = F(T) \prod_{i=1}^{b_T} D(T_i),$$ (9.1)

where $F(T)$ is the *form factor* of the tree T, b_T is the number of subtrees of T, and $D(T_i)$ is the diversity of the ith subtree. Various expressions for $F(T)$ have been proposed by Ceccatto and Huberman (1988), the simplest one being

$$\log_2 F(T) = N_T \log_2(2^{k_T} - 1),$$ (9.2)

where N_T is the number of leaves of T at the depth m. By the recursivity of definition (9.1), N_{T_i} and k_{T_i} must be considered in $F(T_i)$ at the next level, for each i (and analogously for all subtrees T_{ij}, T_{ijk}, ...). The *tree complexity* C_T is finally defined as (Ceccatto & Huberman, 1988)

$$C_T(T) = \log_2 D(T).$$ (9.3)

For the tree in Fig. 9.1, we have: $D(T) = F(T)\prod_{i=1}^{3}D(T_i)$, $F(T) = 3^5$ (since there are $N_T = 5$ leaves and $k_T = 2$ classes for $m = 1$); $D(T_1) = F(T_1)D(T_{11})D(T_{12})$, $F(T_1) = 3^4$, $D(T_2) = F(T_2)D(T_{21})$, $F(T_2) = 1$ (one class), $D(T_3) = F(T_3)D(T_{31})D(T_{32})$, $F(T_3) = 3^3$. A collection of p nonisomorphic trees has complexity

$$C_T\left(\cup_{i=1}^{p} T_i\right) = \sum_{i=1}^{p} C_T(T_i) + \log_2 F\left(\cup_{i=1}^{p} T_i\right).$$

1. We slightly modified the expression of $D(T)$ with respect to Huberman and Hogg (1986) by letting the product include all the b_T subtrees of T and not just the k_T classes.

The expression (9.2) of the form factor $F(T)$ shows that each terminal leaf, having no other subtrees than itself (i.e., $k_T = 1$), has diversity $D = 1$. The complexity is computed up to level n and the limit $n \to \infty$ is taken (possibly, the calculation is repeated for a larger discrimination depth m and the double limit $n \to \infty$, $m \to \infty$ is evaluated).

Self-similar trees are simple since they have just one isomorphism class. The term $2^{k_T} - 1$ in Eq. (9.2) represents the number of all (binary, ternary, ..., n-ary) "interactions" that can take place among k_T distinct classes (which are associated with elements of a physical system) connected to each other through the branches of T. The factor N_T was introduced to privilege trees with a large number of branches. The most complex (i.e., "diverse") trees are isomorphic, up to irrelevant permutations, to one with the following structure: the ith subtree T_i has i branches and $i = 1, 2, \ldots, b$; in turn, the jth subsubtree T_{ij} of T_i has j branches (with $j = 1, 2, \ldots, i$), and so on. For a random tree with an average number k of isomorphism classes and an average branching ratio b, Ceccatto and Huberman (1988) show that

$$C_T \sim m^\beta N(n) \frac{\log_2(2^k - 1)}{1 - k/b},$$

where $N(n)$ is the number of leaves at level n and the exponent β characterizes the dependence of the "complexity per leaf" $C_T/N(n)$ on the discrimination depth m: $\beta \approx 0$ for nearly ordered trees and $\beta \approx 1$ for complex ones.

Other forms for $F(T)$, such as $\log_2 F(T) = R^{-m} \log_2(2^{k_T} - 1)$, with R a ratio between energy barriers, may be employed (Ceccatto & Huberman, 1988) to relate complexity to diffusion on hierarchical structures (Huberman & Kerszberg, 1985; Rammal et al., 1986). Bachas and Huberman (1986) have shown that the decay of correlation functions in such problems is slowest for trees with large diversity. In their view, the problem of characterizing complexity acquires importance more from these connections than from the mere evaluation of an abstract indicator.

The tree complexity C_T clearly refers to the topology of general trees, without any reference to shift dynamical systems or formal languages. Hence, it is quite generally applicable but not sufficiently sharp. In fact, C_T is nonzero as soon as strict self-similarity does not hold. Forbidden words, whether irreducible or not, have the same influence on the value of C_T or have, at most, a weight that depends on the tree level but not on their importance in shaping the language. Moreover, the effect of high-order (i.e., low-lying) levels asymptotically dominates since the number of branches increases with the tree depth. On the contrary, low-order (i.e., short) prohibitions should contribute most, as they affect the whole structure of the tree. Finally, the most diverse trees, with their several nonbifurcating branches, correspond to dynamical systems with periodic attractors, that is, to nontransitive systems. In general, trees arising from shift dynamical systems need not have large diversity. Hence, although some of these incongruences can be healed, a more refined measure of complexity is needed to comprehend adequately the scaling properties of generic hierarchical structures.

9.1.1 Horton–Strahler indices

A very simple topological classification of ramified patterns is given by the indexing scheme devised by Horton (1945) and Strahler (1957) to describe river networks. For any finite tree T with root V, one needs to assign an order k to each vertex, starting with $k = 1$ at the leaves. The order of an internal vertex equals the largest of its immediate descendent's orders $(k_1, k_2, \ldots, k_b)$ if they are distinct, and is set to $k + 1$ if the k_i are all equal to k. The *Strahler number* S_T of the tree is the order of its root. A *path of order* k is a maximal sequence of branches joining a vertex of order $k + 1$ with a vertex of order k: that is, if a branch B leads from a vertex of order 3 to one of order 2 and its successor B' to another vertex of order 2, B' is considered to be a continuation of B and belongs to the same path. The scaling behaviour of infinite trees can be investigated by means of the *overall branching ratio*

$$\beta_k = \frac{p_k}{p_{k+1}},$$

where p_k is the number of paths of order k, and of the *length ratio*

$$\lambda_k = \frac{\langle \ell_{k+1} \rangle}{\langle \ell_k \rangle},$$

where $\langle \ell_k \rangle$ is the average length of the paths of order k (the ℓ_k may take the values of a physical observable, such as the actual length for rivers or the number of particles for fractal aggregates). For a complete binary tree of depth n, the number of order-k paths is $p_k = 2^{n-k}$. In general, for self-similar trees, $\beta_k \sim \beta$, independently of k. The branching ratios β_1, β_2, and β_3, averaged over the ensemble of all distinct binary trees with N leaves taken with the same statistical weight, tend to the value 4 for $N \to \infty$; the tree's Strahler number S_T is of the order of $\log_4 N$, and the expectation value of the top ratio β_{S_T} is smaller than 4 and tends asymptotically to a periodic function of $\log_4 N$, with mean 3.34 and amplitude 0.19 (Yekutieli & Mandelbrot, 1994). Extensive numerical simulations on off-lattice two-dimensional DLA clusters (Yekutieli *et al.*, 1994; see also Vannimenus, 1988, Vannimenus and Viennot, 1989) have shown that the branching ratios β_k with $k > 3$ also tend to a constant for $N \to \infty$, although convergence occurs only for very large values of N ($N > 10^6$). Hence, very large DLA clusters behave, from a purely topological point of view, as self-similar objects.

The evaluation of the branching ratios in the limit of an infinite tree associated with a shift dynamical system[2] can be assimilated to that of the topological entropy K_0. The sensitivity of the β_k to prohibitions (some β_k are larger than 2 in an incomplete binary tree) indicates that they may be used to quantify deviations from self-similarity. The Horton–Strahler ordering, however, is still quite rough in so far as it refers to unlabelled trees (vertices corresponding to irreducible forbidden strings are treated as any other incompletely branching vertex) and ignores metric features altogether.

9.2 Effective-measure and forecasting complexity

Referring to the convergence of the block entropies H_n (see Eq. (5.36)), Grassberger (1986) introduced a quantity C_{EM}, called the *effective-measure complexity* (EMC), which is related to the amount of information necessary in order to obtain an optimal estimate of the conditional probability $P(s_{n+1}|s_1 s_2 \dots s_n)$, given the n-block probabilities $P(s_1 s_2 \dots s_n)$. Consider the difference

$$h_n = H_{n+1} - H_n \tag{9.4}$$

between block entropies at consecutive hierarchical levels (see Ex. 5.25) which represents the additional average information needed to specify symbol s_{n+1}, given the previous n. By the positivity of $-\ln P$ and the subadditivity property (5.42), the function H_n is monotonic nondecreasing and concave in n. Hence, the values of H_n, $h_n \approx dH/dn$, and of the "curvature" $H_n'' = d^2 H/dn^2$ provide the least average information needed for a reliable estimate of $P(s_{n+1}|s_1 s_2 \dots s_n)$. Indicating by

$$\delta_n = h_{n-1} - h_n = 2H_n - H_{n+1} - H_{n-1} \geq 0 \tag{9.5}$$

the second variation of H_n changed of sign, Grassberger (1989b) remarks that this quantity is the average amount of information by which the uncertainty about s_{n+1} decreases when one more symbol in the past is taken into account. The information needed for the whole n-block $S = s_1 s_2 \dots s_n$ is then n times this contribution (i.e., $n\delta_n$). The effective measure complexity is accordingly defined as

$$C_{EM} = \sum_{i=1}^{\infty} i\delta_i \ . \tag{9.6}$$

2. Notice that such trees may have vertices with only one outgoing branch (all others being forbidden by the dynamics), so that the order k remains unchanged in passing from the descedent to the parent. When this happens more than once in cascade, there may be descendents of a common parent having orders k that differ by more than one unit: e.g., 3 and 1. The assignment of the order to the paths must be therefore modified with respect to the original definition, an order-k path being now one that joins an order-k vertex with a higher-order one.

In the case when the convergence of H_n to the large-n limit nK_1 occurs rapidly, C_{EM} is small. In general, one postulates (Szépfalusy & Györgyi, 1986; Misiurewicz & Ziemian, 1987) that

$$H_n \sim nK_1 + C - a\,[\ln(n+1)]^\alpha\, n^\beta e^{-\eta n}\,, \tag{9.7}$$

where C, a, and η are nonnegative constants and α, β are finite. Slow convergence is observed when $\eta = 0$: in such a case, $\beta < 0$ (if also $\beta = 0$, $\alpha < 0$, and so on for even weaker converging functions). Other laws satisfying $H'_n \geq 0$, $H''_n \leq 0$, and $H''_n \to 0$ are possible: for example,

$$H_n \sim nK_1 + C + an^\beta \tag{9.8}$$

with $a > 0$, and $0 < \beta < 1$, or even "multiplicative" corrections to the linear term, such as $H_n \sim a_1 n / \ln(a_2 + n) + C$ (Ebeling & Nicolis, 1991). The entropy differences h_n (9.4) exhibit a quicker convergence to K_1 than H_n/n because of the cancellation of the additive constant C:

$$h_n = K_1 + f(n)\,, \tag{9.9}$$

where the *convergence speed* $f(n) \geq 0$ is often exponentially small in n (Grassberger, 1986). When this verifies, since

$$\sum_{i=1}^{n} i\delta_i = \sum_{i=1}^{n} h_i - nh_{n+1}$$

and the second term on the r.h.s. differs from nK_1 by a quantity of the order of e^{-n}, the EMC can be rewritten as the "integral"

$$C_{EM} \approx \sum_{i=1}^{\infty}(h_i - K_1) \tag{9.10}$$

of the convergence speed $f(n)$ and depends, therefore, on the exponential rate η (9.7).

The alternative expression (Grassberger, 1989b)

$$C'_{EM} = \lim_{n\to\infty} [H_n - n(H_n - H_{n-1})] = \lim_{n\to\infty}(H_n - nK_1) \tag{9.11}$$

brings additional insight into the concept of effective-measure complexity. In fact, C'_{EM} is the intercept with the $n = 0$ axis of the tangent to the curve H_n vs. n at (n, H_n) for $n \to \infty$, so that it represents the minimal information to start with when constructing a model for the system. On the other hand, the r.h.s. in Eq. (9.11) shows that C'_{EM} equals the constant C in the asymptotic expression (9.7) of H_n. If (9.8) holds, $C'_{EM} = \infty$. It must be remarked that $C'_{EM} \neq C_{EM}$: the two quantities differ by $H_1 - K_1 \geq 0$, since $H_n = \sum_{i=1}^{n-1} h_i + H_1$.

The EMC is a lower bound to the amount of information needed to estimate $P(s_{n+1}|s_1 \ldots s_{n-1}s_n)$ "optimally", which Grassberger (1986, 1989b) calls *true-measure complexity* or *forecasting complexity*. The latter, however, is infinite in many simple cases as, e.g., when the sequence depends on a real parameter, which implies an infinity of digits of accuracy (Zambella & Grassberger, 1988). The inequality $C_{EM} \leq C_{SC}$ (see Eq. (8.23)) also holds (Grassberger, 1986).

The EMC is positive and finite for periodic sequences (with period larger than 1) and for any finite-order Markov process, since $h_n = K_1$ once the memory of the process is exhausted. Notice that the value of K_1 must be known with high accuracy in order to have a reliable estimate of the EMC, otherwise the errors accumulate dramatically in the sum (9.10). The EMC is zero only for period-one (constant) and Bernoulli sequences since they are characterized by the equality $h_n = K_1$, for all $n \geq 1$. Numerical investigations (Grassberger, 1986) indicate that the differences h_n tend exponentially to the limit K_1 in most one-dimensional unimodal maps, so that the EMC is finite. At the period-doubling accumulation point, however, $h_n \sim n^{-1}$ and $C_{EM} = \infty$. For the cellular automaton 22, instead, the convergence appears to be exponential (Ex. 6.7), although with a small rate, and C_{EM} is finite.

The positivity of the EMC for periodic signals can be corrected by letting the sum (9.10) start from a cutoff index n_0 (Szépfalusy, 1989), larger than the period p: in this way, also Markov processes with order $k \leq n_0$ have zero EMC. Even so, however, systems with exponentially converging entropy differences h_n remain complex and the actual value of C_{EM} depends on n_0, which is not satisfactory. These problems arise from the "integral" nature of the definition: although the probabilities of all subsequences are taken into account, the EMC is given by the sum over all lengths n and not by some scaling exponent in the large-n limit. The EMC is hence not a genuinely hierarchical measure of complexity. Finally, notice that its definition can be extended by referring it to the convergence properties of the generalized Renyi entropies K_q (see Chapter 6).

9.3   Topological exponents

The number $N(n)$ of different words of length n quantifies the richness of a language $\mathscr{L}$. In the theory of computation, the exponential growth rate of $N(n)$, the topological entropy K_0, is sometimes referred to as a complexity measure. Although questionable in general, this interpretation is meaningful within the hierarchical classification introduced by D'Alessandro and Politi (1990). They proposed, in fact, to attribute a series of complexity measures $C^{(i)}$ to the language $\mathscr{L}$ as the result of increasingly finer investigations, whereby $K_0 = C^{(1)}$ is placed at the bottom of the hierarchy and provides the coarsest information about $\mathscr{L}$: the one concerning its cardinality. This immediately ensures that periodic sequences are simple. In fact, a single check for each input symbol is sufficient to confirm a conjectured periodicity. Conversely, a random sequence has a higher degree of complexity, since an exponentially growing amount of work is needed to verify that all its subsequences are compatible with a random process.

As already pointed out, however, it is desirable to call "complex" languages that statisfy certain grammatical rules, irrespective of a possibly low entropy content. A partial solution to this problem is provided by the order-two exponent $C^{(2)}$, defined as the topological entropy of the set $\mathscr{F}$ of all irreducible forbidden words (IFW) which has been introduced in Chapter 4 (see also Chapter 7). As long as factorial transitive languages $\mathscr{L}$ are considered, the mere knowledge of $\mathscr{F}$ is sufficient to reconstruct the original language $\mathscr{L}$, so that $\mathscr{F}$ can be seen as a *presentation* of $\mathscr{L}$. Since the number $N_f(n)$ of IFWs of length n is at most equal to $bN(n-1)$, it follows that $C^{(2)} \leq C^{(1)}$, so that the passage from $\mathscr{L}$ to $\mathscr{F}$ is *de facto* a compression of information. On the one hand, an eventual equality between the two indicators implies the maximality of the order-two complexity $C^{(2)}$. On the other hand, purely random signals are simple not just because $C^{(2)} = 0$ but, even more, because $\mathscr{F}$ is empty. More generally, subshifts of finite type are simple by the finiteness of $\mathscr{F}$. Indeed, $C^{(2)}$ can be seen to measure the difficulty of approximating the language $\mathscr{L}$ through subshifts of finite type with increasing memory (length of the IFWs).

Still in the class of regular languages, strictly sofic systems may have positive $C^{(2)}$, as illustrated by the language corresponding to the graph in Fig. 9.2. Its IFWs are given by the regular expression $101(1 + 00)^*101$. The subword 101 acting as a suffix and a prefix guarantees that the words are not reducible to one another. In this case, $C^{(1)} \approx 0.6374$, while $C^{(2)} \approx 0.4812$, in agreement with the supposed inequality between the two indicators.

Besides RLs with a nonzero $C^{(2)}$, there are nonRLs which are instead classified as simple. These are the languages generated by one-dimensional (1d) maps with a finite number n_e of extrema, for which $N_f(n)$ cannot exceed n_e (see Section 7.3, where the case $n_e = 1$ has been discussed). Accordingly, $N_f(n)$ is bounded from above by a constant and $C^{(2)} = 0$. In particular, this holds at the (generic) parameter values that yield uncomputable languages (Friedman, 1991). The existence of such nonRELs in unimodal maps is qualitatively understood as follows. The language $\mathscr{L}$ of a 1d map is closely related to the kneading sequence T. In Section 7.3, we have shown how the former is constructed from the latter. Hence, the complexity of $\mathscr{L}$ can be ultimately traced back to that of T. Although not all possible concatenations of symbols arise from legal kneading sequences, there is a sufficiently large (uncountable) number of them to include encodings of uncomputable numbers. The apparent conflict between the order-two simplicity ($C^{(2)} = 0$) of these languages and their high-level classification in Chomsky's scheme is resolved by noticing that $C^{(2)}$ does not refer to the exact, machine-optimized reproduction of a given signal, but rather to a model that yields exactly a finite number of words in the language.

The usefulness of this approach is confirmed by the analysis of two-dimensional

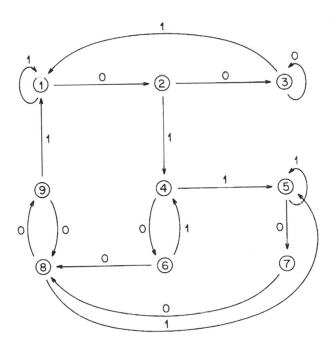

Figure 9.2 Graph of the regular language having a set $\mathscr{F}$ of irreducible forbidden words described by the regular expression $101(1+00)^{*}101$.

maps, which are attributed a higher degree of complexity. As discussed in Chapter 4, their homoclinic tangencies play the role of the extrema in 1d maps in separating the elements of a generating partition. The important difference is the infinity of the number of tangencies which, moreover, form a fractal set. This allowed D'Alessandro and Politi (1990) to show that $N_f(n)$ grows exponentially for generic parameter values with an exponent

$$C^{(2)} = \frac{\lambda_1^2}{\lambda_1 - \lambda_2}$$

where λ_1 and λ_2 are the positive and the negative Lyapunov exponent, respectively. In the limit $\lambda_2 \to -\infty$, the expected value $C^{(2)} = 0$ for 1d maps is recovered. In area-preserving maps, $\lambda_1 = -\lambda_2$ and $C^{(2)} = \lambda_1/2 \leq K_0/2$.

The present approach concentrates, in the first step, on the set of all admissible words (the language $\mathscr{L}$) and, in the second, on the set $\mathscr{F}$ of the IFWs. This suggests continuation of the analysis with the set (which might be indicated with $\mathscr{F}(\mathscr{F})$) of all irreducible prohibitions found in $\mathscr{F}$. At variance with $\mathscr{L}$, however, $\mathscr{F}$ is both nonfactorial and nontransitive: no subsequence of any word in $\mathscr{F}$ is, by definition, an IFW itself. In the language $\mathscr{L}$ of Fig. 9.2, for example, each IFW is identified not only by an even number of consecutive zeros (as in the factorial, transitive component of $\mathscr{F}$), but also by a prefix (101) and a suffix (101). In general, therefore, knowledge of all prohibitions in the language $\mathscr{F}$ does not permit the reconstruction of $\mathscr{F}$ itself unambiguously. As a further example, consider a language $\mathscr{L}$ generated by randomly choosing IFWs of length n from A^* with a given probability distribution, for all n. Knowledge of $\mathcal{F}$ does not help understand $\mathcal{L}$ at all.

Although a general theory that covers all languages is unlikely to exist, the hierarchy $\mathcal{L} \to \mathcal{F} \to \mathcal{F}(\mathcal{F}) \to \ldots$ sketched here can be ascended whenever the main source of diversity in $\mathcal{F}$ consists of a finite number of factorial and transitive components. In such cases, the complexity of $\mathcal{F}$ originates from its own IFWs and an index $C^{(3)}$ can be properly defined as the topological entropy of the language $\mathcal{F}(\mathcal{F})$.

This is, in particular, possible for regular languages. In fact, for any RL accepted by some graph G, there is a procedure to construct the graph G_f accepting the language $\mathcal{F}$ (see Appendix 5) which is, in turn, regular. In general, G_f contains transients (necessary to account for the lack of factoriality and transitivity of $\mathcal{F}$) as well as distinct ergodic components. The complexity measure $C^{(2)}$ is just the largest of the entropies of such components and it is conjectured to be strictly smaller than the topological entropy $K_0 = C^{(1)}$. Since each ergodic component in G_f is a finite graph, the procedure can be iterated, giving rise to a sequence of languages characterized by decreasing topological entropies $C^{(i)}$. We conjecture that, for each RL, there exists a finite k such that $C^{(k+1)} = 0$. This is the number of nested hierarchical levels in the language $\mathcal{L}$ (which may be seen as the "topological depth" of the language).

The elementary CA rule 22 illustrates the above ideas. Consider the regular language $\mathcal{L}$ associated with the spatial configuration obtained in one time step from a random initial condition (Section 7.3). The IFW analysis of $\mathcal{L}$, $\mathcal{F}$, $\mathcal{F}(\mathcal{F})$, etc., reveals the existence of four hierarchical levels with complexity indices $C^{(1)} \approx 0.6508$, $C^{(2)} \approx 0.5297$, $C^{(3)} \approx 0.4991$, and $C^{(4)} \approx 0.1604$. No rigorous results are instead known for the limit set $\Omega^{(\infty)}$ (Eq. 3.22) of this rule. Since, however, the size of the graph associated with the nth image is a rapidly increasing function of n, one expects that both the topological depth $k = k(n)$ and the values $C^{(i)}(n)$ also increase with n. With the approach devised in Section 7.3, it has been possible to determine all IFWs up to length 25. The results, displayed in Fig. 9.3, yield the value $C^{(2)} \approx 0.545$, very close indeed to the best finite-size estimate of the topological entropy $K_0 = C^{(1)} \approx 0.55$, and thus hint at the maximal degree of order-two complexity compatible with K_0. This finding explains the previously observed slow convergence of the thermodynamic functions (Chapter 6) and supports the presence of a phase transition in the entropy spectrum $g(\kappa)$. It is presently unknown, however, whether the equality $C^{(1)} = C^{(2)}$ holds exactly and whether the order-three index $C^{(3)}$ can be determined.

A language for which it is instead possible to prove that $C^{(2)} = C^{(1)}$ is the CFL associated with a one-dimensional discrete random walker which is free to move either up (u) or down (d) everywhere except at a barrier where it can either move up again or stand still (s). The number of admissible n-words is

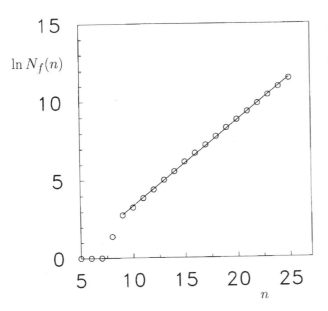

$\ln N_f(n)$

Figure 9.3 Natural logarithm of the number $N_f(n)$ of irreducible forbidden words of length n for the asymptotic spatial configuration of the elementary CA rule 22 vs. n, obtained with the method of the preimages described in Section 7.3. The fitted slope is $C^{(2)} = 0.545$.

exactly $N(n) = 2^n$, since each path can be extended in two ways, independently of the past. Besides us and sd, $\mathscr{F}$ includes all the sequences associated with paths that cross the barrier. From the theory of Brownian motion (see Feller, 1970, for the discrete case), their number is proportional to $N(n)/\sqrt{n}$. Accordingly, $C^{(1)} = C^{(2)} = \ln 2$, while $N_f(n)/N(n)$ vanishes algebraically.

In addition to solving many of the contradictions arising with the previously discussed complexity measures, the topological exponents introduced in this section fulfil a new, natural requirement: namely, that the complexity of the union $\mathscr{L}^u(m) = \bigcup_j^m \mathscr{L}_j$ of a finite number m of languages $\mathscr{L}_j$ be bounded from above by the largest of the single components' complexities. In fact, $K_0(\mathscr{L}^u(m)) \leq \max_j K_0^{(j)}$, where $K_0^{(j)}$ is the topological entropy of $\mathscr{L}_j$, while the number of IFWs may even decrease upon finite union (provided that the single languages share the same alphabet). This is a positive feature since the elementary operation of joining finitely many objects cannot be a source of complexity.

Analogously, it is important to study the complexity of the *product* $\mathscr{L}^p(m)$ of m languages, which can be conveniently defined as follows. Consider the two sequences $\mathscr{S} = \ldots s_1 s_2 \ldots$ and $\mathscr{S}' = \ldots s_1' s_2' \ldots$, with $s_i \in \{0, \ldots, b-1\}$ and $s_i' \in \{0, \ldots, b'-1\}$. Their direct product $\mathscr{S} \otimes \mathscr{S}'$ is the sequence $\ldots (s_1, s_1')(s_2, s_2') \ldots$ of corresponding pairs of symbols which can be turned into an ordinary symbol sequence $\mathscr{T}$ by means of the mapping from (s_i, s_i') to $t_i = b's_i + s_i' \in \{0, \ldots, bb'-1\}$ (where the numerical values of t, s, and s' are identified with the symbols themselves). The extension to m components is straightforward. This operation corresponds to the usual product $\mathbf{F} \times \mathbf{G}$ of maps. Since $K_0(\mathbf{F} \times \mathbf{G}) = K_0(\mathbf{F}) +$

$K_0(\mathbf{G})$ (Petersen, 1989), it follows that $K_0(\mathbf{F}^n) = nK_0(\mathbf{F})$: hence, K_0 is not an exhaustive measure of complexity, as already remarked, because it can be made arbitrarily large by simple iteration of the map $\mathbf{F}$. The order-two indicator $C^{(2)}$ presents a similar property. Each IFW of length n for $\mathscr{S}$ can, in fact, be paired with all *legal* n-blocks in $\mathscr{S}'$ and vice versa to yield the IFWs of $\mathscr{T}$. Hence, $C^{(2)}(\mathscr{L} \otimes \mathscr{L}') = \max\{K_0 + C^{(2)\prime}, K_0' + C^{(2)}\}$. In the m-component case, the maximum is over all sums involving any $m-1$ distinct topological entropies and the index $C^{(2)}$ of the remaining language. Finally, notice that this approach deals with topological features only. Extensions to metric properties are however not precluded.

9.4   Convergence of model predictions

A primary objective in the physical characterization of a symbolic pattern is the evaluation of its thermodynamic properties. Quantities like the generalized entropies K_q (6.11) (or the dimensions D_q (6.25) when they are accessible) measure the abundance of parts in the system and the spread of the values $O(S)$ assumed by the observable $\mathscr{O}$ (P or ε, respectively). They do not provide, however, any direct information about the complexity of the interactions which, as discussed in Chapter 6, reflect the dependence of $O(S)$ on changes in the configuration S. As illustrated by the phenomenon of phase transitions, correlations that decay slowly with $|S|$ are rather revealed by nonanalytic behaviour of the thermodynamic functions and by the corresponding slow convergence of the finite-size estimates. Accordingly, the expectation value of a generic multiplicative observable $\mathscr{O}$ is assumed to scale as

$$\langle O(S) \rangle \sim A \left(1 + an^\beta e^{-\eta n} \right) e^{n\alpha} \tag{9.12}$$

for large $n = |S|$ (cf. Eq. (9.7)). If $\eta > 0$, the system is said to have weak scaling correlations: this notion will be clarified below with the example of a Markov process. The vanishing of η, instead, indicates the existence of long-range interactions and critical phenomena as, for example, in self-avoiding random walks (Section 7.2).

The following two sections focus on the interactions as revealed either by global estimates of averages such as (9.12) or by a detailed model-system comparison at all levels of resolution. For definiteness, we take $O(S) = P(S)$, unless otherwise stated.

9.4.1   Global prediction

A class of models with important conceptual and practical applications to discrete stochastic processes is represented by the Markov approximations.

Recalling the definition (5.11) of an order-k Markov chain, the corresponding predictor $P_0(S)$ for a sequence $S = s_1 \ldots s_n$ is defined by the product

$$P_0(s_1 \ldots s_n) = P(s_1 \ldots s_k)P(s_{k+1}|s_1 \ldots s_k) \cdot \ldots \cdot P(s_n|s_{n-k} \ldots s_{n-1}) . \qquad (9.13)$$

The accuracy of such models in the evaluation of the thermodynamic functions can be investigated by means of the reduced canonical partition function (cf. Eq. (6.11))

$$Z_{\mathrm{can}}(n, q; s_{n-k+1}, \ldots, s_n) = \sum_{\{s_1, \ldots, s_{n-k}\}} P^q(s_1 \ldots s_n)$$

$$= \sum_{\{s_1, \ldots, s_{n-k}\}} P^q(s_1 \ldots s_{n-1})P^q(s_n|s_1 \ldots s_{n-1}) .$$

Making the assumption that the conditional probability actually depends only on the most recent k symbols, the sum over $s_1, \ldots, s_{n-k-1}$ readily yields

$$Z_{\mathrm{can}}(n, q; s_{n-k+1}, \ldots, s_n) = \sum_{s_{n-k}} Z_{\mathrm{can}}(n-1, q; s_{n-k}, \ldots, s_{n-1})P^q(s_n|s_{n-k} \ldots s_{n-1}).$$

This is a linear recursive equation for the reduced partition function which, in vector notation, reads

$$\mathbf{Z}_{\mathrm{can}}(n, \cdots) = \mathscr{T} \mathbf{Z}_{\mathrm{can}}(n-1, \cdots) ,$$

where $\mathscr{T}$ denotes the *transfer operator*. The ordinary partition function $Z_{\mathrm{can}}(n, q)$ (6.11) is recovered by adding together the b^k components (corresponding to all possible configurations of k symbols) of the vector $\mathbf{Z}_{\mathrm{can}}(n, \cdots)$. The explicit matricial form of the transfer operator is

$$T(q)_{s_0 s_1' \ldots s_{k-1}'; s_1 s_2 \ldots s_k} = P^q(s_k|s_0 s_1' \ldots s_{k-1}')\delta_{s_1' s_1} \cdot \ldots \cdot \delta_{s_{k-1}' s_{k-1}} , \qquad (9.14)$$

where the origin of the time lattice has been shifted by $n - k$. The $b^k \times b^k$ *transfer matrix* (9.14) describes the memory-k scaling relation between partition functions at consecutive hierarchical orders (n vs. $n - 1$). Since

$$Z_{\mathrm{can}}(n + 1, q) \sim e^{(1-q)K_q} Z_{\mathrm{can}}(n, q) ,$$

for $n \to \infty$, the entropy K_q is obtained from the largest eigenvalue $\mu_1(q)$ of $\mathscr{T}$: in fact, $\mathbf{Z}_{\mathrm{can}}(n + m, \cdots) \sim \mathscr{T}^n \mathbf{Z}_{\mathrm{can}}(m, \cdots)$, for $n \to \infty$ and sufficiently large m ($m > k$ for an order-k Markov process). This is the same relation as Eq. (7.12), deduced in Section 7.2 for probabilistic regular languages, where node-to-node transitions were considered.

The correlations along the "scaling direction" are responsible for the dependence of the conditional probability $P(s_n|s_1 \ldots s_{n-1})$ on the whole past history: i.e., the b-furcation rates of the observable P are neither always the same at each level, as in a Bernoulli process, nor are they completely unpredictable as in a random tree. Their range and amplitude can be globally quantified by the convergence properties of the finite-size estimate

$$K_q(n) = \frac{1}{1-q} \ln \frac{Z_{\text{can}}(n+1,q)}{Z_{\text{can}}(n,q)} \tag{9.15}$$

to the limit K_q, for $n \to \infty$. For the Bernoulli process, $K_q(n) = K_q$, $\forall\, n$, since $Z_{\text{can}}(n,q) = Z_{\text{can}}(1,q)^n$. For a Markov process of finite order k, the convergence exponent η (Eq. (9.12)) is given by the ratio

$$\eta_M \equiv \ln \frac{\mu_1}{|\mu_2|}$$

between the first two eigenvalues of $\mathcal{T}$. If the memory is infinite, the operator $\mathcal{T}$ can only be approximated. Accordingly, K_q may be compared with the estimates $\hat{K}_q(n)$ obtained from the largest eigenvalue μ_1 of a sequence of reduced operators $\mathcal{T}_k$ corresponding to Markov models of increasing order $k = n - 1$. Obviously, if the signal is Markovian, $\hat{K}_q(n)$ coincides with the asymptotic value as soon as n exceeds the order of the process. For more general signals and $q = 0$, D'Alessandro *et al.* (1990) have conjectured an exponential convergence with a rate $\hat{\eta} = C^{(1)} - C^{(2)}$, where $C^{(1)}$ and $C^{(2)}$ are the first two topological scaling indices defined in Section 8.2.3. In fact, indicating by $N_0(n)$ the number of n-blocks predicted to exist (i.e., with nonzero measure) by an order-$(n-1)$ Markov model, the actual number $N(n)$ of n-blocks is given by

$$N(n) = N_0(n) - N_f(n) \sim e^{n\hat{K}_0(n-1)} - e^{nC^{(2)}} \sim e^{n\hat{K}_0(n)},$$

where $N_f(n)$ is the number of irreducible forbidden words. By expressing $\hat{K}_0(n)$ in terms of $\hat{K}_0(n-1)$ and recalling that $C^{(1)} \equiv K_0$, the exponent $\hat{\eta}$ is obtained from the first-order correction to the scaling.

Whenever $\hat{\eta} > \eta_M$ (i.e., the interactions are sufficiently short-ranged), the convergence rate η of the finite-size estimates (9.15) is expected to coincide with η_M. This occurs indeed for the symbolic signal generated by the Hénon map (3.12) at $(a,b) = (1.4, 0.3)$ where the estimates at $q = 0$ yield $\hat{\eta} = 0.37$ and $\eta \approx 0.25$ (D'Alessandro *et al.*, 1990). If, instead, $\hat{\eta} < \eta_M$, the convergence speed of $K_q(n)$ is given by that of the Markov approximations: $\eta = \hat{\eta}$. This is the scenario illustrated in Ex. 6.7 for the elementary CA 22 and $q = 0$, where $K_0(n)$ turns out to be even more accurate than $\hat{K}_0(n)$.

9.4.2 Detailed prediction

The "microscopic" counterpart of the global approach discussed above requires comparison of suitable model predictions $O_0(S)$ with the actual values $O(S)$ of an observable $\mathcal{O}$ for each sequence S in the limit $|S| \to \infty$. This procedure is the basis of the notion of *scaling dynamics complexity* (Badii, 1990a and b). In conformity with the literal meaning of the term "complex" (see the preface), the system is seen as an arrangement of parts at all levels of resolution, labelled by the sequences S, the mutual interactions of which are accounted for by the effective dependence of $O(s_1 \ldots s_n | s_{-m} \ldots s_{-1} s_0)$ on the "environment" $s_{-m} \ldots s_{-1} s_0$. The observer measures his/her understanding of the system by trying to predict its asymptotic scaling behaviour from nonasymptotic data. The difficulty of the task of course depends on the observable O.

In this view, the numerical value of complexity is *relative* to the observer's ability. It measures the agreement between the object and the model inferred by the subject: in fact, a phenomenon ceases to "look" complex as soon as an "explanation" is found for it. In addition to these considerations, the measure to be defined next incorporates the following (Badii *et al.*, 1991):

1. Complexity lies between "order" and "disorder": in its most restrictive interpretation, this statement requires complexity to vanish just for periodic and random uncorrelated processes.

2. Complexity is an intrinsic, "intensive" property of the pair system-model. The complexity $C(\mathcal{L}^u(m))$ of the union of finitely many objects (languages or trees) $\mathcal{L}_j$ is not given by the sum $\sum_j C(\mathcal{L}_j)$ of the single complexities but rather satisfies $C(\mathcal{L}^u(m)) \leq \sup_j C(\mathcal{L}_j)$. In fact, if system $\mathcal{L}_1$ involves a number of rules which grows, for increasing resolution, more rapidly than that of system $\mathcal{L}_2$, the overall descriptive effort is asymptotically dominated by the properties of $\mathcal{L}_1$ alone[3].

3. Complexity should not increase when direct products of independent languages are taken without supplying any additional structure. Accordingly, it is required that $C(\mathcal{L}^p(m)) = \sup_j C(\mathcal{L}_j)$, where the "product language" $\mathcal{L}^p(m)$ is defined as in Section 9.3.

Consider the tree representing a language $\mathcal{L} \subseteq A^*$ arising from a shift dynamical system. Scaling dynamics complexity deals with the question:"What can be inferred about the composition of the tree at level $n+1$ from the knowledge of its structure at levels 1 to n, in the limit $n \to \infty$?"

This question can be further specialized to topological, metric, or more general properties of the system, which are associated, respectively, with the shape of the tree, with the probabilities $P(S)$ of the words at the vertices, or with the values $O(S)$ of other system observables (as, e.g., the size $\varepsilon(S)$ of the elements in the covering of a fractal set or eddy velocity differences in a turbulent fluid) which may similarly be attached to the vertices.

For each of the $N(n)$ legal sequences $S = s_1 \ldots s_n$ at level n, the observer forms all concatenations of the type $S' = Ss$, with $s \in A$, retaining only those which do not contain any known prohibition: i.e., it must be verified that $s_n s$,

---

3. This remark implies that finite languages (or objects with a finite number of homogeneous, structureless components) have no bearing on complexity.

$s_{n-1}s_n s, \ldots, s_2 \ldots s_n s \in \mathcal{L}$, otherwise S' is rejected *a priori*. Let the number of these concatenations S' be $N_0(n) \geq N(n)$. A prediction $P_0(S')$ is then made for the probability of each S'. All these operations use exclusively the information stored in the available n levels of the tree. A very crude prediction is, for example, $P_0(Ss) = P(S)P(s)$ (independence of s from S). The expected probabilities $P_0(S')$ need to be normalized by requiring their sum to be 1 at each level. Prediction and reality are then compared by means of the relative entropy

$$C_1 = \limsup_{n\to\infty} C_1(n) \equiv \limsup_{n\to\infty} \sum_S P(S) \ln \frac{P(S)}{P_0(S)}, \tag{9.16}$$

where the sum extends over all sequences S that are predicted to exist at level $n = |S|$. The nonnegative quantity C_1 is called *metric complexity* (Badii, 1990a). It vanishes when the predictions coincide with the actual probabilities for all S. This occurs not only for periodic and quasiperiodic signals but also for random, uncorrelated ones: in fact, even the simplest possible predictor, defined by $P_0(s_1 \ldots s_n) = \prod_i P(s_i)$, is exact in that case for all n.

More powerful predictors are able to recognize as simple signals with increasingly longer memory (interaction range). The order-k Markov model (9.13) yields the expression

$$C_1(n) = -H_n + H_k + (n-k)(H_{k+1} - H_k)$$

where H_n is the block entropy (5.36). Hence, $C_1 = 0$ if the signal is Markovian with memory r less than k (so that the model is redundant), and $C_1 = \infty$ if $r > k$ (insufficient model). In the former case, in fact, both $h_k = H_{k+1} - H_k$ (Eq. 9.4) and H_n/n coincide with the metric entropy K_1. In the latter, h_k is strictly larger than H_n/n. By setting $k = n-2$, the model uses all the information provided by the $(n-1)$-blocks to predict the level-n distribution. As a result,

$$C_1(n) = -H_n + 2H_{n-1} - H_{n-2} = h_{n-2} - h_{n-1} \tag{9.17}$$

gives the second variation of the block entropy H_n, changed of sign. Recalling Eq. (9.9), one immediately verifies that $C_1 = 0$ since $C_1(n) \approx -f'(n-1)$, where the convergence speed $f(n)$ tends to zero from above. Such a predictor, hence, is so powerful that no system is metrically complex. This result should not surprise the reader. In fact, $C_1 = 0$ only because the size of the model diverges, for increasing resolution n, at the same speed as the size of the system representation (the tree). In practice, however, one can never fulfil this condition.

With a suitable choice of the model, one can extract information about the strength of the scaling correlations through the finite-size estimates $C_1(n)$. Taking a memory $k = k(n) \sim \gamma_1 \ln n$ permits the testing of a possible exponential convergence of $|H_n - nK_1 - C| = r(n)$ (see Eq. (9.7)): if $r(n) \sim e^{-\eta n}$, $C_1(n)$ diverges for $\gamma_1 < 1/\eta$ and vanishes for $\gamma_1 > 1/\eta$. Analogously, with $k(n) \sim n^{\gamma_2}$ with $0 < \gamma_2 < 1$, one may investigate a possible algebraic law of the form

$r(n) \sim n^{-\beta}$: $C_1 = 0$ (∞) for γ_2 smaller (larger) than $(1+\beta)^{-1}$. Finally, $k(n) \sim \gamma_3 n$, with $0 < \gamma_3 < 1$ gives similar results when $r(n) \sim [\ln(n+1)]^{-\alpha}$.

Definition (9.16) is readily extended to a generic (nonnegative) observable $\mathcal{O}$, provided that the values $O(S)$ (and $O_0(S)$) are normalizable, by introducing the *pseudo-probability*

$$\pi(S) = \frac{O(S)}{\sum_S O(S)}, \tag{9.18}$$

where the sum runs over all sequences S at level n. When $O(S)$ is the size $\varepsilon(S)$ of a subelement of a fractal set, the corresponding complexity $\tilde{C}_1$ could be called *geometric*. With the introduction of $\pi(S)$ and of the corresponding prediction $\pi_0(S)$, the positivity (5.49) of the relative entropy is maintained. This important property permits one to interpret C_1 as a distance in the space of measure-preserving transformations: the true (π) and expected (π_0) distributions, in fact, can be traced back to two piecewise-linear maps $\mathbf{F}$ and $\mathbf{F}_0$ (or $\mathbf{F}_n$ if a sequence of approximations is made), respectively (Badii, 1991). The agreement between the two (observable-dependent) maps is measured by a quantity $C_1^{(\pi)}$, defined as in (9.16) but referred to π and π_0 (as recalled by the superscript). With this interpretation, the finite-n metric complexity $C_1^{(\pi)}(n)$ is reminiscent of the convergence speed $f(N)$ considered in ergodic theory, in which $N = N(n)$ is the number of partition elements at the nth approximation level (see Eq. (5.33)). The quantity $f(N) \leq 1$ is the $\mathbf{F}$-measure of the set of points over which the *first iterates* of the nth approximate map $\mathbf{F}_n$ and of the given map $\mathbf{F}$ differ. In the present approach, the partitions $\mathscr{C}_n$ and $\mathscr{C}$, respectively associated with $\mathbf{F}_n$ and $\mathbf{F}$, need not be the same: the former contains the $N_0(n)$ elements to which $\mathbf{F}_n$ attributes a nonzero mass. Moreover, the two (in general different) invariant measures of $\mathbf{F}_n$ and $\mathbf{F}$ are involved, since one compares the masses contained in the elements of $\mathscr{C}_n$ and $\mathscr{C}$ that carry the same label S. Hence, one evaluates the asymptotic discrepancy between the maps and not one-step differences as in Eq. (5.33). No periodic maps are considered, at variance with the analysis of Katok and Stepin (1967) which, however, can be extended to approximations by more general measure-preserving homeomorphisms (and, in particular, by Markov models), as discussed by Alpern (1978). With this generalization, expression (5.33) is itself a measure of complexity since it quantifies the agreement between model and object. Periodic and finite-order Markov maps can be seen as approximations to the system from "below" and from "above", respectively: both are dense in the group G of all automorphisms, according to suitable topologies (Halmos, 1956; Alpern, 1978). If any of them gives good agreement, no complexity subsists; if not, models in other classes of transformations (e.g., minimal or Markov with increasing memory) must be tried. This view is closely related to the so-called realization problem in ergodic theory, one of the main results of which is that every ergodic MPT on a Lebesgue space can be realized as (i.e., is metrically isomorphic to) a minimal, uniquely ergodic homeomorphism of a compact metric space (Petersen, 1989).

When referred to a generic pseudo-probability π (Eq. 9.18), the metric complexity $C_1^{(\pi)}$ can be nonzero even in connection with a Markov predictor with memory $k = n - 2$. In particular, expression (9.17) does not apply. In order to see this, let us define a random tree through the relation $\pi(Ss) = a_s(S)\pi(S)$, in which S is an n-block and $s \in A$ a symbol. The arbitrary branching rates $a_s(S)$ depend, in general, on the parent node S. At variance with true probabilities, the π need not satisfy conditions (5.4) and (5.5) and the sum $\sum_s a_s(S)$ need not be unity for every S. The only constraint is, in fact, $\sum_S \pi(S) = 1$ at each level $n = |S|$.

Examples:

[9.1] In a Cantor set covered by segments of lengths $\varepsilon(0) = 1/2$ and $\varepsilon(1) = 1/3$ at the first level and $\varepsilon(00) = 4/9$, $\varepsilon(01) = 1/27$, $\varepsilon(10) = 1/9$, and $\varepsilon(11) = 1/27$ at the second, one has $\pi(0) = 3/5 < \pi(00) = 12/17$: $a_0(0) = 20/17$ is larger than 1! This situation is by no means exceptional and does not descend by any inconsistency of the model (indeed, $\sum_s \varepsilon(Ss) \leq \varepsilon(S)$ for all S, as it should).   □

[9.2] Consider a random tree with ordinary probabilities $P(S)$ and let $\tilde{h}_S = -\sum_s a_s(S) \ln a_s(S) \in [0, \ln b]$ denote the entropy of the b-furcation at node S. Then,

$$H_{n+1} = H_n + \sum_{S:|S|=n} P(S)\tilde{h}_S = \sum_{i=0}^{n} \langle \tilde{h} \rangle_i \,, \tag{9.19}$$

where $\langle \tilde{h} \rangle_i = \sum_{S:|S|=i} P(S)\tilde{h}_S$ and $\langle \tilde{h} \rangle_0 = H_1$. The sequence $\{\langle \tilde{h} \rangle_i\}$ is a stochastic process, with its own probability measure $m(\tilde{h})$, which governs the scaling properties of the Ps. In this sense, one may speak of "scaling dynamics" (Feigenbaum, 1988) when considering the law that relates the values of an observable at different resolution levels. For independent increments $\langle \tilde{h} \rangle_i$ one finds, by the law of the iterated logarithm (Feller, 1970), $H_n \sim nK_1 + O(\sqrt{n \ln \ln n})$ and, hence, $C_1 = 0$ by Eq. (9.17).   □

In the case of pseudo-probabilities, sums like $\sum_{sSs'} \pi(sSs') \ln \pi(S)$ do not simplify to $\sum_S \pi(S) \ln \pi(S)$ as it occurs, instead, in the derivation of Eq. (9.17). Hence, the values $C_1^{(\pi)}(n)$ are expected to oscillate in quite an unpredictable way for arbitrary choices of the rates $a_s(S)$. Different subsequences $C_1^{(\pi)}(n_k)$, with $\lim_{k\to\infty} n_k = \infty$, may converge to different limits. This phenomenon is reminiscent of chaotic renormalization group transformations arising in disordered systems (Section 6.3).

A purely topological characterization of the complexity of the tree is obtained by setting all nonzero probabilities to a fixed value: $P(S) = 1/N(n)$ and $P_0(S) = 1/N_0(n)$ for all S, where $N(n)$ is the number of legal n-blocks and $N_0(n)$ the number of predicted n blocks. By substituting these values into Eq. (9.16), one obtains the *topological complexity*

$$C_0 = \limsup_{n \to \infty} C_0(n) = \limsup_{n \to \infty} \ln \frac{N_0(n)}{N(n)}. \tag{9.20}$$

As illustrated at the beginning of this subsection, $N_0(n) \geq N(n)$, so that $C_0 \geq 0$. The topological complexity is identically zero if all predicted orbits exist, past some length n_0, as happens for the subshifts of finite type. Moreover, since $N_0(n) \leq bN(n-1)$, we have the bound $C_0 \leq \ln b - K_0(n-1)$, where $K_0(n-1)$ is the finite-size estimate (9.15) of K_0. Every extension $S' = Ss$ of S which does not contain any shorter forbidden word and is found to be forbidden itself is, by definition, an IFW. Therefore, $N_0(n) = N(n) + N_f(n)$ and interesting situations occur when $N_f(n)$ diverges at the same speed as $N(n)$. This has been numerically observed, for example, in the limit set of the elementary cellular automaton 22 (Fig. 9.3). Since, instead, $N_f(n)$ is bounded from above by a constant in one-dimensional maps with a finite number of monotonic pieces and by $e^{nK_0/2}$ in two-dimensional maps, $C_0(n)$ vanishes in those cases.

If $N(n)$ and $N_f(n)$ diverge with the same exponential rate (i.e., $C^{(1)} = C^{(2)}$), $C_0(n)$ is usually expected to vanish algebraically, as shown with an example in the previous section. The result $C_0 = 0$ for trees with a finite maximal branching number b should cause no surprise since the predictor described above again makes use of *all* the IFWs up to length $n-1$, so that model and tree size have the same leading behaviour. As already remarked, such predictors are mainly theoretical tools which cannot be implemented in practice when n becomes very large (e.g., $n > 20$). Using a smaller or more slowly increasing amount of information, on the other hand, may yield a positive C_0. These considerations are completely analogous to those made in the metric case. Finally, notice that the definition of C_0 does not rely on a positive topological entropy K_0. In fact, a strictly positive C_0 may be conjectured to occur, irrespectively of the model, in peculiar limit cases in which $N(n)$ and $N_f(n)$ diverge with the same, arbitrarily slow, speed.

Topological and metric complexity $C_0^{(\pi)}$ and $C_1^{(\pi)}$ are two special cases of a complexity function $C_q^{(\pi)}$ which may be defined in analogy with the generalized dimensions and entropies. In order to guarantee the positivity of the relative entropy (9.16), it is useful to introduce the second-order pseudo-probability $\tilde{\pi}^{(q)}(S) = \pi^q(S)/\sum_S \pi^q(S)$ and the analogous quantity $\tilde{\pi}_0^{(q)}$ for the predictor, so that

$$C_q^{(\pi)} = \limsup_{n \to \infty} \sum_S \tilde{\pi}^{(q)}(S) \ln \frac{\tilde{\pi}^{(q)}(S)}{\tilde{\pi}_0^{(q)}(S)}. \tag{9.21}$$

Clearly, when $\pi(S) = P(S)$, C_q reduces to C_0 (C_1) when $q \to 0$ ($q \to 1$).

The scaling dynamics complexity (9.21) is a relative quantity which depends on both the system and the model. It is also easy to verify that it fulfils the three requirements listed at the beginning of this subsection. First of all, unpredictability is referred to the class of all equivalent signals compatible with the source (and to the scaling properties of the associated observables) rather than to the specific conformation of yet unexplored neighbourhoods of a given sequence[4]. Hence, "ordered" (e.g., periodic or quasiperiodic) signals are as simple as disordered (random, delta-correlated) ones, even for the simplest predictor. The zone between the regions denoted by the words "order" and "disorder", in an imaginary one-dimensional representation of all shift dynamical systems, may therefore be associated with complexity and be progressively reduced by improving the models as one's "understanding" of the system deepens.

The second point is most easily illustrated for the topological complexity. The expression $R(n) = \ln\left[N_0(n) + N_0'(n)\right] / \left[N(n) + N'(n)\right]$, where primed and unprimed quantities refer to different systems, asymptotically either coincides with one of the two complexities C_0 or C_0' (when both $N_0(n) > N_0'(n)$ and $N(n) > N'(n)$ or vice versa), or it takes the mixed form $R(n) = \ln\left[N_0(n)/N'(n)\right]$ (the other being symmetric). In the latter case, $R(n) < \ln\left[N_0(n)/N(n)\right] = C_0(n)$ by definition and the statement is verified.

The validity of the third property is evident: the factorization of both the number of admissible words and of the word probabilities of a product language $\mathscr{L}^p(m)$ into the product of the single components' contributions is realized even by the lowest-order predictor.

Finally, notice that the scaling dynamics complexity is not expected to increase when the language is translated by using a code (Sections 8.1 and 8.2) that compresses the information. On the contrary, we may think that the complexity of the signal is transferred to the code. In the ideal case, when the resulting signal is a completely random sequence, the inverse code can be interpreted as a perfect model and its size used, in principle, as a complexity measure. This is the case of the language $(1 + 01)^*$. Once the two codewords 1 and 01 have been identified and renamed 1 and 0, respectively, one obtains a full shift and no prohibitions occur. The rule "00 is forbidden" is thus incorporated into the code and is, therefore, automatically accounted for by the model. A more general example is provided by the roof map displayed in Fig. 4.4, where the tree, after the recoding $(1, 01, 001) \rightarrow (a, b, c)$, has ca as the only IFW. Moreover, words like 10, 100, 010, etc., appear only as subwords inside concatenations (as, e.g., ab, ac, bb, and bc). The whole language can then be reconstructed from a model based on a "smaller" and more regular tree over three symbols. In the case of the Hénon map (3.12) at $(a, b) = (1.4, 0.3)$, the compression of information and

---

4. The two views may be seen as mutually "orthogonal". In the latter, one has the usual unpredictability of local behaviour, either in space or time, which is accounted for by the metric entropy.

the increased efficiency of a Markov model after recoding has been illustrated in Section 8.2, Fig. 8.1. The property of being nonincreasing under (information-compressing) coding for a complexity measure may be added to the previously discussed three primary properties.

Finally, let us illustrate in more detail the concept of scaling dynamics, introduced by Feigenbaum (1988) and briefly mentioned in Ex. 9.2, in connection with the present approach to complexity. Using the order-n Markov predictor defined by

$$\pi_0(sSs') = \pi(s'|S)\pi(sS) , \tag{9.22}$$

where s and s' are symbols and S an n-block, the ratio π/π_0, appearing in the definition of $C_1^{(\pi)}$ (Eq. 9.21), takes the form $\pi(s'|sS)/\pi(s'|S)$. Complexity hence deals with the dependence of conditional probabilities on symbols (here s) lying farther than n steps away from the current position. In the thermodynamic picture, $-\ln \pi(S) = U(S)$ is the potential energy of the spin chain S, and such dependence can be interpreted as the change in the energy difference $U(Ss') - U(S)$ that occurs when interaction with a further spin at the left extreme of the chain is taken into account. The rules that govern the asymptotic behaviour of $\pi(S)$ for each symbolic path S constitute the scaling dynamics. In the next section, we shall address the question of how to produce dynamical equations of this kind of "motion".

With reference to a binary alphabet ($s \in \{0,1\}$), let $\rho(m) = \sum_{i=1}^{m} s_i$, with i running along the sequence sSs' and $m \leq n + 1$. The values of $\rho(m)$ and s' represent the number of particles in the first m sites and in the last site of the chain, respectively. Then, comparison with the tempering condition (6.8) shows that the expression

$$\ln\left[\pi(s'|sS)/\pi(s'|S)\right] = U(sSs') - U(sS) - U(Ss') + U(S) \tag{9.23}$$

yields the variation $W_n' = W_{\rho(n+1),s'} - W_{\rho(n),s'}$ of the interaction between two groups of particles disposed on n and 1 lattice sites, upon changing n to $n + 1$. The complexity $C_1^{(\pi)}$ is the average of W_n' over all sequences sSs'. Both $\rho(n)$ and the distance r between the two groups (recall Eq. (6.8)) depend on the specific sequence sSs'. In fact, r takes all values between 0 and n (as, e.g., for $sSs' = ss_1 \ldots 11$ and $sSs' = 100 \ldots 01$) and is, on average, proportional to n. Hence, $\langle W_n' \rangle$ may be nonzero in the limit $n \to \infty$ if the convergence exponent γ is sufficiently low and if groups of particles with large $\rho(n)$ and small r have a larger weight. In fact, the tempering condition breaks for $\gamma \leq 1$ in one-dimensional lattices. Such behaviour implies that the thermodynamic description is no more appropriate and witnesses the unpredictability of the scaling dynamics, a stronger condition than just the occurrence of a phase transition (related to interactions decaying with an algebraic exponent $1 < \alpha < 2$). In other words, one may say that the statistical description of the object is as "complex" as the object itself.

9.5   Scaling function

The existence of a range of local exponents (given by the finite-volume fluctuations of the energy) and the lack of self-similarity (corresponding to long-ranged correlations) are the primary challenging aspects of the thermodynamics of shift dynamical systems. Because of these phenomena, which are ubiquitous in nature, model predictions based on a limited amount of information fail to be accurate beyond a certain resolution level. The hierarchical organization of the values of the observables on a tree, which renders the observer's task straightforward in the self-similar case, is however still fundamental in the study of generic systems. With the introduction of pseudo-probabilities π, as in Eq. (9.18), the problem becomes equivalent to the investigation of the scaling properties of a fractal measure (which need not satisfy the consistency conditions (5.4, 5.5)). While the thermodynamic functions defined in Chapter 6 provide a global macroscopic characterization of the measure, the detailed microscopic information is carried by the *local scaling ratios*

$$\sigma(s_1,\ldots,s_{n+1}) = \pi(s_1,\ldots,s_{n+1})/\pi(s_1,\ldots,s_n) \tag{9.24}$$

between the "masses" at two consecutive vertices of the tree. When self-similarity holds, the descendant-to-parent ratio (9.24) depends uniquely on the present symbol s_{n+1}: the measure splits in the same b rates at each stage of the construction as, e.g., in the Bernoulli process (Ex. 6.1). Nevertheless, the energy $E = -\ln \pi(S)/|S|$ takes values in a real interval, in the limit $|S| \to \infty$, as a result of the combinations of the b local rates. A much more interesting and physically relevant situation occurs when $\sigma(S)$ actually depends on the entire history S. In such a case, the scaling ratios $\sigma(S)$ themselves may vary in a continuous range when $|S| \to \infty$. As a consequence, not only phase transitions are expected to arise but the whole thermodynamic construction becomes questionable if the conditions for the existence of the thermodynamic limit are no longer fulfilled.

In order to visualize the interactions along each possible path on the hierarchical tree, Feigenbaum (1979b, 1980) proposed to represent the local rates $\sigma(S)$ as a function $\sigma(t)^5$ of the unit interval, called the *scaling function*, by mapping each sequence S to a set $I_S \in [0,1]$. At the nth resolution level, $\sigma(t)$ is

----

5. We use the same symbol σ, as long as no confusion arises.

approximated by a piecewise constant function $\sigma_n(t)$ which takes the value $\sigma(S)$ (Eq. (9.24)) in the interval I_S. Defining

$$t_n(s_1 \dots s_n) = \sum_{i=1}^{n} s_{n-i+1} b^{-i} , \qquad (9.25)$$

one obtains the b^n adjacent intervals $I_S = [t_n(S), t_n(S) + b^{-n})$. For $n = 2$ and 3 and $b = 2$, this yields $t_2(00) < t_2(10) < t_2(01) < t_2(11)$ and $t_3(000) < t_3(100) < \dots < t_3(111)$. The sequences are in an inverse alphabetical order, so that $\sigma_3(010)$ and $\sigma_3(110)$, for example, refine $\sigma_2(10)$ in the interval $[t_2(10), t_2(10) + 1/4)$, when passing from $n = 2$ to $n = 3$. In other words, the change in the conditional probability of 0 after 1 upon further extension of the memory by one symbol in the past is investigated. In a thermodynamic context, this is a change in the energy upon a spin flip at a next-nn site.

The aim of the scaling function representation is to order the hierarchical cascade consisting of all values of the observable $\mathcal{O}$ with a clever choice of the symbolic dynamics, in such a way as to let the interactions depend in a weaker and weaker way on distant symbols. If this falling off is sufficiently rapid, the nth approximant $\sigma_n(t)$ converges to a limit function $\sigma(t)$. Formally,

$$\sigma(t) = \lim_{n \to \infty} \sigma_n \left(b^{-n} \lfloor b^n t \rfloor \right) . \qquad (9.26)$$

The eventual convergence crucially depends on the symbolic ordering and, of course, on the scaling properties of $\mathcal{O}$[6].

Examples:

[9.3] For the Bernoulli shift, either $\sigma(S) = p_0$ or $\sigma(S) = p_1$, depending on whether $s_n = 0$ or 1, respectively. Hence, $\sigma(t)$ also takes these two values, with a discontinuity at $t = 1/2$. A finite number of values for $\sigma(t)$ is found for every self-similar cascade. □

[9.4] The attractor of the logistic map (3.11) at the period-doubling accumulation point (Ex. 3.3) is a Cantor set which, at the kth resolution level, can be covered by 2^k segments with extrema x_i and x_{i+2^k}, where $x_n = f^n(0)$ and $i = 1, 2, 3, \dots, 2^k$. Let $\varepsilon_i^{(k)}$ denote the length of the ith segment, so that its offsprings have lengths $\varepsilon_i^{(k+1)}$ and $\varepsilon_{i+2^k}^{(k+1)}$. Since the logistic map has an order-reversing branch ($f'(x) < 0$, for $x > 0$), some care is necessary in the symbolic assignments. Suppose that, at $k = 1$, the label $s = 1$ is attributed to the longer segment and $s = 0$ to the shorter (the former, $\overline{x_2 x_4}$, with length $\varepsilon_2^{(1)}$, being the leftmost one). For consistency, symbol 1 should always be reserved for the larger

6. Notice that σ (Eq. (9.24)) can be defined by considering the dependence of π on one more symbol (s_0) in the "past" rather than in the "future" (s_{n+1}), since the reading direction along the signal is irrelevant. In fact, either both or none of the two corresponding scaling functions converge. The two limits may, however, differ for a generic shift dynamical system. They obviously coincide for the spatial configurations of symmetric cellular automata.

of the two descendants at each further splitting. Hence, the two offsprings $\overline{x_2 x_6}$ and $\overline{x_8 x_4}$ (with lengths $\varepsilon_2^{(2)} < \varepsilon_4^{(2)}$) of $\overline{x_2 x_4}$ are labelled by 10 and 11, respectively. Analogously, $\overline{x_3 x_7}$ (longer than $\overline{x_5 x_1}$ and lying to its left) has label 01 and $\overline{x_5 x_1}$ 00. Calling a pair of adjacent intervals direct (D) if its left member is the longer, and inverse (I) otherwise, the metasymbols D and I form an ordinary tree in which I always labels the left branch at each bifurcation.

Although this attractor is not self-similar, it is well approximated by a self-similar two-scale Cantor set. In fact, $\sigma(t)$ is nearly constant for $t \in [0, 1/4)$, where it is close to the value $\sigma(0^+) = \alpha^{-2}$ (with $\alpha = -2.502\ldots$ as in Ex. 3.3), and for $t \in [1/4, 1/2)$, where $\sigma(t) \approx \sigma(1/2^-) = -\alpha^{-1}$, and satisfies the relation $\sigma(t + 1/2) = \sigma(t)$. The function $\sigma(t)$ exhibits a discontinuity at every diadic rational point $t_{ik} = i 2^{-k}$. The next approximation hence consists of a four-scale Cantor set, with rates given by the mean value of $\sigma(t)$ in the t-intervals $[0, 1/8)$, $[1/8, 1/4)$, $[1/4, 3/8)$, and $[3/8, 1/2)$: this model is formally equivalent to the Markov chain of Ex. 6.2. At the period-doubling chaotic threshold, the ratios $\sigma(s_1, \ldots, s_{n+1})$ of Eq. (9.24) with $\pi(S) \propto \varepsilon(S)$ satisfy the Hölder condition (Aurell, 1987; Przytycki & Tangerman, 1996)

$$|\sigma(s_1, \ldots, s_n, 1) - \sigma(s_1, \ldots, s_n, 0)| \le 2^{-n\gamma(s_1, \ldots, s_n)},$$

where $\gamma_1 \le \gamma(s_1, \ldots, s_n) \le 2\gamma_1$ and $\gamma_1 = \ln|\alpha| / \ln 2$.   □

Each rate $\sigma(S)$ involves, through the ratio (9.24), values of the observable measured at possibly distant points on the invariant set but with a close symbolic link, as shown in the example above. Feigenbaum (1988) stresses that obtaining the "equations of motion" for σ itself in a general (chaotic) case is equivalent to a change of variables that permits computation of the salient features of the motion, rather than merely exhibiting the real-space trajectory. The high irregularity of the motion in a Bernoulli shift is thereby mapped to an extremely simple σ. Analogously, a moderately turbulent fluid or a fractal aggregate may turn out to possess straightforward scaling properties by a suitable symbolic ordering of its components.

The scaling function hence constitutes a further characterization of complexity. The range $R(\sigma)$ of $\sigma(t)$ contains all local asymptotic scaling rates of the system. At the lowest level of a classification based on $\sigma(t)$, one finds a finite $R(\sigma)$, which corresponds to self-similar processes; at the next, a countable infinity of rates; then, a Cantor-like $R(\sigma)$; and, finally, a continuum of rates. It is also clear that $\sigma(t)$ is closely related to the transfer matrix $\mathbf{T}(q)$ defined in Eq. (9.14), the entries of which are just powers of the ratios (9.24) for the observable of interest. All previously discussed convergence issues are, hence, obtainable from $\sigma(t)$.

Notice that the scaling function approach is sharper than the macroscopic thermodynamic description based on quantities like $g(\kappa)$, K_q, and similar ones.

Consider, in fact, a four-scale Cantor set with a Hamiltonian $\mathcal{H}(S) = -\ln \varepsilon(S)$ of the form (6.5), where the coefficients $p_{\alpha|\beta} = \sigma_{\alpha\beta}$ are arbitrary. Its free energy D_q can be shown to depend on two of the rates (σ_{01} and σ_{10}) only through their product and is, therefore, the same for all sets with the same value of $\sigma_{01}\sigma_{10}$ (Feigenbaum, 1987). These sets can be therefore distinguished through their $\sigma(t)$ but not through D_q.

The relationship between the scaling function $\sigma(t)$ and the metric complexity C_1 (or, more generally, $C_q^{(\pi)}$; see Eqs. (9.16 and 9.21)) can be illustrated by defining a new ordering index $t_n^{(\pi)}$ in terms of the usual t_n of Eq. (9.25). Let

$$t_n^{(\pi)}(S) = \sum_{S':t_n(S')<t_n(S)} \pi(S'), \qquad (9.27)$$

where the sum is taken over all sequences S' with a smaller t_n-index than the one of the actual sequence S. In this way, for $b = 2$ one obtains the intervals $[0, \pi(0))$ and $[\pi(0), 1]$ at level 1, $[0, \pi(00))$, $[\pi(00), \pi(00) + \pi(10))$, ..., and $[\pi(00) + \pi(10) + \pi(01), 1]$ at level 2, and so forth. The values of the new scaling function $\sigma^{(\pi)}(t^{(\pi)})$ are still evaluated as in Eq. (9.24). Since the width of each interval $I(S)$ is given by the (pseudo)-probability $\pi(S)$, forbidden sequences simply do not appear in $\sigma^{(\pi)}$. This procedure is illustrated in Fig. 9.4(a) (Badii et al., 1991) for the ternary alphabet $A = \{1, 2, 3\}$ and the prohibition "31": the superscript (π) has been omitted for simplicity and the shorthand notation σ_S has been used for the ratio $\sigma(S)$ of Eq. (9.24), so that σ_{12} is the conditional probability of 2 given 1 (to be compared with the unconditioned probability σ_2 of 2). This representation permits one to write a compact expression for what can be considered as the "hardest" type of complexity in the present approach. Clearly, the most selective choice is obtained with the best possible predictor in the Markov class which does not reduce to the mere definition of conditional probability: namely, the one given by Eq. (9.22). Recalling Eq. (9.24), the metric complexity $C_1^{(\pi)}$ (9.16) (or, analogously, the generalized complexity $C_q^{(\pi)}$ (9.21)) can be written as

$$C_1^{(\pi)} = \limsup_{n \to \infty} \left\langle \ln \frac{\sigma_{n+1}^{(\pi)}}{\sigma_n^{(\pi)}} \right\rangle, \qquad (9.28)$$

where the average is given by the area (with sign) between the logarithms of two successive approximations to the scaling function in the representation of Fig. 9.4(a). If $\sigma_n \to \sigma$ for $n \to \infty$ rapidly enough for all t or from different directions for different ts, so that mutual cancellation of terms occurs, $C_1^{(\pi)} = 0$. This relates complexity to the convergence properties of the scaling function or, in other words, to the unpredictability of the scaling dynamics.

Although nonconvergence of $\sigma_n(t)$ (which is possible only for pseudo-probabilities) is not easily detected in mathematical models, it is believed to be quite common (Feigenbaum, 1993; Tresser, 1993). As already remarked

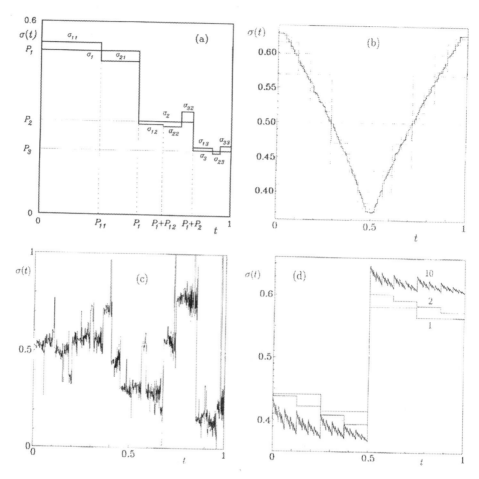

Figure 9.4 Generalized scaling functions: (a) illustration of the coding procedure for $\sigma(t)$ with the ternary alphabet $A = \{1, 2, 3\}$ and the forbidden word "31"; (b) first 6 approximations $\sigma_n(t)$ to the probability scaling function of the Lorenz equations (3.2) at "standard" parameter values; (c) approximation $\sigma_{10}(t)$ for the logistic map (3.11) at $a = 1.85$; (d) $\sigma_n(t)$ for the filtered baker map, at $r_0 = 0.3$, $r_1 = 0.5$, and $\gamma = 0.35$, and levels $n = 1, 2,$ and 10.

in Section 9.4, the unpredictability of the scaling dynamics lies beyond the appearance of phase transitions in the associated thermodynamic formalism. Indeed, one can have the latter phenomenon without the former, as illustrated in Fig. 9.4(b) for the probability scaling function of the Lorenz system (3.2) at "standard" parameter values ($\sigma = 10$, $r = 28$, $b = 8/3$) and with a binary generating partition (Badii *et al.*, 1991 and 1994). Notice the exponential convergence to a smooth limit curve (up to statistical fluctuations). The continuous range for $\sigma(t)$ indicates the existence of a phase transition. Indeed, the system is close to an intermittency in the vicinity of the two fixed points $\mathbf{x}_{\pm} \equiv (\pm\sqrt{b(r-1)}, \pm\sqrt{b(r-1)}, r-1)$. The persistence of the trajectories in their

neighbourhoods and the scarceness of jumps between the two folds of the attractor are reflected in the high probability of the sequences $00\ldots0$, $11\ldots1$, and
in the low probability of 01 and 10, respectively. This explains the form of the
curve $\sigma(t)$, high at the extrema and low in the middle.

An example of an irregular but converging probability scaling function is
shown in Fig. 9.4(c) for the logistic map (3.11) at $a = 1.85$. Notice that a
few rates approach the value 1: they correspond to unconditioned symbol-to-
symbol transitions and have a counterpart in prohibitions which are represented
by (asymptotic) zeros of $\sigma(t)$. Most of the rates are, at this resolution, around
$\sigma = 0.5$ and, thus, indicate maximal uncertainty in the time evolution.

Finally, we give an example of an interval-length scaling function which is
explicitly computable by iterating a one-dimensional "scaling map". Consider
the filtered baker map of Ex. 6.8 with $p_0 = p_1 = 1/2$ and denote by σ_{n+1} the
ratio $\sigma_{n+1}(Ss) = \varepsilon(Ss)/\varepsilon(S)$ between two interval lengths at orders $n = |S|$ and
$n + 1$. A straightforward procedure, analogous to that of Paoli et al. (1989a),
yields the recurrence

$$\sigma_{n+1} = r(n+1) + \gamma \left[1 - \frac{r(n)}{\sigma_n} \right] \frac{r(n+1) - c(\gamma)}{r(n) - c(\gamma)} , \qquad (9.29)$$

where $r(k) = r_0$ or r_1 depending on whether $s_k = 0$ or 1, respectively, and
$c(\gamma) - 2/(3 - \gamma)$. For each sequence S, the value of the scaling function $\sigma_n(t)$ is
obtained by iterating Eq. (9.29) for n steps with initial condition $\sigma_1 = \gamma + r(1)d(\gamma)$,
with $d(\gamma) = (1 - \gamma)(1 - \gamma/2)$, and by supplying the proper sequence $\{r(k)\}$ of
rates. Thus, $\sigma_n(t)$ is the outcome of a stochastic nonlinear map. The sequences
$000\ldots0$ and $111\ldots1$ yield, respectively, the two pairs (r_0, γ) and (r_1, γ) of fixed
points for $\sigma(0)$ and $\sigma(1)$. The former fixed point of each pair has eigenvalue
$\mu_s = \gamma/r_s$ (with $s = 0$ or 1) and the second $\mu'_s = 1/\mu_s$. Hence, $\sigma = r_s$ is attractive
for $\gamma < r_s$, whatever s is. More generally, the solution $\sigma_n(t) = r(n)$ is attractive
for $\gamma < r_0 < r_1$ (weak filtering) and any t, so that $\sigma(t)$ consists of two horizontal
segments at heights r_0 and r_1. A more interesting behaviour occurs in the
intermediate region $r_0 < \gamma < r_1$. At the extrema of the t-interval, one observes
that $\sigma_n(0) \rightarrow \gamma$ while $\sigma_n(1)$ still tends to r_1. Along generic values of t, $\sigma(t)$ is
a fractal curve with a continuous range. The plots of three approximations
$\sigma_n(t)$ to the scaling function at $r_0 = 0.3$, $r_1 = 0.5$, and $\gamma = 0.35$, are shown in
Fig. 9.4(d) for $n = 1$, 2, and 10. Pseudo-probabilities $\pi(S) = \varepsilon(S)/\sum_S \varepsilon(S)$ and
the order index t (Eq. (9.25)) have been used.

The fluctuations of $\sigma(t)$ can be estimated by evaluating $\sigma_n(t)$ for two sequences
S and S' with close t-values. If the distance between the t is of the order of
2^{-n}, these sequences do not differ in the n rightmost symbols. Hence, the
corresponding change in σ is obtained by iterating Eq. (9.29) with two close
initial conditions σ_n and σ'_n at "time" n, and using the same rates r_s during
the subsequent evolution. If the map is globally stable (i.e., for any t), the

final difference will be exponentially small in n. The appropriate tool for such an investigation is hence the Lyapunov exponent λ of map (9.29). Although its average value $\langle\lambda\rangle$ is negative for any choice of the parameters, its value increases with increasing γ. The local Lyapunov exponent (i.e., computed over a finite number of steps) can be positive for certain sequences S. Indeed, the support of the associated spectrum $h(\lambda)$ (see Eq. (6.32)) includes an interval of positive values (Paoli $et\ al.$, 1989a). This reflects the occurrence of long-range interactions for $\gamma > r_0$, which have been discussed in Chapter 6. The scaling behaviour of this system may therefore be called "prechaotic". Although this is sufficient to yield phase transitions, the convergence of $\sigma_n(t)$ for almost every t (negativity of $\langle\lambda\rangle$) still implies simplicity from the point of view of the metric complexity in its most restrictive interpretation: i.e., when this is referred to the maximal-order Markov model (9.22) at each level n as in Eq. (9.28).

It is therefore advisable to introduce some weaker notion of complexity which is still directly related to the scaling function. One is suggested by the apparent fractality of the curve $\sigma(t)$ vs. t for certain systems, which can be established with the help of the following theorem (Falconer, 1990) for Hölder-continuous functions.

Theorem 9.1 $Let\ f : [0, 1] \rightarrow \mathbb{R}\ be\ a\ continuous\ function\ such\ that$

$$|f(t) - f(t')| \leq c|t - t'|^{2-d} \tag{9.30}$$

$for\ all\ 0 \leq t, t' \leq 1,\ where\ c > 0\ and\ 1 \leq d \leq 2.\ Then,\ the\ Hausdorff\ measure$ $m_d(\varepsilon)\ with\ index\ d\ (see\ Eq.\ 5.52)\ is\ finite\ for\ \varepsilon \rightarrow 0\ and\ the\ Hausdorff\ dimension$ $d_h\ of\ the\ graph\ (t, f(t))\ satisfies\ d_h \leq D_0 \leq d.\ This\ result\ remains\ true\ if\ (9.30)$ $holds\ when\ |t - t'| < \delta\ for\ some\ \delta > 0.$
$Suppose\ that\ there\ are\ numbers\ c > 0,\ \delta_0 > 0\ and\ 1 \leq d \leq 2\ with\ the\ following$ $property:\ for\ each\ t \in [0, 1]\ and\ 0 < \delta \leq \delta_0\ there\ exists\ t'\ such\ that\ |t - t'| \leq \delta$ and

$$|f(t) - f(t')| \geq c\delta^{2-d} . \tag{9.31}$$

$Then,\ d \leq D_0.$

While Eq. (9.30) represents a criterion for the regularity of the function f, Eq. (9.31) requires f to have sufficiently large variability for it to be a fractal curve.

Accordingly, one may define a complexity $C_f = D_0(\sigma) - 1$, related to the fractality of the scaling-function σ, where $D_0(\sigma)$ is the box-dimension D_0 of $\sigma(t)$. Clearly, $C_f = 0$ if σ is a smooth function and $C_f \in (0, 1]$ if σ is fractal as in Fig. 9.4(d). If no convergence of σ_n to a limit curve takes place, C_f is undefined.

Chapter 10

Summary and perspectives

The term "complex" is being used more and more frequently in science, often in a vague sense akin to "complication", and referred to any problem to which standard, well-established methods of mathematical analysis cannot be immediately applied. The spontaneous, legitimate reaction of the careful investigator to this attitude can be summarized by the questions: "Why study complexity?", "What is complexity?".

In the first part of the book, we have illustrated several examples from various disciplines in which complexity purportedly arises, trying, on the one hand, to exclude phenomena which do not really call for new concepts or mathematical tools and, on the other, to find common features in the remaining cases which could be of guidance for a sound and sufficiently general formulation of the problem. While amply answering the former question, the observed variety of apparently complex behaviour renders the task of formalizing complexity, i.e., of answering the latter question, quite hard. This is the subject of the main body of the book.

Aware of the difficulty of developing a formalism which is powerful enough to yield meaningful answers in all cases of interest, we have presented a critical comparison among various approaches, with the help of selected examples, stressing their complementarity. Since "complexity" has become a widely misused term, invoked in connection with all kinds of obstacles encountered in the advancement of scientific knowledge, it is useful to recall the context to which we have continuously referred. From Chapter 4 on, the object of our study has been a symbolic signal extracted from the original process. When dealing with continuous systems, it has been assumed that no relevant information is lost in the discretization procedure upon which our approach relies. This need not always be the case. The best example is perhaps DNA which is already in the form of a sequence of discrete units: the nucleotides. When these are represented by mere symbols, however, are we guaranteed that the "life program" is indeed entirely coded by the symbol sequences alone and not also by the chemical-physical interaction between nucleotides and environment?

Generating partitions permit the recovery of the whole information about a continuous dynamical system by considering increasingly long sequences, but it is not known whether such partitions can always be found. Consideration of sequences with infinitely many symbols or a computational complexity theory for real numbers might, therefore, become necessary in some cases. Let us, for instance, recall hierarchical trees with a continuum of branches, which arise in the study of spin glasses and may prove useful also in optimization problems concerning the spatial structure of proteins or, more generally, of biomolecules.

In the second part of the book, several available tools for the analysis of spatio-temporal patterns have been reviewed and reinterpreted in the spirit of complexity. In particular, the application of information theory, statistical mechanics, and formal language theory has permitted us to split the object system into parts having various degrees of internal homogeneity and to investigate their mutual interactions. The progressive coarse-graining of the system and the identification of its internal correlations proved to be fundamental steps in the characterization of complexity.

All issues we were confronted with could be traced back to the problem of inferring an appropriate model from a given data set. Unfortunately, as we have learned from the theory of computation, the identification of the optimal (minimal) representation of an infinite symbolic signal is not in general possible since, loosely speaking, there are by far "fewer" models than signals. Because of this fundamental limitation and of the disparity of motivations for the construction of models, the proposed definitions of complexity sometimes reflect quite particular, subjective points of view. We have reviewed the most promising ones, with the purpose of promoting the personal investigation of the motivated reader.

Two main topics have been omitted: multidimensional patterns and quantum mechanical systems. The former are expected to exhibit new features, as indeed happens in statistical mechanics, where the occurrence of phase transitions in systems with short-range forces and the critical indices depend on the lattice dimension. Some of the measures discussed in the book, however, apply to patterns of arbitrary dimension, provided that the sequences are replaced by arbitrary connected blocks containing the same number of symbols. In particular, we recall definitions based on entropy, enumeration of legal blocks, and comparisons between observables at different levels of resolution. The difficulty of a practical implementation of these techniques in generic systems is closely related to those encountered in proofs of the existence of the thermodynamic limit, which require careful consideration of all possible subdomains of the lattice. A major simplification could result from the mapping of a multidimensional pattern to a one-dimensional sequence (e.g., by following a space-filling, self-avoiding walk). Unfortunately, most of the parametrizations produce nonstationarity and artificial long-range correlations, even when these are absent in the original system. While the former invalidate a statistical analysis, the latter alter the asymptotic behaviour in the thermodynamic limit: that is, the transformation might even increase the complexity of the system!

These considerations are tied in with our continuous reference to infinite signals. This choice, inherited from statistical mechanics, where it permits the understanding of the salient features of phase transitions, profitably applies to shift dynamical systems, where Hamiltonian functions can be constructed and spectra of local entropies are deduced from a large fluctuation theory. The associated study of the scaling properties of some observables (undertaken in the last chapter) could be extended to generic formal languages, thus eluding conflicts with undecidable questions which are expected when dealing with infinite sequences (e.g., in the assessment of membership of a given language class).

As for quantum mechanics, most of the tools we presented can be straightforwardly carried over to the study of wavefunctions, whether their evolution is given by a continuous Schrödinger equation or by a discrete tight-binding approximation. Their utility has been especially demostrated for wavefunctions exhibiting typical features of fractal measures. The problem of quantum measurement, instead, recalls the above-mentioned question of the information loss associated with the encoding procedure. The most important aspect of quantum evolution related to complexity is, however, the new research field of quantum computation. Although quantum computers are still far from experimental realizations, there is evidence that quantum mechanics, with its possible simultaneous support for infinitely many pure states, offers tools beyond the capabilities of classical computers. We recall, for instance, the recent proof that the prime factors of a generic integer number can be estimated in polynomial time. As research in this area is still in its infancy, we have preferred not to address it in the book.

What does one learn, in conclusion, from a study of complexity? As long as the outcome of the investigation is a collection of numbers or functions, one achieves a classification of patterns of different origin and possibly defines the conditions (either necessary or sufficient) under which various forms of complexity can arise and evolve. Even more important than giving a sharp definition of complexity *per se*, however, is establishing general criteria for the identification of analogies among seemingly unrelated fields and for the inference of appropriate models. The concept of complexity, in fact, is inseparably connected with the observer's analytical skills. While part of the present scientific research still deals with the fundamental forces acting among the elementary constituents of matter, new challenging problems are put forward by the need for models which, abstracting from the irrelevant interactions, permit extraction of the key features of a given (biological, physical, chemical) system. To what extent can one disregard the "microscopic" details? It is obvious that, as soon as correct encodings of different systems turn out to belong to the same class, the understanding of one of them can be exported to the others.

The natural extension of the study of complexity that has been carried out so far seems, therefore, to point inevitably to a theory of model inference. Not only is such a theory missing, however, but the task of constructing a complete and mathematically consistent one definitely looks too ambitious, as the failure of similar projects in the past (e.g., Hilbert's program and its end decreed by Gödel's theorem) suggests. Indeed, we have presented just a few examples of inference methods, extensively using only Markov models, because of their accuracy and generality. Further progress is presumably possible with a careful definition of the objectives. After all, no theory can be required to apply to all conceivable situations. On the other hand, our mind's ability in finding analogies and dealing with subtle forms of correlations is still awaiting to be transferred into faithful models of reality.

Appendix 1

The Lorenz model

A fluid system in a Rayleigh–Bénard experiment is described by Eq. (2.1) where the force is given by $\mathbf{q} = -(1 - \alpha(T - T_0))g\hat{\mathbf{z}}$, α being the expansion coefficient, T_0 the temperature of the bottom plate, g the acceleration of gravity, and $\hat{\mathbf{z}}$ the unit vector in the z-direction. By introducing the temperature and pressure fluctuations $\theta = T - T_0 + \beta z$ and $\tilde{p} = p - p_0 + g\rho_0 z(1 + \alpha\beta z/2)$, respectively, Eq. (2.1) is transformed to

$$\frac{\partial \mathbf{v}}{\partial t} + \mathbf{v} \cdot \nabla\mathbf{v} = \sigma\left(-\nabla\tilde{p} + \theta\hat{\mathbf{z}} + \nabla^2\mathbf{v}\right)$$
$$\nabla\mathbf{v} = 0 \tag{A1.1}$$
$$\frac{\partial \theta}{\partial t} + \mathbf{v} \cdot \nabla\theta = R_a\hat{\mathbf{z}} \cdot \mathbf{v} + \nabla^2\theta \, ,$$

where space and time have been suitably rescaled, $\sigma = v/\kappa$ is the Prandtl number, and $R_a = g\alpha\beta d^4/\kappa v$ is the Rayleigh number. As long as R_a is not much larger than its value at the onset of convection, the system presents a translational invariance along the roll-axis (y-axis). In these conditions, equations (A1.1) can be simplified to

$$\frac{\partial}{\partial t}\nabla^2\psi + \frac{\partial\psi}{\partial x}\frac{\partial}{\partial z}\nabla^2\psi - \frac{\partial\psi}{\partial z}\frac{\partial}{\partial x}\nabla^2\psi = \sigma\left(\nabla^4\psi + \frac{\partial\theta}{\partial x}\right)$$
$$\frac{\partial\theta}{\partial t} + \frac{\partial\psi}{\partial x}\frac{\partial\theta}{\partial z} - \frac{\partial\psi}{\partial z}\frac{\partial\theta}{\partial x} = R_a\frac{\partial\psi}{\partial x} + \nabla^2\theta \, , \tag{A1.2}$$

where ψ is the *stream function*, defined by

$$v_x = -\frac{\partial\psi}{\partial z} \quad \text{and} \quad v_z = \frac{\partial\psi}{\partial x}$$

in terms of the x and z components of the velocity $\mathbf{v}$. The obvious requirement that no flow of matter occurs across the horizontal plates (defined by $z = \pm 1/2$) implies $v_z|_{z=\pm 1/2} = 0$. Furthermore, the *stress-free* condition $\partial v_x/\partial z|_{z=\pm 1/2} = 0$ is often imposed. In terms of ψ, these two relations are equivalent to $\psi = \partial^2\psi/\partial z^2 = 0$ at the boundaries. Finally, perfect conductivity is assumed at the plates, so that $\theta|_{z=\pm 1/2} = 0$. A solution of Eq. (A1.2) can be sought by expanding the field ψ in Fourier series as

$$\psi(\mathbf{x}) = i \sum_{\mathbf{k} \neq 0} X_{\mathbf{k}} e^{i\mathbf{k}\cdot\mathbf{x}} , \tag{A1.3}$$

where the reality condition $X_{\mathbf{k}} = -X^*_{-\mathbf{k}}$ must be satisfied. A similar expansion is taken for θ. Insertion in the equations of motion (A1.2) yields an open hierarchy of differential equations for the Fourier coefficients. This can be profitably truncated by restricting the sums to the lowest-order modes. The approximation is effective if the amplitudes of the neglected modes remain sufficiently small at all times. The most severe reduction that still retains nonlinear contributions is obtained by setting

$$\psi(x, z, t) = X(t)\cos(\pi z)\sin(kx)$$
$$\theta(x, z, t) = Y(t)\cos(\pi z)\cos(kx) + Z(t)\sin(2\pi z) , \tag{A1.4}$$

which clearly satisfies the boundary conditions. After discarding higher harmonics generated by the nonlinear couplings, this substitution yields the Lorenz model

$$\dot{x} = \sigma(y - x)$$
$$\dot{y} = -y + rx - xz$$
$$\dot{z} = -bz + xy ,$$

where x, y, and z are respectively proportional to X, Y, and Z, time has been rescaled, and

$$r = \frac{k^2}{(\pi^2 + k^2)^3} R_a \quad , \quad b = \frac{4\pi^2}{\pi^2 + k^2} .$$

The horseshoe map

The action of Smale's horseshoe map **G** (Smale, 1963, 1967) is sketched in Fig. A2.1. The unit square $R = [0, 1] \times [0, 1]$ is stretched in the y (unstable) direction and contracted in the x (stable) direction to yield a long vertical strip (middle picture) which is then bent as shown in the figure. The horizontal strips H_0 and H_1 are mapped inside R to the vertical strips V_0 and V_1, respectively. One further application of **G** yields two vertical substrips inside each of the regions V_0 and V_1. After n iterations, 2^n thin rectangles are left in R. Vice versa, n backward iterations yield 2^n horizontal strips. The map is piecewise linear inside R; the smooth nonlinearity, responsible for the folding process, is neglected in this model. In fact, reinjections of points mapped outside the

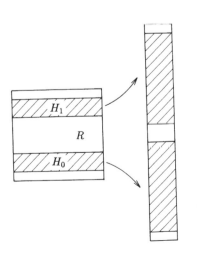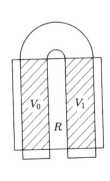

Figure A2.1 Illustration of the action of the horseshoe map. After an intermediate stretching step, the image rectangle is folded over the original unit square R.

square are not considered, although they are allowed in a generic map $\mathbf{F}$. The largest invariant set of $\mathbf{G}$ is the intersection $\Lambda = \bigcap_{i=-\infty}^{\infty} \mathbf{G}^i(R)$ of two mutually orthogonal families of straight lines. The largest connected subset of Λ is a point and every point of Λ is a limit point. This is one of the possible definitions of a *Cantor set*.

Mathematical definitions

Measure and integration

A collection $\tilde{\mathscr{B}}$ of subsets of a set X is a σ-algebra if:

1. $X \in \tilde{\mathscr{B}}$;
2. $B \in \tilde{\mathscr{B}}$ implies $X \setminus B \in \tilde{\mathscr{B}}$;
3. $B_k \in \tilde{\mathscr{B}}$, for $k \geq 1$, implies $\bigcup_{k=1}^{\infty} B_k \in \tilde{\mathscr{B}}$.

The pair $(X, \tilde{\mathscr{B}})$ is called a *measurable space*. A *finite measure* on $(X, \tilde{\mathscr{B}})$ is a function $m : \tilde{\mathscr{B}} \to \mathbb{R}^+$ such that:

1. $m(\emptyset) = 0$;
2. $m(B_k) < \infty$ for all $B_k \in \tilde{\mathscr{B}}$;
3. $m\left(\bigcup_{k=1}^{\infty} B_k\right) = \sum_{k=1}^{\infty} m(B_k)$, whenever $\{B_k\}_1^{\infty}$ is a sequence of members of $\tilde{\mathscr{B}}$ that are pairwise disjoint subsets of X.

The elements of $\tilde{\mathscr{B}}$ are called measurable sets because a measure is defined for each of them by the properties listed above. The triple $(X, \tilde{\mathscr{B}}, m)$ is a *finite measure space*. If, in addition, $m(X) = 1$, $(X, \tilde{\mathscr{B}}, m)$ is a *probability space* and m a *probability measure* on $(X, \tilde{\mathscr{B}})$. In $X = \mathbb{R}$, the *Borel measure* is defined by $m([a,b]) = |b - a|$. In the *product space* $(\mathbb{R}^d, \tilde{\mathscr{B}}^d, m^d)$, $\tilde{\mathscr{B}}^d$ is the smallest σ-algebra containing all sets of the form $B^d = B_1 \times \cdots \times B_d$ (with $B_i \in \tilde{\mathscr{B}}$) and $m^d(B^d) = \prod m(B_k) = \prod |b_k - a_k|$. The *complete* Borel measure on $\mathbb{R}^d$ (i.e., such that all subsets of zero-measure sets also have measure zero) is called the *Lebesgue measure* and is indicated with m_L. A point $\mathbf{x} \in X$ is an *atom* of the measure m if $m(\mathbf{x}) > 0$.

A function $f : X \to \mathbb{R}$ is *measurable* if $f^{-1}(I) \in \tilde{\mathscr{B}}$ for every interval $I \subset \mathbb{R}$. Two functions $f, g : X \to \mathbb{R}$ are a.e. (almost everywhere) equal if $m(\{\mathbf{x} : f(\mathbf{x}) \neq g(\mathbf{x})\}) = 0$. A function f is called *simple* if it can be decomposed as $f(\mathbf{x}) = \sum_{k=0}^{b-1} \beta_k \chi_{B_k}(\mathbf{x})$, where $\beta_k \in \mathbb{R}$, B_k are disjoint subsets of X, and χ_B is the characteristic function of set B ($\chi_B(\mathbf{x}) = 1$ if $\mathbf{x} \in B$ and $\chi_B(\mathbf{x}) = 0$ otherwise). Every simple function f is measurable and its Lebesgue integral is defined by $\int_X f(\mathbf{x}) m(d^d x) = \sum_{k=0}^{b-1} \beta_k m(B_k)$, where $m(d^d x) \equiv dm = m(dx_1 \cdots dx_d)$. The Lebesgue integral of an arbitrary, non-negative, bounded, measurable function f is defined as

$$\int_X f(\mathbf{x}) dm = \lim_{n \to \infty} \int_X g_n(\mathbf{x}) dm ,$$

where $\{g_n\}$ is a sequence of simple functions converging uniformly to f. This definition can be easily extended to unbounded measurable functions with arbitrary sign (Walters, 1985). A function with a finite integral is called *integrable*.

Baire category

A set A is *nowhere dense* if the complement of its closure $\overline{A}$ is dense. Equivalently, $\overline{A}$ contains no nonempty open set. A set A is of *first category* or *meagre* if it is the union of a countable collection of nowhere dense sets. A set that is not of first category is said to be of *second category* or *nonmeagre*. Finally, the complement of a set of the first category is called *residual*. Sets of the first category can be considered as "thin" sets. For further properties of sets involving category theory, see Royden (1989).

Appendix 4

Lyapunov exponents, entropy, and dimension

Generalized dimensions and entropies of chaotic dynamical systems are closely related to expansion and contraction properties of phase space, which are accounted for by the Lyapunov exponents (see Section 3.3). While referring the reader to the specialized literature for an exhaustive discussion of this subject (Eckmann & Ruelle, 1985), we recall a few fundamental relations that are often employed in this book.

Consider a trajectory $\omega_{0n} = \{x_0, x_1, \ldots, x_n\}$ of an MPT $(\mathbf{F}, X, \tilde{\mathscr{B}}, m)$. The probability $P(\varepsilon, n; x_0)$ of finding an orbit within a distance ε from ω_{0n} can be rewritten as $P(\varepsilon; x_0)P(\omega_{1n}|x_0)$, where $P(\varepsilon; x_0)$ is the probability of observing a point x in a domain $B_\varepsilon(x_0)$ of size ε around x_0 and $P(\omega_{1n}|x_0)$ is the conditional probability for an orbit originating at x_0 to pass within a distance ε from $\omega_{1n} = \{x_1, \ldots, x_n\}$ in the subsequent n time steps. Hence, it is natural to assume that

$$P(\varepsilon, n; x) \sim \varepsilon^{\alpha(x)} e^{-\kappa(x)n}, \text{ for } \varepsilon \to 0 \text{ and } n \to \infty, \qquad (A4.1)$$

where the symbol $\sim$ indicates the leading asymptotic behaviour. Since the first term on the r.h.s. corresponds to $P(\varepsilon; x)$, the index $\alpha(x)$, already introduced in Section 5.1, can be interpreted as the *"local dimension"* at x and characterizes the scaling behaviour of the invariant measure m when the resolution is indefinitely sharpened. Analogously, the exponent $\kappa(x)$ represents the "local" entropy of the given portion of trajectory. Both $\alpha(x)$ and $\kappa(x)$ are by no means the "pointwise" dimension and entropy at x but are defined only for each finite choice of ε and n in the coarse-graining procedure, through Eq. (A4.1). For m-almost every point x, $\alpha(x) = D_1$ and $\kappa(x) = K_1$ in the infinite-resolution limit (Shannon–McMillan–

Breiman theorem, Section 6.2). In the following, we drop the **x**-dependence and refer to typical values. The importance of the fluctuations of the local variables around them is extensively discussed in Chapter 6 (see also Grassberger *et al.*, 1988).

For strange attractors that are locally the product of continua and Cantor sets, one may consider the scaling of the probability along each eigendirection of the linearized map separately by setting

$$P(\varepsilon_1,\ldots,\varepsilon_d;\mathbf{x}) \sim \prod_{i=1}^{d} \varepsilon_i^{D_1^{(i)}} , \tag{A4.2}$$

where the reference domain $B(\mathbf{x})$ is now an ellipsoid with axes $\{\varepsilon_i\}$ and d indicates the dimension of X. The exponents $D_1^{(i)}$, called *partial information dimensions* (Grassberger *et al.*, 1988), satisfy

$$D_1 = \sum_{i=1}^{d} D_1^{(i)} . \tag{A4.3}$$

In n time steps, the ith axis ε_i is mapped to $\varepsilon_i' = \varepsilon_i e^{n\lambda_i}$, where λ_i is the ith Lyapunov exponent. Assuming the hyperbolicity of the map $\mathbf{F}$ (i.e., a constant sign for each λ_i along the trajectory), expanding and contracting directions can be treated separately. The conditional probability $P(\omega_{1n}|\mathbf{x}_0)$ decreases with n because some trajectories leave the "ε-neighbourhood" $C(\varepsilon,n)$ of ω_{1n} along the unstable directions. All trajectories belonging to $C(\varepsilon,n)$ have started off within a distance ε_i'' from $\mathbf{x}_0$ along the ith direction, where $\varepsilon_i'' = \varepsilon e^{-n\lambda_i}$, if $\lambda_i > 0$, and $\varepsilon_i'' = \varepsilon$, if $\lambda_i \leq 0$. The mass in the ellipsoid defined by these axes scales as

$$P(\varepsilon,n;\mathbf{x}_0) \sim \varepsilon^{\Sigma_i D_1^{(i)}} e^{-n\Sigma_i^+ D_1^{(i)}\lambda_i} , \tag{A4.4}$$

where the superscript + in the sum indicates that only the positive Lyapunov exponents are included. Recalling Eq. (A4.1), one obtains

$$K_1 \approx \kappa = \sum_i^+ D_1^{(i)}\lambda_i \tag{A4.5}$$

and verifies Eq. (A4.3) at the same time. In the limit $n \to \infty$, this is an exact relation which links the Kolmogorov–Sinai entropy with the average Lyapunov exponents and the partial dimensions. For hyperbolic attractors, $D_1^{(i)} = 1$ along each unstable direction, so that Eq. (A4.5) reduces to

$$K_1 = \sum_i^+ \lambda_i . \tag{A4.6}$$

This result is subjected to the two assumptions of uniform hyperbolicity and factorization of the measure m into the product of independent measures along mutually transverse subspaces. When the latter is relaxed, the sum of the positive Lyapunov exponents provides an upper bound to the metric entropy K_1 (Pesin, 1977).

Finally, consider a sphere $B_\varepsilon(\mathbf{x})$ of radius ε centred at $\mathbf{x}$ and its nth image B', an ellipsoid with axes $\{\varepsilon_i' = \varepsilon e^{n\lambda_i(\mathbf{x})} : i = 1, \ldots, d\}$. The mass $m(B) = P(\varepsilon; \mathbf{x})$ in $B_\varepsilon(\mathbf{x})$ is entirely transferred to B' by the measure-preservation property of $\mathbf{F}$. Hence, the obvious relation

$$m(B) \sim \varepsilon^{D_1} \sim \varepsilon^{\Sigma_i D_1^{(i)}} e^{n\Sigma_i D_1^{(i)} \lambda_i} \sim m(B') \qquad (A4.7)$$

yields

$$\sum_{i=1}^{d} D_1^{(i)} \lambda_i = 0 . \qquad (A4.8)$$

Let j denote the largest integer such that $\sum_{i=1}^{j} \lambda_i \geq 0$, where $\lambda_1 \geq \lambda_2 \geq \ldots \geq \lambda_d$. The index j (≥ 1 for chaotic maps) is always well defined for invertible transformations $\mathbf{F}$, since $\sum_i^d \lambda_i \leq 0$. With the assumption that $D_1^{(i)} = 1$, for $i \leq j$, and $D_1^{(i)} = 0$, for $i > j + 1$, the fractality (if any) is attributed to the partial dimension in the direction $(j+1)$ only. Its value can be determined from Eq. (A4.8). The result, known as the *Kaplan–Yorke relation* (Kaplan & Yorke, 1979; Young, 1982; Pesin, 1984), reads

$$D_1 = j + \frac{\sum_{i=1}^{j} \lambda_i}{|\lambda_{j+1}|} . \qquad (A4.9)$$

Appendix 5

Forbidden words in regular languages

Given a regular language $\mathscr{L}$ with graph G, it is possible to construct automatically a graph G_f that generates the irreducible forbidden words (IFWs) of $\mathscr{L}$. For simplicity, we treat the binary case $b = 2$. If a given test word w is forbidden (but not necessarily an IFW), its initial $n - 1$ symbols yield a path on G which ends in a node with only one outgoing arc. Not each path of this kind, however, corresponds to a forbidden word, since it might appear somewhere else on G followed by both possible extensions 0 and 1. Therefore, in order to recognize w as forbidden, all paths on G which share the same prefixes as w must be taken into account. Additional checks are obviously needed to test irreducibility.

Let us search all IFWs ending with symbol s. To this aim, we separate the nodes of G into the two sets Q, containing the nodes q_i $(i = 1, \ldots, I)$ that are not followed by symbol s, and P, containing the remaining nodes p_j $(j = 1, \ldots, J)$[1]. Let then consider the (nondeterministic) graph obtained from G by reversing all arrows, and its equivalent deterministic version G', which includes possible transient parts (see Chapter 7 for an illustration of this transformation), and denote by r_k its nodes. Notice that $Q \cup P \subseteq R = \{r_k\}$ since, in general, there are r_k which belong neither to Q nor to P.

All possible backward paths on G correspond to forward paths on G' with initial conditions in $Q \cup P$. In order to follow them simultaneously, keeping the information about Q and P, we introduce metasymbols of the type $\mathbf{r} = (R'; R'')$ where R' and R'' are suitable subsets of R. The first of such metasymbols, $\mathbf{r}_0 = (Q; P)$, represents the initial condition in G_f (i.e., the set of all nodes in

---

1. Here and in the following, the dependence of this regrouping on s is understood in order to keep the notation as simple as possible

G). Its image in G_f upon emission of symbol t is denoted by $\mathbf{r}_1' = (Q_1^t; P_1^t)$, where Q_1^t (P_1^t) is the set of all r_k reached from Q (P) through an arc labelled by t. The sequence $T = t_1 t_2 \ldots t_n$ on the graphs G' and G_f is the inverse of the test sequence $S = s_1 s_2 \ldots s_n$ on G. The nodes $(Q_n; P_n)$ of G_f correspond to the metasymbols and are all obtained in a finite number of iterations with this procedure. Since G' is deterministic, the cardinalities $N_n(Q)$ and $N_n(P)$ of the sets Q_n and P_n, respectively, are nonincreasing functions of n. As long as $N_n(P)$ is nonzero, there is at least one path in the original graph G which corresponds to the inverse S of the T-sequence and terminates at a P-node. Therefore, Ss is allowed. As soon as $N_n(P) = 0$, with $N_n(Q) > 0$, instead, the only possible end nodes in G for S are in Q and the concatenation Ss is forbidden. Moreover, not only its prefix S is allowed but also its longest suffix $s_2 s_3 \ldots s_n s$ is. Otherwise, it would have been recognized as forbidden at some previous step in the procedure.

All IFWs ending with symbol s can, hence, be obtained from all paths on the graph G_f that start with node $\mathbf{r}_0$ and terminate at the "accepting" nodes, which are characterized by $P = \emptyset$. The existence of a construction method for the graph G_f represents the proof that the IFW language $\mathscr{F}$ of a regular language $\mathscr{L}$ is regular itself.

References

The following list, while not an exhaustive set of references for the topics treated in this book, includes all the articles and books mentioned in the text and in the additional bibliographical suggestions. The first author's name of books and survey articles has been set in bold type to facilitate the search.

1. **Abraham**, N. B., Albano, A. M., Passamante, A., and Rapp, P. E., eds., (1989). *Measures of complexity and chaos*, Plenum, New York.

2. Adler, R. L., (1991). Geodesic flows, interval maps, and symbolic dynamics, in Bedford *et al.* (1991), p. 93.

3. Adler, R. L., and Weiss, B., (1967). Entropy, a complete metric invariant for automorphisms of the torus, *Proc. Natl. Acad. Sci. USA* **57**, 1573.

4. Adler, R. L., Konheim, A. G., and McAndrew, M. H., (1965). Topological entropy, *Trans. Am. Math. Soc.* **114**, 309, Providence, RI.

5. Agnes, C., and Rasetti, M., (1991). Undecidability of the word problem and chaos in symbolic dynamics, *Nuovo Cimento B* **106**, 879.

6. **Aharony**, A., and Feder, J., eds., (1989). *Fractals in physics: essays in honor of Benoit B. Mandelbrot*, North-Holland, Amsterdam.

7. Aho, A. V., Hopcroft, J. E., and Ullman, J. D., (1974). *The design and analysis of computer algorithms*, Addison-Wesley, Reading, MA.

8. Akhmanov, S. A., Vorontsov, M. A., and Ivanov, V. Yu., (1988). Large-scale transverse nonlinear interactions in laser beams; new types of nonlinear waves; onset of "optical turbulence", *JETP Lett.* **47**, 707.

9. **Alekseev**, V. M., and Yakobson, M. V., (1981). Symbolic dynamics and hyperbolic dynamic systems, *Phys. Rep.* **75**, 287.

10. Alexander, J. C., Yorke, J. A., You, Z., and Kan, I., (1992). Riddled basins, *Int. J. Bifurc. and Chaos* **2**, 795.

11. Alpern, S., (1978). Approximation to and by measure preserving homeomorphisms, *J. London Math. Soc.* **18**, 305.

12. Anderson, P. W., (1972). More is different, *Science* **177**, 393.

13. Anderson, P. W., and Stein, D. L., (1987). Broken symmetry, emergent properties, dissipative structures, life: are they related?, in Yates (1987), p. 445.

14. Anselmet, F., Gagne, Y., Hopfinger, E. J., and Antonia, R., (1984). High-order velocity

structure functions in turbulent shear flows, *J. Fluid Mech.* **140**, 63.

15. **Arecchi**, F. T., (1994). Optical morphogenesis: pattern formation and competition in nonlinear optics, *Nuovo Cim. A* **107**, 1111.

16. Arecchi, F. T., Meucci, R., Puccioni, G., and Tredicce, J., (1982). Experimental evidence of subharmonic bifurcations, multistability, and turbulence in a Q-switched gas laser, *Phys. Rev. Lett.* **49**, 1217.

17. Argoul, F., Arneodo, A., and Richetti, P., (1991). Symbolic dynamics in the Belousov-Zhabotinskii reaction: from Rössler's intuition to experimental evidence for Šil'nikov's homoclinic chaos, in *Chaotic Hierarchy*, edited by G. Baier and M. Klein, World Scientific, Singapore, p. 79.

18. Arneodo, A., Argoul, F., Bacry, E., Muzy, J. F., and Tabard, M., (1992a). Golden mean arithmetic in the fractal branching of diffusion-limited aggregates, *Phys. Rev. Lett.* **68**, 3456.

19. Arneodo, A., Argoul, F., Muzy, J. F., and Tabard, M., (1992b). Structural five-fold symmetry in the fractal morphology of diffusion-limited aggregates, *Physica A* **188**, 217.

20. Arnold, V. I., (1963). Proof of A. N. Kolmogorov's theorem on the preservation of quasiperiodic motions under small perturbations of the Hamiltonian, *Russ. Math. Surv.* **18**, 9.

21. **Arnold**, V. I., (1983). *Geometrical methods in the theory of ordinary differential equations*, Springer, New York.

22. **Arnold**, V. I., and Avez, A., (1968). *Ergodic problems of classical mechanics*, Benjamin, New York.

23. **Arrowsmith**, D. K., and Place, C. M., (1990). *An introduction to dynamical systems*, Cambridge University Press, Cambridge.

24. **Ash**, R. B., (1965). *Information theory*, Interscience, New York.

25. **Ashcroft**, N. W., and Mermin, N. D., (1976). *Solid state physics*, Holt, Rinehart, and Winston, New York.

26. **Atlan**, H., (1979). *Entre le cristal et la fumée*, Seuil, Paris.

27. Atlan, H., (1987). Self creation of meaning, *Physica Scripta* **36**, 563.

28. Atlan, H., (1990). Ends and meaning in machine-like systems, *Communication and Cognition* **23**, 143.

29. Atlan, H., and Koppel, M., (1990). The cellular computer DNA: program or data, *Bull. of Mathematical Biology* **52**, 335.

30. Aubry, S., Godrèche, C., and Luck, J. M., (1988). Scaling properties of a structure intermediate between quasiperiodic and random, *J. Stat. Phys.* **51**, 1033.

31. Auerbach, D., (1989). Dynamical complexity of strange sets, in Abraham *et al.* (1989), p. 203.

32. Auerbach, D., and Procaccia, I., (1990). Grammatical complexity of strange sets, *Phys. Rev. A* **41**, 6602.

33. Aurell, E., (1987). Feigenbaum attractor as a spin system, *Phys. Rev. A* **35**, 4016.

34. Avron, J. E., and Simon, B., (1981). Transient and recurrent spectrum, *J. Func. Anal.* **43**, 1.

35. Bachas, C. P., and Huberman, B. A., (1986). Complexity and the relaxation of hierarchical structures, *Phys. Rev. Lett.* **57**, 1965.

36. **Badii**, R., (1989). Conservation laws and thermodynamic formalism for dissipative dynamical systems, *Riv. Nuovo Cim.* **12**, No. 3, 1.

37. Badii, R., (1990a). Complexity as unpredictability of the scaling dynamics, *Europhys. Lett.* **13**, 599.

38. Badii, R., (1990b). Unfolding complexity in nonlinear dynamical systems, in Abraham *et al.* (1989), p. 313.

39. Badii, R., (1991). Quantitative characterization of complexity and predictability, *Phys. Lett. A* **160**, 372.

40. Badii, R., (1992). Complexity and unpredictable scaling of hierarchical structures, in *Chaotic Dynamics, Theory and Practice*, edited by T. Bountis, Plenum, New York, p. 1.

41. Badii, R., and Broggi, G., (1988). Measurement of the dimension spectrum $f(\alpha)$: fixed-mass approach, *Phys. Lett. A* **131**, 339.

42. Badii, R., and Politi, A., (1984). Hausdorff dimension and uniformity factor of strange attractors, *Phys. Rev. Lett.* **52**, 1661.

43. Badii, R., and Politi, A., (1985). Statistical

description of chaotic attractors: the dimension function, *J. Stat. Phys.* **40**, 725.

44. Badii, R., Broggi, G., Derighetti, B., Ravani, M., Ciliberto, S., Politi, A., and Rubio, M. A., (1988). Dimension increase in filtered chaotic signals, *Phys. Rev. Lett.* **60**, 979.

45. Badii, R., Finardi, M., and Broggi, G., (1991). Unfolding complexity and modelling asymptotic scaling behaviour, in *Chaos, Order, and Patterns*, edited by R. Artuso, P. Cvitanović, and G. Casati, Plenum, New York, p. 259.

46. Badii, R., Finardi, M., Broggi, G., and Sepúlveda, M. A., (1992). Hierarchical resolution of power spectra, *Physica D* **58**, 304.

47. Badii, R., Brun, E., Finardi, M., Flepp, L., Holzner, R., Parisi, J., Reyl, C., and Simonet, J., (1994). Progress in the analysis of experimental chaos through periodic orbits, *Rev. Mod. Phys.* **66**, 1389.

48. Baker, A., (1990). *A concise introduction to the theory of numbers*, Cambridge University Press, Cambridge.

49. Bartuccelli, M., Gibbon, J. D., Constantin, P., Doering, C. R., and Gisselfält, M., (1990). On the possibility of soft and hard turbulence in the complex Ginzburg Landau equation, *Physica D* **44**, 421.

50. Bates, J. E., and Shepard, H. K., (1993). Measuring complexity using information fluctuation, *Phys. Lett. A* **172**, 416.

51. Batchelor, G. K., (1982). *The theory of homogeneous turbulence*, Cambridge University Press, Cambridge.

52. Batchelor, G. K., (1991). *An introduction to fluid dynamics*, 2nd printing, Cambridge University Press, Cambridge.

53. Baxter, R. J., (1982). *Exactly solved models in statistical mechanics*, Academic Press, London.

54. Beck, C., and Schlögl, F., (1993). *Thermodynamics of chaotic systems*, Cambridge University Press, Cambridge.

55. Bedford, T., Keane, M., and Series, C., eds., (1991). *Ergodic theory, symbolic dynamics and hyperbolic spaces*, Oxford University Press, Oxford.

56. Behringer, R. P., (1985). Rayleigh-Bénard convection and turbulence in liquid helium, *Rev. Mod. Phys.* **57**, 657.

57. Bell, G. I., (1990). The human genome: an introduction, in Bell and Marr (1990), p. 3.

58. Bell, G. I., and Marr, T. G., eds., (1990). *Computers and DNA*, SFI Studies in the Sciences of Complexity, vol. VII, Addison-Wesley, Redwood City, CA.

59. Bell, T. C., Cleary, J. G., and Witten, I. H., (1990). *Text compression*, Prentice-Hall, Englewood Cliffs, NJ.

60. Benedicks, M., and Carleson, L., (1991). The dynamics of the Hénon map, *Ann. of Math.* **133**, 73.

61. Bennett, C. H., (1988). Dissipation, information, computational complexity and the definition of organization, in *Emerging Syntheses in Science*, edited by D. Pines, Addison-Wesley, Reading, MA, p. 215.

62. Bennett, C. H., (1990). How to define complexity in physics and why, in Zurek (1990), p. 137.

63. Benzi, R., Paladin, G., Parisi, G., and Vulpiani, A., (1984). On the multifractal nature of fully developed turbulence and chaotic systems, *J. Phys. A* **17**, 3521.

64. Bergé, P., Pomeau, Y., and Vidal, C., (1986). *Order within chaos*, Wiley, New York.

65. Berlekamp, E. R., Conway, J. H., and Guy, R. K., (1982). *Winning ways for your mathematical plays*, Academic Press, London.

66. Billingsley, P., (1965). *Ergodic theory and information*, Wiley, New York.

67. Blanchard, F., and Hansel, G., (1986). Systems codes, *Theor. Computer Science* **44**, 17.

68. Blum, L., (1990). Lectures on a theory of computation and complexity over the reals (or an arbitrary ring), in Jen (1990), p. 1.

69. Blum, L., and Smale, S., (1993). The Gödel incompleteness theorem and decidability over a ring, in *From Topology to Computation*, edited by M. W. Hirsch, J. E. Marsden, and M. Shub, Springer, New York, p. 321.

70. Bohr, T., and Tél, T., (1988). The thermodynamics of fractals, in *Directions in Chaos*, Vol. 2, edited by B.-L. Hao, World Scientific, Singapore, p. 194.

71. Bombieri, E., and Taylor, J. E., (1986). Which distributions of matter diffract? An initial investigation, *Journal de Physique*, Colloque C3, Suppl. 7, **47**, 19.

72. Bombieri, E., and Taylor, J. E., (1987). Quasicrystals, tilings, and algebraic number theory: some preliminary connections, *Contemp. Math.* **64**, 241.

73. Borštnik, B., Pumpernik, D., and Lukman, D., (1993). Analysis of apparent $1/f^\alpha$ spectrum in DNA sequences, *Europhys. Lett.* **23**, 389.

74. Bouchaud, J.-P., and Mézard, M., (1994). Self induced quenched disorder: a model for the spin glass transition, *J. Phys. I* (France) **4**, 1109.

75. Bowen, R., (1975). Equilibrium states and the ergodic theory of Anosov diffeomorphisms, *Lect. Notes in Math.* **470**, Springer, New York.

76. Bowen, R., (1978). On axiom A diffeomorphisms, *CBMS Regional Conference Series in Mathematics* **35**, American Mathematical Society, Providence, RI.

77. Brandstater, A., and Swinney, H. L., (1987). Strange attractors in weakly turbulent Couette-Taylor flow, *Phys. Rev. A* **35**, 2207.

78. Bray, A. J., and Moore, M. A., (1987). Chaotic nature of the spin-glass phase, *Phys. Rev. Lett.* **58**, 57.

79. Broggi, G., (1988). *Numerical characterization of experimental chaotic signals*, Doctoral Thesis, University of Zurich, Switzerland.

80. Brudno, A. A., (1983). Entropy and the complexity of the trajectories of a dynamical system, *Trans. Moscow Math. Soc.* **44**, 127.

81. Busse, F. H., (1967). The stability of finite amplitude cellular convection and its relation to an extremum principle, *J. Fluid Mech.* **30**, 625.

82. **Busse**, F. H., (1978). Nonlinear properties of thermal convection, *Rep. Prog. Phys.* **41**, 1929.

83. Caianiello, E. R., (1987). A thermodynamic approach to self-organizing systems, in Yates (1987), p. 475.

84. **Callen**, H. B., (1985). *Thermodynamics and an introduction to thermostatistics*, 2nd edition, Wiley, Singapore.

85. **Carrol**, J., and Long, D., (1989). *Theory of finite automata with an introduction to formal languages*, Prentice-Hall, Englewood Cliffs, NJ.

86. Casartelli, M., (1990). Partitions, rational partitions, and characterization of complexity, *Complex Systems* **4**, 491.

87. Casti, J. L., (1986). On system complexity: identification, measurement, and management, in *Complexity, Language, and Life: Mathematical Approaches*, edited by J. L. Casti and A. Karlqvist, Springer, Berlin, p. 146.

88. Cates, M. E., and Witten, T. A., (1987). Diffusion near absorbing fractals: harmonic measure exponents for polymers, *Phys. Rev. A* **35**, 1809.

89. Ceccatto, H. A., and Huberman, B. A., (1988). The complexity of hierarchical systems, *Physica Scripta* **37**, 145.

90. Chaitin, G. J., (1966). On the length of programs for computing binary sequences, *J. Assoc. Comp. Math.* **13**, 547.

91. **Chaitin**, G. J., (1990a). *Algorithmic information theory*, 3rd edition, Cambridge University Press, Cambridge.

92. **Chaitin**, G. J., (1990b). *Information, randomness and incompleteness*, 3rd edition, World Scientific, Singapore.

93. Chandra, P., Ioffe, L. B., and Sherrington, D., (1995). Possible glassiness in a periodic long-range Josephson array, *Phys. Rev. Lett.* **75**, 713.

94. **Chandrasekar**, S., (1961). *Hydrodynamic and hydromagnetic stability*, Clarendon Press, Oxford.

95. Chhabra, A. B., and Sreenivasan, K. R., (1991). Probabilistic multifractals and negative dimensions, in *New Perspectives in Turbulence*, edited by L. Sirovich, Springer, New York, p. 271.

96. Chomsky, N., (1956). Three models for the description of language, *IRE Trans. on Information Theory* **2**, 113.

97. Chomsky, N., (1959). On certain formal properties of grammars, *Information and Control* **2**, 137.

98. Christiansen, F., and Politi, A., (1995). A generating partition for the standard map, *Phys. Rev. E* **51**, 3811.

99. Ciliberto, S., and Bigazzi, P., (1988). Spatiotemporal intermittency in Rayleigh-Bénard convection, *Phys. Rev. Lett.* **60** 286.

100. Ciliberto, S., Douady, S., and Fauve, S., (1991a). Investigating space-time chaos in Faraday instability by means of the

fluctuations of the driving acceleration, *Europhys. Lett.* **15**, 23.

101. Ciliberto, S., Pampaloni, E., and Perez Garcia, C., (1991b). The role of defects in the transition between different symmetries in convective patterns, *J. Stat. Phys.* **64**, 1045.

102. **Collet**, P., and Eckmann, J.-P., (1980). *Iterated maps on the interval as dynamical systems*, Birkhäuser, Boston.

103. Constantin, P., and Procaccia, I., (1991). Fractal geometry of isoscalar surfaces in turbulence: theory and experiments, *Phys. Rev. Lett.* **67**, 1739.

104. Cook, C. M., Rosenfeld, A., and Aronson, A. R., (1976). Grammatical inference by hill climbing, *Informational Sciences* **10**, 59.

105. **Cornfeld**, I. P., Fomin, S. V., and Sinai, Ya. G., (1982). *Ergodic theory*, Springer, New York.

106. Courbage, M., and Hamdan, D., (1995). Unpredictability in some nonchaotic dynamical systems, *Phys. Rev. Lett.* **74**, 5166.

107. Coven, E. M., and Hedlund, G. A., (1973). Sequences with minimal block growth, *Math. Systems Theory* **7**, 138.

108. Coven, E. M., and Paul, M. E., (1975). Sofic systems, *Israel J. of Math.* **20**, 165.

109. **Cover**, T. M., and Thomas, J. A., (1991). *Elements of information theory*, Wiley, New York.

110. **Crisanti**, A., Paladin, G., and Vulpiani, A., (1993). *Products of random matrices in statistical physics*, Springer, Berlin.

111. Crisanti, A., Falcioni, M., Mantica, G., and Vulpiani, A., (1994). Applying algorithmic complexity to define chaos in the motion of complex systems, *Phys. Rev. E* **50**, 1959.

112. **Cross**, M. C., and Hohenberg, P. C., (1993). Pattern formation outside of equilibrium, *Rev. Mod. Phys.* **65**, 851.

113. Crutchfield, J. P., and Young, K., (1989). Inferring statistical complexity, *Phys. Rev. Lett.* **63**, 105.

114. Crutchfield, J. P., and Young, K., (1990). Computation at the onset of chaos, in Zurek (1990), p. 223.

115. Culik II, K., Hurd, L. P., and Yu, S., (1990). Computation theoretic aspects of cellular automata, *Physica D* **45**, 357.

116. **Cvitanović**, P., ed., (1984). *Universality in chaos*, Adam Hilger, Bristol.

117. D'Alessandro, G., and Firth, W. J., (1991). Spontaneous hexagon formation in a nonlinear optical medium with feedback mirror, *Phys. Rev. Lett.* **66**, 2597.

118. D'Alessandro, G., and Politi, A., (1990). Hierarchical approach to complexity with applications to dynamical systems, *Phys. Rev. Lett.* **64**, 1609.

119. D'Alessandro, G., Grassberger, P., Isola, S., and Politi, A., (1990). On the topology of the Hénon map, *J. Phys. A* **23**, 5285.

120. Damgaard, P. H., (1992). Stability and instability of renormalization group flows, *Int. J. Mod. Phys. A* **7**, 6933.

121. **Davies**, P., ed., (1993). *The new physics*, 2nd printing, Cambridge University Press, Cambridge.

122. **de Gennes**, P. G., (1979). *Scaling concepts in polymer physics*, Cornell University Press, Ithaca, NY.

123. Dekking, F. M., and Keane, M., (1978). Mixing properties of substitutions, *Z. Wahrsch. verw. Geb.* **42**, 23.

124. De Luca, A., and Varricchio, S., (1988). On the factors of the Thue-Morse word on three symbols, *Information Processing Letters* **27**, 281.

125. **Denker**, M., Grillenberger, C., and Sigmund, K., (1976). *Ergodic theory on compact spaces*, Springer, Berlin.

126. Derrida, B., (1981). Random-energy model: an exactly solvable model of disordered systems, *Phys. Rev. B* **24**, 2613.

127. Derrida, B., and Flyvbjerg, H., (1985). A new real-space renormalization method and its Julia set, *J. Phys. A* **18**, L313.

128. Derrida, B., and Spohn, H., (1988). Polymers on disordered trees, spin glasses and traveling waves, *J. Stat. Phys.* **51**, 817.

129. Derrida, B., Eckmann, J.-P., and Erzan, A., (1983). Renormalization groups with periodic and aperiodic orbits, *J. Phys. A* **16**, 893.

130. **Devaney**, R. L., (1989). *An introduction to chaotic dynamical systems*, 2nd edition, Addison Wesley, Redwood City, CA.

131. **Dietterich**, T. G., London, R., Clarkson, K., and Dromey, R., (1982). Learning and inductive inference, in *The Handbook of*

Artificial Intelligence, edited by P. Cohen and E. Feigenbaum, Kaufman, Los Altos, CA, p. 323.

132. **Domb**, C., and Green, M. S., eds., (1972 ff.). *Phase transitions and critical phenomena*, Vols. 1-7 (Vols. 8 ff. edited by C. Domb and J. L. Lebowitz), Academic Press, London and New York.

133. **Doolittle**, R. F., ed., (1990). *Methods in enzymology*, Vol. 183, Academic Press, San Diego, CA.

134. **Drazin**, P. G., and King, G. P., eds., (1992). *Interpretation of time series from nonlinear systems*, North-Holland, Amsterdam.

135. Dumont, J. M., (1990). Summation formulae for substitutions on a finite alphabet, in Luck *et al.* (1990), p. 185.

136. **Dutta**, P., and Horn, P. M., (1981). Low-frequency fluctuations in solids: $1/f$ noise, *Rev. Mod. Phys.* **53**, 497.

137. **Dyachenko**, S., Newell, A. C., Pushkarev, A., and Zakharov, V. E., (1992). Optical turbulence: weak turbulence, condensates, and collapsing filaments in the nonlinear Schrödinger equation, *Physica D* **57**, 96.

138. **Dyson**, F. J., (1969). Existence of a phase transition in a one-dimensional Ising ferromagnet and Non-existence of spontaneous magnetization in a one-dimensional Ising ferromagnet, *Commun. Math. Phys.* **12**, 91 and 212.

139. Ebeling, W., and Nicolis, G., (1991). Entropy of symbolic sequences: the role of correlations, *Europhys. Lett.* **14**, 191.

140. **Eckmann**, J.-P., and Ruelle, D., (1985). Ergodic theory of chaos and strange attractors, *Rev. Mod. Phys.* **57**, 617.

141. Eckmann, J.-P., Meakin, P., Procaccia, I., and Zeitak, R., (1989). Growth and form of noise-reduced diffusion limited aggregation, *Phys. Rev. A* **39**, 3185.

142. **Edgar**, G. A., (1990). *Measure, topology, and fractal geometry*, Springer, New York.

143. Edwards, S. F., and Anderson, P. W., (1975). Theory of spin glasses, *J. Phys. F* **5**, 965.

144. Eggers, J., and Großmann, S., (1991). Does deterministic chaos imply intermittency in fully developed turbulence?, *Phys. Fluids A* **3**, 1958.

145. Ehrenfeucht, A., and Rozenberg, G., (1981). On the subword complexity of D0L

languages with a constant distribution, *Information Processing Letters* **27**, 108.

146. Eigen, M., (1986). The physics of molecular evolution, *Chemica Scripta B* **26**, 13.

147. Elliott, R. J., (1961). Phenomenological discussion of magnetic ordering in the heavy rare-earth metals, *Phys. Rev.* **124**, 346.

148. Ernst, M. H., (1986). Kinetics of clustering in irreversible aggregation, in *Fractals in Physics*, edited by L. Pietronero and E. Tosatti, North-Holland, Amsterdam, p. 289.

149. Fahner, G., and Grassberger, P., (1987). Entropy estimates for dynamical systems, *Complex Systems* **1**, 1093.

150. **Falconer**, D. S., (1989). *Introduction to quantitative genetics*, 3rd edition, Longman, Harlow.

151. **Falconer**, K. J., (1990). *Fractal geometry: mathematical foundations and applications*, Wiley, New York.

152. **Family**, F., and Landau, D. P., eds., (1984). *Kinetics of aggregation and gelation*, North-Holland, Amsterdam.

153. Farmer, J. D., Ott, E., and Yorke, J. A., (1983). The dimension of chaotic attractors, *Physica D* **7**, 153.

154. **Farmer**, J. D., Toffoli, T., and Wolfram, S., eds., (1984). *Cellular automata*, North-Holland, Amsterdam.

155. Feigenbaum, M. J., (1978). Quantitative universality for a class of nonlinear transformations, *J. Stat. Phys.* **19**, 25.

156. Feigenbaum, M. J., (1979a). The universal metric properties of nonlinear transformations, *J. Stat. Phys.* **21**, 669.

157. Feigenbaum, M. J., (1979b). The onset spectrum of turbulence, *Phys. Lett. A* **74**, 375.

158. Feigenbaum, M. J., (1980). The transition to aperiodic behaviour in turbulent systems, *Commun. Math. Phys.* **77**, 65.

159. Feigenbaum, M. J., (1987). Some characterizations of strange sets, *J. Stat. Phys.* **46**, 919 and 925.

160. Feigenbaum, M. J., (1988). Presentation functions, fixed points and a theory of scaling function dynamics, *J. Stat. Phys.* **52**, 527.

161. Feigenbaum, M. J., (1993). Private communication.

162. Feigenbaum, M. J., Kadanoff, L. P., and

Shenker, S. J., (1982). Quasiperiodicity in dissipative sytems: a renormalization group analysis, *Physica D* **5**, 370.

163. Felderhof, B. U., and Fisher, M. E., (1970). Phase transitions in one-dimensional cluster-interaction fluids: (II) simple logarithmic model, *Ann. of Phys.* **58**, 268.

164. **Feller**, W., (1970). *An introduction to probability theory and its applications*, Vol. 1, 3rd edition, Wiley, New York.

165. Feudel, U., Pikovsky, A., and Politi, A., (1996). Renormalization of correlations and spectra of a strange nonchaotic attractor, *J Phys. A*, **29**, 5297.

166. **Field**, R. J., and Burger, M., eds., (1985). *Oscillations and traveling waves in chemical systems*, Wiley, New York.

167. **Fischer**, K. H., and Hertz, J. A., (1991). *Spin glasses*, Cambridge University Press, Cambridge.

168. **Fisher**, M. E., (1967a). The theory of equilibrium critical phenomena, *Rep. Prog. Phys.* **30**, 615.

169. Fisher, M. E., (1967b). The theory of condensation and the critical point, *Physics* **3**, 255.

170. Fisher, M. E., (1972). On discontinuity of the pressure, *Commun. Math. Phys.* **26**, 6.

171. **Fisher**, M. E., (1983). Scaling, universality, and renormalization group theory, in *Critical Phenomena*, edited by F. J. W. Hahne, Lect. Notes in Phys. **186**, Berlin, Springer, p. 1.

172. Fisher, M. E., and Milton, G. W., (1986). Classifying first-order phase transitions, *Physica A* **138**, 22.

173. Fitch, W. M., (1986). Unresolved problems in DNA sequence analysis, in Miura (1986), p. 1.

174. Flepp, L., Holzner, R., Brun, E., Finardi, M., and Badii, R., (1991). Model identification by periodic-orbit analysis for NMR-laser chaos, *Phys. Rev. Lett.* **67**, 2244.

175. **Fletcher**, C. A. J., (1984). *Computational Galerkin methods*, Springer, New York.

176. Fogedby, H. C., (1992). On the phase space approach to complexity, *J. Stat. Phys.* **69**, 411.

177. **Friedlander**, S. K., (1977). *Smoke, dust, and haze*, Wiley Interscience, New York.

178. Friedman, E. J., (1991). Structure and uncomputability in one-dimensional maps, *Complex Systems* **5**, 335.

179. Frisch, U., and Orszag, S. A., (1990). Turbulence: challenges for theory and experiment, *Physics Today* **43**, 24.

180. Frisch, U., and Parisi, G., (1985). On the singularity structure of fully-developed turbulence, appendix to U. Frisch, Fully-Developed Turbulence and Intermittency, in *Turbulence and Predictability in Geophisical Fluid Dynamics and Climate Dynamics*, edited by M. Ghil, R. Benzi and G. Parisi, North-Holland, New York, p. 84.

181. Frisch, U., Sulem, P. L., and Nelkin, M., (1978). On the singularity structure of fully developed turbulence, *J. Fluid Mech.* **87**, 719.

182. Frisch, U., Hasslacher, B., Pomeau, Y., (1986). Lattice-gas automata for the Navier-Stokes equation, *Phys. Rev. Lett.* **56**, 1505.

183. Fu, Y., and Anderson, P. W., (1986). Application of statistical mechanics to NP-complete problems in combinatorial optimisation, *J. Phys. A* **19**, 1605.

184. Fujisaka, H., (1992). A contribution to statistical nonlinear dynamics, in *From Phase Transitions to Chaos*, edited by G. Györgyi, I. Kondor, L. Sasvári, and T. Tél, World Scientific, Singapore, p. 434.

185. Furstenberg, H., (1967). Disjointness in ergodic theory, minimal sets, and a problem in Diophantine approximation, *Math. Systems Theory* **1**, 1.

186. **Garey**, M. R., and Johnson, D. S., (1979). *Computers and intractability: a guide to the theory of NP-completeness*, Freeman, San Francisco.

187. Gelfand, M. S., (1993). Genetic Language: metaphor or analogy?, *Biosystems* **30**, 277.

188. Gibbs, J. W., (1873). Private communication.

189. **Gibbs**, J. W., (1948). *The collected works of J. Willard Gibbs*, Longmans, New York.

190. Gilman, R. H., (1987). Classes of linear automata, *Ergod. Theory and Dynam. Syst.* **7**, 105.

191. Girsanov, I. V., (1958). Spectra of dynamical systems generated by stationary Gaussian processes, *Dokl. Akad. Nauk SSSR* **119**, 851.

192. Gledzer, E. B., (1973). Systems of hydrodynamic type admitting two integrals

of motion (in Russian), *Dokl. Akad. Nauk SSSR* **209**, 1046.

193. Gödel, K., (1931). Über formal unentscheidbare Sätze der Principia Mathematica und verwandter Systeme I, *Monatshefte für Mathematik und Physik* **38**, 173.

194. Godrèche, C., and Luck, J. M., (1990). Multifractal analysis in reciprocal space and the nature of Fourier transform of self-similar structures, *J. Phys. A* **23**, 3769.

195. Gold, E. M., (1967). Language identification in the limit, *Inform. Contr.* **10**, 447.

196. **Goldbeter**, A., ed., (1989). *Cell to cell signalling: from experiments to theoretical models*, Academic Press, London.

197. Gollub, J. P., (1991). Nonlinear waves: dynamics and transport, *Physica D* **51**, 501.

198. Goodwin, B. C., (1990). Structuralism in biology, *Sci. Progress* (Oxford) **74**, 227.

199. Goren, G., Procaccia, I., Rasenat, S., and Steinberg, V., (1989). Interactions and dynamics of topological defects: theory and experiments near the onset of weak turbulence, *Phys. Rev. Lett.* **63**, 1237.

200. Grassberger, P., (1981). On the Hausdorff dimension of fractal attractors, *J. Stat. Phys.* **26**, 173.

201. Grassberger, P., (1986). Toward a quantitative theory of self-generated complexity, *Int. J. Theor. Phys.* **25**, 907.

202. Grassberger, P., (1989a). Estimating the information content of symbol sequences and efficient codes, *IEEE Trans. Inform. Theory* **35**, 669.

203. Grassberger, P., (1989b). Randomness, information and complexity, in *Proc. 5th Mexican Summer School on Statistical Mechanics*, edited by F. Ramos-Gomez, World Scientific, Singapore, p. 59.

204. Grassberger, P., and Kantz, H., (1985). Generating partitions for the dissipative Hénon map, *Phys. Lett. A* **113**, 235.

205. Grassberger, P., and Procaccia, I., (1983). Characterization of strange attractors, *Phys. Rev. Lett.* **50**, 346.

206. Grassberger, P., Badii, R., and Politi, A., (1988). Scaling laws for hyperbolic and non-hyperbolic attractors, *J. Stat. Phys.* **51**, 135.

207. **Grassberger**, P., Schreiber, T., and

Schaffrath, C., (1991). Nonlinear time-sequence analysis, *Int. J. Bifurc. and Chaos* **1**, 521.

208. Grebogi, C., Ott, E., Pelikan, S., and Yorke, J.A., (1984). Strange attractors that are not chaotic, *Physica D* **13**, 261.

209. Greene, J. M., (1979). A method for determining a stochastic transition, *J. Math. Phys.* **20**, 1183.

210. Griffiths, R. B., (1972). Rigorous results and theorems, in Domb and Green (1972), Vol. 1, p. 7.

211. Grillenberger, C., (1973). Constructions of strictly ergodic systems: (I) given entropy, (II) *K*-systems, *Z. Wahrsch. verw. Geb.* **25**, 323 and 335.

212. Großmann, S., and Horner, R., (1985). Long time tail correlations in discrete chaotic dynamics, *Z. Phys. B* **60**, 79.

213. Großmann, S., and Lohse, D., (1994). Universality in fully developed turbulence, *Phys. Rev. E* **50**, 2784.

214. **Grünbaum**, B., and Shephard, G. C., (1987). *Tilings and patterns*, Freeman, New York.

215. **Guckenheimer**, J., and Holmes, P., (1986). *Nonlinear oscillations, dynamical systems, and bifurcations of vector fields*, 2nd edition, Springer, New York.

216. Günther, R., Schapiro, B., and Wagner, P., (1992). Physical complexity and Zipf's law, *Int. J. Theor. Phys.* **31**, 525.

217. Gurevich, B. M., (1961). The entropy of horocycle flows, *Dokl. Akad. Nauk SSSR* **136**, 768; *Soviet Math. Dokl.* **2**, 124.

218. Gutowitz, H. A., (1990). A hierarchical classification of cellular automata, *Physica D* **45**, 136.

219. Gutowitz, H. A., (1991a). Transients, cycles, and complexity in cellular automata, *Phys. Rev. A* **44**, 7881.

220. **Gutowitz**, H. A., ed., (1991b). *Cellular automata: theory and experiment*, MIT Press, Cambridge, MA.

221. Gutowitz, H. A., Victor, J. D., and Knight, B. W., (1987). Local structure theory for cellular automata, *Physica D* **28**, 18.

222. Haken, H., (1975). Analogy between higher instabilities in fluids and lasers, *Phys. Lett. A* **53**, 77.

223. Halmos, P. R., (1944). Approximation theories for measure preserving

transformations, *Trans. Am. Math. Soc.* **55**, 1.

224. **Halmos**, P. R., (1956). *Lectures on ergodic theory*, The mathematical society of Japan, Tokyo.

225. Halsey, T. C., Jensen, M. H., Kadanoff, L. P., Procaccia, I., and Shraiman, B., (1986). Fractal measures and their singularities: the characterization of strange sets, *Phys. Rev. A* **33**, 1141.

226. **Hamming**, R. W., (1986). *Coding and information theory*, 2nd edition, Prentice-Hall, Englewood Cliffs, NJ.

227. **Hao**, B.-L., (1989). *Elementary symbolic dynamics*, World Scientific, Singapore.

228. **Hardy**, G. H., and Wright, H. M., (1938). *Theory of numbers*, Oxford University Press, Oxford.

229. **Harrison**, R. G., and Uppal, J. S., eds., (1993). *Nonlinear dynamics and spatial complexity in optical systems*, SUSSP, Edinburgh, and IOP, Bristol.

230. Hedlund, G. A., (1969). Endomorphisms and automorphisms of the shift dynamical system, *Math. Systems Theory* **3**, 320.

231. Hennequin, D., and Glorieux, P., (1991). Symbolic dynamics in a passive Q-switching laser, *Europhys. Lett.* **14**, 237.

232. Hénon, M., (1976). A two-dimensional mapping with a strange attractor, *Commun. Math. Phys.* **50**, 69.

233. **Herken**, R., ed., (1994). *The universal Turing machine: a half-century survey*, 2nd edition, Springer, Wien.

234. **Hofstadter**, D., (1979). *Gödel, Escher, Bach: an eternal golden braid*, Basic Books, New York.

235. Hohenberg, P. C., and Shraiman, B. I., (1988). Chaotic behaviour of an extended system, *Physica D* **37**, 109.

236. Holschneider, M., (1994). Fractal wavelet dimensions and localization, *Commun. Math. Phys.* **160**, 457.

237. **Hopcroft**, J. E., and Ullman, J. D., (1979). *Introduction to automata theory, languages, and computation*, Addison-Wesley, Reading, MA.

238. Horton, R. E., (1945). Erosional development of streams and their drainage basins: hydrophysical approach to quantitative morphology, *Bull. Geol. Soc. Am.* **56**, 275.

239. **Howie**, J. M., (1991). *Automata and languages*, Oxford University Press, Oxford.

240. **Huang**, K., (1987). *Statistical mechanics*, 2nd edition, Wiley, Singapore.

241. Huberman, B. A., and Hogg, T., (1986). Complexity and adaptation, *Physica D* **22**, 376.

242. Huberman, B. A., and Kerszberg, M., (1985). Ultradiffusion: the relaxation of hierarchical systems, *J. Phys. A* **18**, L331.

243. Huffman, D. A., (1952). A method for the construction of minimum redundancy codes, *Proc. IRE* **40**, 1098.

244. Hurd, L. P., (1990a). Recursive cellular automata invariant sets, *Complex Systems* **4**, 119.

245. Hurd, L. P., (1990b). Nonrecursive cellular automata invariant sets, *Complex Systems* **4**, 131.

246. Hurd, L. P., Kari, J., and Culik, K., (1992). The topological entropy of cellular automata is uncomputable, *Ergod. Theory and Dynam. Sys.* **12**, 255.

247. **Hurewicz**, W., and Wallman, H., (1974). *Dimension theory*, 9th printing, Princeton University Press, Princeton, NJ.

248. Hurley, M., (1990). Attractors in cellular automata, *Ergod. Theory and Dynam. Syst.* **10**, 131.

249. Jahnke, W., Skaggs, W. E., and Winfree, A. T., (1989). Chemical vortex dynamics in the Belousov-Zhabotinsky reaction and in the two-variable Oregonator model, *J. Phys. Chem.* **93**, 740.

250. **Jen**, E., ed., (1990). *Lectures in complex systems*, SFI Studies in the Sciences of Complexity, Vol. II, Addison-Wesley, Redwood City, CA.

251. Jensen, M. H., Paladin, G., and Vulpiani, A., (1991). Intermittency in a cascade model for three-dimensional turbulence, *Phys. Rev. A* **43**, 798.

252. Joets, A., and Ribotta, R., (1991). Localized bifurcations and defect instabilities in the convection of a nematic liquid crystal, *J. Stat. Phys.* **64**, 981.

253. **Jullien**, R., and Botet, R., eds., (1987). *Aggregation and fractal aggregates*, World Scientific, Singapore.

254. Kadanoff, L. P., (1966). Scaling laws for Ising models near T_c, Physics **2**, 263.

255. Kadanoff, L. P., (1976a). Scaling, universality and operator algebras, in Domb and Green (1976), Vol. 5a, p. 1.

256. Kadanoff, L. P., (1976b). Notes on Migdal's recursion formulas, *Ann. Phys.* (NY) **100**, 359.

257. Kadanoff, L. P., (1991). Complex structures from simple systems, *Physics Today* **44**, Vol. 3, p. 9.

258. Kakutani, S., (1973). Examples of ergodic measure preserving transformations which are weakly mixing but not strongly mixing, in *Recent Advances in Topological Dynamics*, Lect. Notes in Math. **318**, Springer, New York, p. 143.

259. **Kakutani**, S., (1986). *Selected papers*, edited by R. R. Kallman, Birkhäuser, Boston.

260. Kakutani, S., (1986a). Ergodic theory of shift transformations, reprinted in Kakutani (1986), p. 268.

261. Kakutani, S., (1986b). Strictly ergodic dynamical systems, reprinted in Kakutani (1986), p. 319.

262. Kaminger, F. P., (1970). The noncomputability of the channel capacity of context-sensitive languages, *Information and Control* **17**, 175.

263. Kaplan, J. L., and Yorke, J. A., (1979). Chaotic behaviour of multidimensional difference equations, in *Functional Differential Equations and Approximations of Fixed Points*, edited by H. O. Peitgen and H. O. Walther, Lect. Notes in Math. **730**, Springer, New York, p. 228.

264. Katok, A. B., and Stepin, A. M., (1966). Approximation of ergodic dynamical systems by periodic transformations, *Soviet Math. Dokl.* **7**, 1638.

265. Katok, A. B., and Stepin, A. M., (1967). Approximations in ergodic theory, *Russ. Math. Surv.* **22**, 77.

266. Keane, M. S., (1991). Ergodic theory and subshifts of finite type, in Bedford *et al.* (1991), p. 35.

267. Kesten, H., (1966). On a conjecture of Erdös and Szüsz related to uniform distribution mod 1, *Acta Aritmetica* **12**, 193.

268. Ketzmerick, R., Petschel, G., and Geisel, T., (1992). Slow decay of temporal correlations in quantum systems with Cantor spectra, *Phys. Rev. Lett.* **69**, 695.

269. **Khinchin**, A. I., (1957). *Information theory (mathematical foundations)*, Dover, New York.

270. **Kimura**, M., (1983). *The neutral theory of molecular evolution*, Cambridge University Press, Cambridge.

271. Klimontovich, Y. L., (1987). Entropy evolution in self-organization processes: H-theorem and S-theorem, *Physica A* **142**, 390.

272. Klir, G. J., (1985). Complexity: some general observations, *Systems Research* **2**, 131.

273. Kirkpatrick, S., Gelatt Jr, C. D., and Vecchi, M. P., (1983). Optimization by simulated annealing, *Science* **220**, 671.

274. Kolář, M., Iochum, B., and Raymond, L., (1993). Structure factor of 1D systems (superlattices) based on two-letter substitution rules: I. δ (Bragg) peaks, *J. Phys. A* **26**, 7343.

275. Kolb, M., Botet, R., and Jullien, R., (1983). Scaling of kinetically growing clusters, *Phys. Rev. Lett.* **51**, 1123.

276. Kolmogorov, A. N., (1941). The local structure of turbulence in an incompressible flow with very large Reynolds numbers, *Dokl. Akad. Nauk SSSR* **30**, 301.

277. Kolmogorov, A. N., (1954). On conservation of conditionally periodic motions under small perturbations of the Hamiltonian, *Dokl. Akad. Nauk SSSR* **98**, 527.

278. Kolmogorov, A. N., (1962). A refinement of previous hypotheses concerning the local structure of turbulence in a viscous incompressible fluid at high Reynolds number, *J. Fluid Mech.* **13**, 82.

279. Kolmogorov, A. N., (1965). Three approaches to the quantitative definition of information, *Problems of Information Transmission* **1**, 4.

280. Koppel, M., (1987). Complexity, Depth, and Sophistication, *Complex Systems* **1**, 1087.

281. Koppel, M., (1994). Structure, in Herken (1994), p. 435.

282. Koppel, M., and Atlan, H., (1991). An almost machine-independent theory of program length complexity, sophistication, and induction, *Information Sciences* **56**, 23.

283. Kraichnan, R. H., and Chen, S., (1989). Is there a statistical mechanics of turbulence?, *Physica D* **37**, 160.

284. Kreisberg, N., McCormick, W. D., and Swinney, H. L., (1991). Experimental demonstration of subtleties in subharmonic intermittency, *Physica D* **50**, 463.

285. Krieger, W., (1972). On unique ergodicity, in *Proc. Sixth Berkeley Symp.*, Vol. 2, University of California Press, Berkeley and Los Angeles, p. 327.

286. Kuich, W., (1970). On the entropy of context-free languages, *Information and Control* **16**, 173.

287. **Kuramoto**, Y., (1984). *Chemical oscillations, waves, and turbulence*, Springer, Berlin.

288. Lakdawala, P., (1996). Computational complexity of symbolic dynamics at the onset of chaos, *Phys. Rev. E*, **53**, 4477.

289. Landauer, R., (1987). Role of relative stability in self-repair and self-maintenance, in Yates (1987), p. 435.

290. Landweber, P., (1964). Decision problems on phrase structure grammars, *IEEE Trans. Electron. Comput.* **13**, 354.

291. Langton, C. G., (1986). Studying artificial life with cellular automata, *Physica D* **22**, 120.

292. Lari, K., and Young, S. J., (1990). The estimation of stochastic context-free grammars using the inside-outside algorithm, *Computer Speech and Language* **4**, 35.

293. **Lasota**, A., and Mackey, M. C., (1985). *Probabilistic properties of deterministic systems*, Cambridge University Press, Cambridge.

294. Lempel, A., and Ziv, J., (1976). On the complexity of finite sequences, *IEEE Trans. Inform. Theory* **22**, 75.

295. Levine, D., and Steinhardt, P. J., (1986). Quasicrystals. I: Definition and structure, *Phys. Rev. B* **34**, 596.

296. Levine, L., (1992). Regular language invariance under one-dimensional cellular automaton rules, *Complex Systems* **6**, 163.

297. **Lewin**, B., (1994). *Genes V*, Oxford University Press, New York.

298. **Li**, M., and Vitányi, P. M. B., (1997). *An introduction to Kolmogorov complexity and its applications*, 2nd edition, Springer, New York.

299. Li, W., and Kaneko, K., (1992). Long-range correlation and partial $1/f^\lambda$ spectrum in a noncoding DNA sequence, *Europhys. Lett.* **17**, 655.

300. Libchaber, A., Fauve, S., and Laroche, C., (1983). Two-parameter study of routes to chaos, *Physica D* **7**, 73.

301. **Lichtenberg**, A. J., and Lieberman, M. A., (1992). *Regular and chaotic dynamics*, 2nd edition, Springer, New York.

302. Lind, D. A., (1984). Applications of ergodic theory and sofic systems to cellular automata, *Physica D* **10**, 36.

303. **Lind**, D., and Marcus, B., (1995). *An introduction to symbolic dynamics and coding*, Cambridge University Press, Cambridge.

304. Lindenmayer, A., (1968). Mathematical models for cellular interactions and development, *J. Theor. Biol.* **18**, 280.

305. Lindgren, K., and Nordahl, M. G., (1988). Complexity measures and cellular automata, *Complex Systems* **2**, 409.

306. Lindgren, K., and Nordahl, M. G., (1990). Universal computation in simple one-dimensional cellular automata, *Complex Systems* **4**, 299.

307. Lloyd, S., and Pagels, H., (1988). Complexity as thermodynamic depth, *Ann. of Phys.* **188**, 186.

308. Löfgren, L., (1977). Complexity of descriptions of systems: a foundational study, *Int. J. General Systems* **3**, 197.

309. **Lorentzen**, L., and Waadeland, H., (1992). *Continuous fractions with applications*, North-Holland, Amsterdam.

310. Lorenz, E. N., (1963). Deterministic non periodic flow, *J. Atmos. Sci.* **20**, 130.

311. **Lothaire**, M., (1983). *Combinatorics on words*, Addison-Wesley, Reading, MA.

312. **Luck**, J. M., Moussa, P., and Waldschmidt, M., eds., (1990). *Number theory and physics*, Springer, Berlin.

313. Luck, J. M., Godrèche, C., Janner, A., and Janssen, T., (1993). The nature of the atomic surfaces of quasiperiodic self-similar structures, *J. Phys. A* **26**, 1951.

314. **Lumley**, J. L., (1970). *Stochastic tools in turbulence*, Academic Press, New York.

315. Lyubimov, D. V., and Zaks, M. A., (1983). Two mechanisms of the transition to chaos in finite-dimensional models of convection, *Physica D* **9**, 52.

316. MacKay, R. S., and Tresser, C., (1988). Boundary of topological chaos for bimodal maps of the interval, *J. London Math. Soc.* **37**, 164.

317. McKay, S. R., Berker, A. N., and Kirkpatrick, S., (1982). Spin-glass behaviour in frustrated Ising models with chaotic renormalization group trajectories, *Phys. Rev. Lett.* **48**, 767.

318. **Madras**, N., and Slade, G., (1993). *The self-avoiding walk*, Birkhäuser, Boston, MA.

319. Mandelbrot, B. B., (1974). Intermittent turbulence in self-similar cascades: divergence of high moments and dimensions of the carrier, *J. Fluid Mech.* **62**, 331.

320. **Mandelbrot**, B. B., (1982). *The fractal geometry of nature*, Freeman, San Francisco.

321. Mandelbrot, B. B., (1989). A class of multinomial multifractal measures with negative (latent) values for the "dimension" $f(\alpha)$, in *Fractals' Physical Origin and Properties*, edited by L. Pietronero, Plenum, New York, p. 3.

322. Manneville, P., (1980). Intermittency, self-similarity and $1/f$ spectrum in dissipative dynamical systems, *J. de Physique* **41**, 1235.

323. **Manneville**, P., (1990). *Dissipative structures and weak turbulence*, Academic Press, Boston.

324. Manneville, P., and Pomeau, Y., (1980). Different ways to turbulence in dissipative dynamical systems, *Physica D* **1**, 219.

325. Marinari, E., Parisi, G., and Ritort, F., (1994a). Replica field theory for deterministic models: binary sequences with low autocorrelation, *J. Phys. A* **27**, 7615.

326. Marinari, E., Parisi, G., and Ritort, F., (1994b). Replica field theory for deterministic models (II): a non random spin glass with glassy behaviour, *J. Phys. A* **27**, 7647.

327. Martiel, J.-L., and Goldbeter, A., (1987). A model based on receptor desensitization for cyclic AMP signaling in Dictyostelium cells, *Biophys. J.* **52**, 807.

328. Martin, J. C., (1973). Minimal flows arising from substitutions of non-constant length, *Math. Systems Theory* **7**, 73.

329. Martin-Löf, P., (1966). The definition of random sequences, *Information and Control* **9**, 602.

330. Masand, B., Wilensky, U., Massar, J.-P., and Redner, S., (1992). An extension of the two-dimensional self-avoiding random walk series on the square lattice, *J. Phys. A* **25**, L365.

331. **Mayer-Kress**, G., ed., (1989). *Dimensions and entropies in chaotic systems*, Springer, Berlin.

332. **McEliece**, R. J., Ash, R. B., and Ash, C., (1989). *Introduction to discrete mathematics*, McGraw-Hill, Singapore.

333. Meakin, P., (1983). Formation of fractal clusters and networks by irreversible diffusion-limited aggregation, *Phys. Rev. Lett.* **51**, 1119.

334. **Meakin**, P., (1990). Fractal structures, *Prog. Solid State Chem.* **20**, 135.

335. **Meakin**, P., (1991). Fractal aggregates in geophysics, *Reviews of Geophysics* **29**, 317.

336. Meneveau, C., and Sreenivasan, K. R., (1991). The multifractal nature of turbulent energy dissipation, *J. Fluid Mech.* **224**, 429.

337. Metropolis, N., Rosenbluth, A. W., Rosenbluth, M. N., Teller, A. H., and Teller, E., (1953). Equation of state calculations by fast computing machines, *J. Chem. Phys.* **21**, 1087.

338. **Mézard**, M., Parisi, G., and Virasoro, M., (1986). *Spin glass theory and beyond*, World Scientific, Singapore.

339. Migdal, A. A., (1976). Phase transitions in gauge and spin-lattice systems, *Sov. Phys. JETP* **42**, 743.

340. Milnor, J., (1985). On the concept of attractor, *Commun. Math. Phys.* **99**, 177.

341. **Minsky**, M. L., ed., (1988). *Semantic Information Processing*, 6th printing, MIT Press, Cambridge, MA, p. 425.

342. Misiurewicz, M., and Ziemian, K., (1987). Rate of convergence for computing entropy of some one-dimensional maps, in *Proceedings of the Conference on Ergodic Theory and Related Topics II*, edited by H. Mickel, Teubner, Stuttgart, p. 147.

343. **Miura**, R. M., ed., (1986). *Some mathematical questions in biology*, Lectures on Mathematics in the Life Sciences, Vol.

17, American Mathematical Society, Providence, RI.

344. **Monin**, A. S., and Yaglom, A. M., (1971, 1975). *Statistical fluid mechanics*, Vols. 1 and 2, MIT Press, Cambridge, MA.

345. Moore, C., (1990). Unpredictability and undecidability in dynamical systems, *Phys. Rev. Lett.* **64**, 2354.

346. Moore, C., (1991). Generalized shifts: unpredictability and undecidability in dynamical systems, Nonlinearity **4**, 199.

347. **Mori**, H., Hata, H., Horita, T., and Kobayashi, T., (1989). Statistical mechanics of dynamical systems, *Prog. Theor. Phys. Suppl.* **99**, 1.

348. Moser, J. K., (1962). On invariant curves of area-preserving mappings of an annulus, *Nachr. Akad. Wiss.*, Göttingen, Math. Phys. Kl. **2**, 1.

349. Mozes, A., (1992). A zero entropy, mixing of all orders tiling system, in *Symbolic dynamics and its applications, Contemporary Mathematics* **135**, edited by P. Walters, American Mathematical Society, Providence, RI, p. 319.

350. **Müller**, B., Reinhardt, J., and Strickland, M. T., (1995). *Neural networks: an introduction*, 2nd edition, Springer, Berlin.

351. Müller, S. C., and Hess, B., (1989). Spiral order in chemical reactions, in Goldbeter (1989), p. 503.

352. **Müller**, S. C., and Plesser, T., eds., (1992). *Spatio-temporal organization in nonequilibrium systems*, Projekt Verlag, Dortmund.

353. Nauenberg, M., and Rudnick, J., (1981). Universality and the power spectrum at the onset of chaos, *Phys. Rev. B* **24**, 493.

354. Nerode, A., (1958). Linear automaton transformations, *Proc. AMS* **9**, 541.

355. **Newell**, A. C., and Moloney, J. V., (1992). *Nonlinear optics*, Addison-Wesley, Redwood City, CA.

356. Newhouse, S. E., (1974). Diffeomorphisms with infinitely many sinks, *Topology* **13**, 9.

357. Newhouse, S. E., (1980). Lectures on dynamical systems, in *Dynamical Systems*, Progress in Mathematics **8**, Birkhäuser, Boston, p. 1.

358. Niemeijer, Th., and van Leeuwen, J. M. J., (1976). Renormalization theory for Ising-like spin systems, in Domb and Green (1976), Vol. **6**, p. 425.

359. Nienhuis, B., (1982). Exact critical point and critical exponents of $O(n)$ models in two dimensions, *Phys. Rev. Lett.* **49**, 1062.

360. Niwa, T., (1978). Time correlation functions of a one-dimensional infinite system, *J. Stat. Phys.* **18**, 309.

361. Novikov, E. A., (1971). Intermittency and scale similarity in the structure of a turbulent flow, *Prikl. Math. Mech.* **35**, 266.

362. Novikov, E. A., and Sedov, Yu. Bu., (1979). Stochastic properties of a four-vortex system, *Sov. Phys. JETP* **48**, 440.

363. Oono, Y., and Osikawa, M., (1980). Chaos in nonlinear difference equations, *Prog. Theor. Phys.* **64**, 54.

364. Oseledec, V. I., (1968). A multiplicative ergodic theorem: Lyapunov characteristic numbers for dynamical systems, *Trans. Moscow Math. Soc.* **19**, 197.

365. Ostlund, S., Rand, D., Sethna, J., and Siggia, E., (1983). Universal properties of the transition from quasiperiodicity to chaos in dissipative systems, *Physica D* **8**, 303.

366. **Ott**, E., (1993). *Chaos in dynamical systems*, Cambridge University Press, Cambridge.

367. Packard, N. H., Crutchfield, J. P., Farmer, J. D., and Shaw, R. S., (1980). Geometry from a time series, *Phys. Rev. Lett.* **45**, 712.

368. **Palmer**, R. G., (1982). Broken ergodicity, *Adv. Phys.* **31**, 669.

369. Pampaloni, E., Ramazza, P. L., Residori, S., and Arecchi, F. T., (1995). Two-dimensional crystals and quasicrystals in nonlinear optics, *Phys. Rev. Lett.* **74**, 258.

370. Paoli, P., Politi, A., and Badii, R., (1989a). Long-range order in the scaling behaviour of hyperbolic dynamical systems, *Physica D* **36**, 263.

371. Paoli, P., Politi, A., Broggi, G., Ravani, M., and Badii, R., (1989b). Phase transitions in filtered chaotic signals, *Phys. Rev. Lett.* **62**, 2429.

372. **Papoulis**, A., (1984). *Probability, random variables and stochastic processes*, 2nd edition, McGraw-Hill, Singapore.

373. Parasjuk, O. S., (1953). Horocycle flows on surfaces of constant negative curvature, *Uspekhi Mat. Nauk* **8**, 125.

374. **Patashinskii**, A. Z., and Pokrovskii, V. L.,

(1979). *Fluctuation theory of phase transitions*, Pergamon, Oxford.

375. **Peitgen**, H.-O., and Richter, P. H., (1986). *The beauty of fractals: images of complex dynamical systems*, Springer, Berlin.

376. Penrose, R., (1974). The rôle of aesthetics in pure and applied mathematical research, *Bull. Inst. Math. Applications* **10**, 266.

377. **Penrose**, R., (1989). *The emperor's new mind*, Oxford University Press, Oxford.

378. Pesin, Ya. B., (1977). Characteristic Lyapunov eponents and smooth ergodic theory, *Russ. Math. Surv.* **32**, 55.

379. Pesin, Ya. B., (1984). On the notion of the dimension with respect to a dynamical system, *Ergod. Theory and Dynam. Syst.* **4**, 405.

380. **Petersen**, K., (1989). *Ergodic theory*, Cambridge University Press, Cambridge.

381. Pikovsky, A. S., and Feudel, U., (1994). Correlations and spectra of strange non-chaotic attractors, *J. Phys. A* **27**, 5209.

382. Pikovsky, A. S., and Grassberger, P., (1991). Symmetry breaking bifurcation for coupled chaotic attractors, *J. Phys. A* **24**, 4587.

383. Politi, A., (1994). Symbolic encoding in dynamical systems, in *From Statistical Physics to Statistical Inference and Back*, edited by P. Grassberger and J.-P Nadal, Kluwer, Dordrecht, p. 293.

384. Politi, A., Oppo, G. L., and Badii, R., (1986). Coexistence of conservative and dissipative behaviour in reversible systems, *Phys. Rev. A* **33**, 4055.

385. Politi, A., Badii, R., and Grassberger, P., (1988). On the geometric structure of nonhyperbolic attractors, *J. Phys. A* **21**, L736.

386. Procaccia, I., and Zeitak, R., (1988). Shape of fractal growth patterns: exactly solvable models and stability considerations, *Phys. Rev. Lett.* **60**, 2511.

387. Procaccia, I., Thomae, S., and Tresser, C., (1987). First-return maps as a unified renormalization scheme for dynamical systems, *Phys. Rev. A* **35**, 1884.

388. Przytycki, F., and Tangerman, F., (1996). Cantor sets in the line: scaling functions and the smoothness of the shift-map, *Nonlinearity* **9**, 403.

389. **Queffélec**, M., (1987). *Substitution dynamical systems: spectral analysis*, Lect. Notes in Math. **1294**, Springer, Berlin.

390. Rabaud, M., Michalland, S., and Couder, Y., (1990). Dynamical regimes of directional viscous fingering: spatiotemporal chaos and wave propagation, *Phys. Rev. Lett.* **64**, 184.

391. Radons, G., (1995). Thermodynamic analysis of inhomogeneous random walks: localization and phase transitions, *Phys. Rev. Lett.* **75**, 4719.

392. **Rammal**, R., Toulouse, G., and Virasoro, M. A., (1986). Ultrametricity for physicists, *Rev. Mod. Phys.* **58**, 765.

393. Rehberg, I., Rasenat, S., and Steinberg, V., (1989). Traveling waves and defect-initiated turbulence in electroconvecting nematics, *Phys. Rev. Lett.* **62**, 756.

394. Reichert, P., and Schilling, R., (1985). Glass-like properties of a chain of particles with anharmonic and competing interactions, *Phys. Rev. B* **32**, 5731.

395. **Renyi**, A., (1970). *Probability theory*, North-Holland, Amsterdam.

396. Rissanen, J., (1986). Complexity of strings in the class of Markov sources, *IEEE Trans. Inform. Theory* **32**, 526.

397. **Rissanen**, J., (1989). *Stochastic complexity in statistical inquiry*, World Scientific, Singapore.

398. **Roberts**, J. A. G., and Quispel, G. R. W., (1992). Chaos and time-reversal symmetry, *Phys. Rep.* **216**, 63.

399. Ross, J., Müller, S. C., and Vidal, C., (1988). Chemical waves, *Science* **240**, 460.

400. **Royden**, H. L., (1989). *Real analysis*, Macmillan, Singapore.

401. **Rozenberg**, G., and Salomaa, A., (1990). *The mathematical theory of L systems*, Academic Press, New York.

402. Rubio, M. A., Gluckman, B. J., Dougherty, A., and Gollub, J. P., (1991). Streams with moving contact lines: complex dynamics due to contact-angle hysteresis, *Phys. Rev. A* **43**, 811.

403. **Ruelle**, D., (1974). *Statistical mechanics: rigorous results*, 2nd printing, Benjamin, New York.

404. **Ruelle**, D., (1978). *Thermodynamic formalism*, vol. 5 of Encyclopedia of Mathematics and its Applications, Addison-Wesley, Reading, MA.

405. Ruelle, D., (1981). Small random perturbations of dynamical systems and the definition of attractors, *Commun. Math. Phys.* **82**, 137.

406. **Ruelle**, D., (1991). *Chance and chaos*, Princeton University Press, Princeton, NJ.

407. Ruelle, D., and Takens, F., (1971). On the nature of turbulence, *Commun. Math. Phys.* **20**, 167.

408. Sakakibara, Y., Brown, M., Hughey, R., Mian, I. S., Sjölander, K., Underwood, R. C., and Haussler, D., (1994). Stochastic context-free grammars for tRNA modeling, *Nucleic Acids Research* **22**, 5112.

409. Saparin, P., Witt, A., Kurths, J., and Anishchenko, V., (1994). The renormalized entropy—an appropriate complexity measure?, *Chaos, Solitons and Fractals* **4**, 1907.

410. Sauer, T., Yorke, J. A., and Casdagli, M., (1991). Embedology, *J. Stat. Phys.* **65**, 579.

411. **Schuster**, H. G., (1988). *Deterministic chaos*, VCH, Physik-Verlag, Weinheim.

412. Schwartzbauer, T., (1972). Entropy and approximation of measure-preserving transformations, *Pacific J. of Math.* **43**, 753.

413. **Seneta**, E., (1981). *Non-negative matrices and Markov chains*, 2nd edition, Springer, New York.

414. Shannon, C. E., (1951). Prediction and entropy of printed English, *Bell Sys. Tech. J.* **30**, 50.

415. **Shannon**, C. E., and Weaver, W. W., (1949). *The mathematical theory of communication*, University of Illinois Press, Urbana, IL.

416. Shechtman, D. S., Blech, I., Gratias, D., and Cahn, J. W., (1984). Metallic phase with long-range orientational order and no translational symmetry, *Phys. Rev. Lett.* **53**, 1951.

417. Shenker, S. J., (1982). Scaling behaviour in a map of a circle onto itself: empirical results, *Physica D* **5**, 405.

418. Sherrington, D., and Kirkpatrick, S., (1975). Solvable model of a spin glass, *Phys. Rev. Lett.* **35**, 1792.

419. Shirvani, M., and Rogers, T. D., (1988). Ergodic endomorphisms of compact abelian groups, *Commun. Math. Phys.* **118**, 401.

420. Shirvani, M., and Rogers, T. D., (1991). On ergodic one-dimensional cellular automata, *Commun. Math. Phys.* **136**, 599.

421. **Simon**, B., (1993). *The statistical mechanics of lattice gases*, Princeton University Press, Princeton, NJ.

422. Simon, H. A., (1962). The architecture of complexity, *Proc. Am. Philos. Soc.* **106**, 467.

423. Simonet, J., Brun, E., and Badii, R., (1995). Transition to chaos in a laser system with delayed feedback, *Phys. Rev. E* **52**, 2294.

424. Sinai, Ya. G., (1972). Gibbs measures in ergodic theory, *Russ. Math. Surv.* **27**, 21.

425. **Sinai**, Ya. G., (1994). *Topics in ergodic theory*, Princeton University Press, Princeton, NJ.

426. Sirovich, L., (1989). Chaotic dynamics of coherent structures, *Physica D* **37**, 126.

427. Skinner, G. S., and Swinney, H. L., (1991). Periodic to quasiperiodic transition of chemical spiral rotation, *Physica D* **48**, 1.

428. Smale, S., (1963). Diffeomorphisms with many periodic points, in *Differential and Combinatorial Topology*, edited by S. S. Cairns, Princeton University Press, Princeton, p. 63.

429. Smale, S., (1967). Differentiable dynamical systems, *Bull. Amer. Math. Soc.* **73**, 747.

430. Solomonoff, R. J., (1964). A formal theory of inductive inference, *Inform. Contr.* **7**, 1 and 224.

431. Sommerer, J. C., and Ott, E., (1993a). Particles floating on a moving fluid: a dynamically comprehensible physical fractal, *Science* **259**, 335.

432. Sommerer, J. C., and Ott, E., (1993b). A physical system with qualitatively uncertain dynamics, *Nature* **365**, 138.

433. Sreenivasan, K. R., (1991). On local isotropy of passive scalars in turbulent shear flows, *Proc. R. Soc. London A* **434**, 165.

434. Staden, R., (1990). Finding protein coding regions in genomic sequences, in Doolittle (1990), p. 163.

435. **Stanley**, H. E., and Ostrowsky, N., eds., (1986). *On growth and form: fractal and non-fractal patterns in physics*, Martinus Nijhoff Publishers, Dordrecht.

436. **Stauffer**, D., and Aharony, A., (1992). *Introduction to percolation theory*, Taylor and Francis, London.

437. Stavans, J., Heslot, F., and Libchaber, A.,

(1985). Fixed winding number and the quasiperiodic route to chaos in a convective fluid, *Phys. Rev. Lett.* **55**, 596.

438. **Stein**, D. L., ed., (1989). *Lectures in the sciences of complexity*, Addison-Wesley, Reading, MA.

439. Stein, D. L., and Palmer, R. G., (1988). Nature of the glass transition, *Phys. Rev. B* **38**, 12035.

440. Steinberg, V., Ahlers, G., and Cannell, D. S., (1985). Pattern formation and wave-number selection by Rayleigh-Bénard convection in a cylindrical container, *Physica Scripta* **32**, 534.

441. **Steinhardt**, P. J., and Ostlund, S., eds., (1987). *The physics of quasicrystals*, World Scientific, Singapore.

442. Strahler, A. N., (1957). Quantitative analysis of watershed geomorphology, *Trans. Am. Geophys. Union* **38**, 913.

443. **Stryer**, L., (1988). *Biochemistry*, 3rd edition, Freeman, New York.

444. Švrakić, N. M., Kertész, J., and Selke, W., (1982). Hierarchical lattice with competing interactions: an example of a nonlinear map, *J. Phys. A* **15**, L427.

445. Szépfalusy, P., (1989). Characterization of chaos and complexity by properties of dynamical entropies, *Physica Scripta T* **25**, 226.

446. Szépfalusy, P., and Györgyi, G., (1986). Entropy decay as a measure of stochasticity in chaotic systems, *Phys. Rev. A* **33**, 2852.

447. Takens, F., (1981). Detecting strange attractors in turbulence, in *Dynamical Systems and Turbulence*, edited by D. A. Rand and L.-S. Young, Lect. Notes in Math. **898**, Springer, New York, p. 366.

448. Tam, W. Y., and Swinney, H. L., (1990). Spatiotemporal patterns in a one-dimensional open reaction-diffusion system, *Physica D* **46**, 10.

449. Tavaré, S., (1986). Some probabilistic and statistical problems in the analysis of DNA sequences, in Miura (1986), p. 57.

450. Tél, T., (1983). Invariant curves, attractors, and phase diagram of a piecewise linear map with chaos, *J. Stat. Phys.* **33**, 195.

451. **Temam**, R., (1988). *Infinite-dimensional dynamical systems in mechanics and physics*, Springer, New York.

452. **Tennekes**, H., and Lumley, J. L., (1990). *A first course in turbulence*, MIT Press, Cambridge, MA.

453. **Toffoli**, T., and Margolus, N., (1987). *Cellular automata machines: a new environment for modelling*, MIT Press, Cambridge, MA.

454. Toulouse, G., (1977). Theory of the frustration effect in spin glasses: I, *Commun. Phys.* **2**, 115.

455. Tresser, C., (1993). Private communication.

456. Tufillaro, N. B., Ramshankar, R., and Gollub, J. P., (1989). Order-disorder transition in capillary ripples, *Phys. Rev. Lett.* **62**, 422.

457. Turing, A. M., (1936). On computable numbers with an application to the Entscheidungsproblem, *Proc. London Math. Soc.* **2**, 230.

458. Turing, A. M., (1952). The chemical basis of morphogenesis, *Philos. Trans. R. Soc. London B* **237**, 37.

459. Ulam, S. M., and von Neumann, J., (1947). On combinations of stochastic and deterministic processes, *Bull. Am. Math. Soc.* **53**, 1120.

460. Uspenskii, V. A., Semenov, A. L., and Shen, A. K., (1990). Can an individual sequence of zeros and ones be random?, *Russ. Math. Surv.* **45**, 121.

461. **Uzunov**, D. I., (1993). *Theory of critical phenomena*, World Scientific, Singapore.

462. **van Kampen**, N. G., (1981). *Stochastic processes in physics and chemistry*, North-Holland, Amsterdam.

463. Vannimenus, J., (1988). On the shape of trees: tools to describe ramified patterns, in *Universalities in Condensed Matter*, edited by R. Jullien, L. Peliti, R. Rammal, and N. Boccara, Springer, Berlin, p. 118.

464. Vannimenus, J., and Viennot, X. G., (1989). Combinatorial tools for the analysis of ramified patterns, *J. Stat. Phys.* **54**, 1529.

465. Vastano, J. A., Russo, T., and Swinney, H. L., (1990). Bifurcation to spatially induced chaos in a reaction-diffusion system, *Physica D* **46**, 23.

466. **von Neumann**, J., (1966). *Theory of self-reproducing automata*, edited by A. W. Burks, Univ. of Illinois Press, Champaign, IL.

467. Voss, R. F., (1992). Evolution of long-range fractal correlations and $1/f$ noise in DNA base sequences, *Phys. Rev. Lett.* **68**, 3805.

468. Walters, P., (1985). *An introduction to ergodic theory*, 2nd edition, Springer, New York.

469. Wang, X.-J., (1989a). Abnormal fluctuations and thermodynamic phase transitions in dynamical systems, *Phys. Rev. A* **39**, 3214.

470. Wang, X.-J., (1989b). Statistical physics of temporal intermittency, *Phys. Rev. A* **40**, 6647.

471. Weaver, W. W., (1968). Science and complexity, *Am. Sci.* **36**, 536.

472. Weiss, B., (1973). Subshifts of finite-type and sofic systems, *Monatshefte für Math.* **77**, 462.

473. Weiss, C. O., and Brock, J., (1986). Evidence for Lorenz-type chaos in a laser, *Phys. Rev. Lett.* **57**, 2804.

474. Weissman, M. B., (1988). $1/f$ noise and other slow, nonexponential kinetics in condensed matter, *Rev. Mod. Phys.* **60**, 537.

475. Weitz, D. A., and Oliveira, M., (1984). Fractal structures formed by kinetic aggregation of aqueous gold colloids, *Phys. Rev. Lett.* **52**, 1433.

476. Wharton, R. M., (1974). Approximate language identification, *Inform. Contr.* **26**, 236,

477. White, D. B., (1988). The planforms and onset of convection with a temperature-dependent viscosity, *J. Fluid Mech.* **191**, 247.

478. Widom, B., (1965). Surface tension and molecular correlations near the critical point, *J. Chem. Phys.* **43**, 3892 and 3898.

479. Wiener, N., (1948). *Cybernetics, or control and communication in the animal and the machine*, MIT Press, Cambridge, MA.

480. Wiggins, S., (1988). *Global bifurcations and chaos*, Applied Mathematical Sciences **73**, Springer, New York.

481. Wightman, A. S., (1979). Introduction to *Convexity in the theory of lattice gases*, by R. B. Israel, Princeton University Press, Princeton, NJ, p. ix-lxxxv.

482. Wilson, K. G., (1971). Renormalization group and critical phenomena: (I) renormalization group and the Kadanoff scaling picture, (II) phase-space cell analysis of critical behaviour, *Phys. Rev. B* **4**, 3174 and 3184.

483. Wilson, K. G., and Kogut, J., (1974). The renormalization group and the ε-expansion, *Phys. Rep. C* **12**, 75.

484. Winfree, A. T., Winfree, E. M., and Seifert, H., (1985). Organizing centers in a cellular excitable medium, *Physica D* **17**, 109.

485. Witten, T. A., and Sander, L. M., (1981). Diffusion-limited aggregation, a kinetic critical phenomenon, *Phys. Rev. Lett.* **47**, 1400.

486. Wolfram, S., (1984). Computation theory of cellular automata, *Commun. Math. Phys.* **96**, 15.

487. Wolfram, S., (1986). *Theory and applications of cellular automata*, World Scientific, Singapore.

488. Yates, F. E., ed., (1987). *Self-organizing systems: the emergence of order*, Plenum, New York.

489. Yekutieli, I., and Mandelbrot, B. B., (1994). Horton-Strahler ordering of random binary trees, *J. Phys. A* **27**, 285.

490. Yekutieli, I., Mandelbrot, B. B., and Kaufman, H., (1994). Self-similarity of the branching structure in very large DLA clusters and other branching fractals, *J. Phys. A* **27**, 275.

491. Young, L.-S., (1982). Dimension, entropy and Lyapunov exponents, *Ergod. Theory and Dynam. Syst.* **2**, 109.

492. Yu, L., Ott, E., and Chen, Q., (1990). Transition to chaos for random dynamical systems, *Phys. Rev. Lett.* **65**, 2935.

493. Zambella, D., and Grassberger, P., (1988). Complexity of forecasting in a class of simple models, *Complex Systems* **2**, 269.

494. Zamolodchikov, A. B., (1986). "Irreversibility" of the flux of the renormalization group in a 2D field theory, *JETP Letters* **43**, 730.

495. Zhang, Y. C., (1991). Complexity and $1/f$ noise. A phase space approach, *J. Phys. I (Paris)* **1**, 971.

496. Zhong, F., Ecke, R., and Steinberg, V., (1991). Asymmetric modes in the transition to vortex structure in rotating Rayleigh–Bénard convection, *Phys. Rev. Lett.* **67**, 2473.

497. Ziv, J., and Lempel, A., (1978). Compression

of individual sequences by variable rate coding, *IEEE Trans. Inform. Theory* **24**, 530.

498. Zuckerkandl, E., (1992). Revisiting junk DNA, *J. Mol. Evol.* **34**, 259.

499. **Zurek**, W. H., ed., (1990). *Complexity, entropy, and the physics of information*, Addison-Wesley, Redwood City, CA.

500. Zvonkin, A. K., and Levin, L. A., (1970). The complexity of finite objects and the development of the concepts of information and randomness by means of the theory of algorithms, *Russ. Math. Surv.* **25**, 83.

Index

The following convention has been adopted: bold-face pages refer to primary entries (e.g., definitions), roman to normal entries, and italics to minor entries.

Printed in the United States
By Bookmasters